T0358410

Aerial view of the silver–gold heap leaching Coeur Rochester Mine in Nevada. (Courtesy of Coeur d'Alene Mines Corporation.)

Solution Mining

Solution Mining

Leaching and Fluid Recovery of Materials

Second Edition

Robert W. Bartlett

Routledge
Taylor & Francis Group

LONDON AND NEW YORK

Published by Routledge
2 Park Square, Milton Park, Abingdon, Oxfordshire OX14 4RN
711 Third Avenue, New York, NY 10017

First issued in paperback 2015

Routledge is an imprint of the Taylor and Francis Group, an informa business

---

British Library Cataloguing in Publication Data

Bartlett, Robert W.
    Solution mining : leaching and fluid recovery of minerals.
    – 2nd ed.
    1. Solution mining   2. Leaching
    I. Title
    622.2'2

Publisher's Note
The publisher has gone to great lengths to ensure the quality of this reprint but
points out that some imperfections in the original may be apparent

ISBN 13: 978-1-138-99638-0 (pbk)
ISBN 13: 978-90-5699-633-8 (hbk)

To

Milt and *Dad*

for teaching extractive metallurgy science
and mineral engineering practice,
and pointing the way

CONTENTS

SYMBOLS

A	area, generally cross-section (horizontal) of ore heap or dump
A_0	acid concentration outside ore fragment (rock) in eqn 5.14
A_c	acid concentration inside ore fragment at core interface, eqn 5.14
A_p	average surface area of reacting mineral grain in reaction zone, eqn 12.17
b	aquifer (ore zone) thickness, eqns 11.2 through 11.8 and 11.11
B	stoichiometric factor, mols acid or oxidant consumed per mol copper extracted in eqns 5.13 and 12.5
c	concentration of diffusing species in solution
c_c	concentration of diffusing species in solution at core interface in the shrinking core model, eqn 5.12
C_{Cl^-}	concentration of chloride ion tracer, eqn 7.13
c_i	initial concentration of dissolved species in ore fragment
c_0	concentration of diffusing species in solution outside rock
c_p	peak well effluent concentration in eqn 10.8
c_z	avg concentration of diffusing species at reaction zone, eqn 12.17
c_{O_2}	concentration of dissolved oxygen, eqn 12.29
$c_{t,r}$	concentration of diffusing species in ore fragment at radius r and time t, eqn 12.3
c_p	peak concentration from a production well before decline, eqn 10.8

$c_{g(x)}$	concentration of gaseous oxygen inside heap at level x
$CF_{(t)}$	cash flow at time t, eqn 10.13
C_{op}	operating unit cost, eqn 10.14
C_p	heap (dump) heat capacity; subscript indicates phase; eqn 12.30
$\mathbf{C^*}$	pulse tracer dimensionless exit concentration, eqn 7.13
$d_{a(avg)}$	average particle size for ore passing sieve $d_{a(H)}$ but larger than the next sieve $d_{a(L)}$; see eqns 2.8 and 2.9
d_c	diameter of capillaries
$d_{m(avg)}$	volume mean diameter of the reacting mineral grain, e.g., chalcopyrite; see eqn 12.4
d_p	effective ore fragment (particle) diameter
d_p^*	critical mean particle diameter effecting permeability in eqn 7.17
d_{80}	particle diameter such that 80% of the mass of particles is smaller; applies to ore particles (and mineral grains in chap. 9)
d_{B-K}	Blake–Kozeny critical ore fragment size, eqn 7.15
D	diffusion coefficient of diffusing species in solution filled pores
D_{eff}	effective diffusion coefficient in porous ore fragment corrected for porosity and tortuosity, $D_{eff}=D\varepsilon/\tau$.
D_A	dispersion coefficient, chap. 7
D_g	gas diffusivity; eqn 8.1
E_{act}	Arrhenius activation energy for mineral surface oxidation rate in eqn 12.4
E^0	half cell potential (volts) for an electrochemical reaction at standard conditions; eqn 5.37
E_h	half cell potential for an electrochemical reaction at non-standard concentrations; eqn 5.37
$\mathscr{F}$	Faraday constant; see eqn 5.37
F or (F_t)	fraction of mineral extracted, usually as a function of time
F_{t,r_0}	F_t for ore fragment with radius, r_0
$\mathbf{F_t}$	fraction of total mineral extracted in an aggregation of ore particles of different diameter, i.e., the heap extraction; see for example eqns 2.3 and 12.10
g	gravitational constant; used in eqns 7.8 and 8.3
G	ore grade: mass fraction of extractable metal (e.g., copper) in ore; see eqns 12.7, 12.21, 12.22, and chap. 5

G_{Au}^{T}	total gold grade in ore, e.g., by fire assay, oz/tonne
G_{Au}^{CN}	direct cyanide extractable gold grade in ore, oz/tonne
G_{Au}^{r}	refractory gold grade in ore, oz/tonne
G_{Au}^{o}	economic cut-off gold grade in ore, oz/tonne
G_{Au}^{*}	critical gold grade defined by eqn 9.10, oz/tonne
h	hydrostatic head of the solution during leaching
h_{avg}	average head in the drain blanket (saturated zone above the leach pad liner), eqn 4.1
h_c	capillary rise accompanying percolation leaching, eqn 7.1
h_f	length of fan wells in fan pattern flooded leaching; see Fig. 11.8
h_0	solution head in the leach pad drain pipe, eqn 4.1
h_w	hydrostatic head at a well
H_{GW}	initial groundwater level in flooded leaching
H	height of heap or dump; as subscript the top of the heap
H_c	distance between horizontal fractures in Fig. 11.6
i	annual financial discount (interest) rate expressed as a fraction
i	index of time increments, t, in finite difference equations
I	total increments of time in finite difference model, eqn 5.29
j	index of ore fragment spherical shell increments, r, in finite difference equations, chap. 12; also, index of vertical stacked volume units in eqn 5.27 and Fig. 5.12
J	total increments of ore fragment concentric shells in finite difference equations, chap. 12; also total stacked volume units in heap height, chap. 5
k_0	proportionality constant in eqn 12.4
k_1	mineral surface oxidation reaction rate constant, eqn 12.4
k_i	intrinsic permeability of the fragmented ore heap or dump
k_z	mineral surface oxidation rate constant for reaction zone model, eqn 12.17
K	hydraulic conductivity; see eqn 7.7
K_{SAT}	hydraulic conductivity at saturation
K_{DB}	lateral hydraulic conductivity of the drain blanket (saturated zone above the leach pad liner), eqn 4.1

K_h	Henry's law constant used in eqn 12.19
K_L	seepage hydraulic conductivity of leach pad liner
K_m	empirical rate constant for mineral oxidation kinetics; eqn 9.5
K_s	hydraulic conductivity of skin zone around a modified well, eqn 11.10
L	drainage lateral path length along the leach pad liner, eqn 4.1
L	air flow travel length under an ore heap, either in the drain blanket (eqn 8.8) or in a forced air injection pipe (eqn 8.13)
L_h	length of cubic unit volume in a vertical ore column, eqns 5.27 and 5.28
MW_{Cu}	molecular weight of copper
MWG	molecular weight of gas, e.g. air being depleted in oxygen; see eqn 8.4
M_{Cl^-}	mass of chloride ion pulse in tracer experiment, eqn 7.13
n	summing index in eqns 1.2 and 1.3
n	number of moles in eqns 5.6 and 12.17–12.21; number of electrons in eqn 5.37
N_p	numerical density of reacting mineral grains in the reaction zone model, eqn 12.17
N_y	mass fraction of fragmented ore with rock size, $r_0 = y$; see eqns 2.3 and 2.4
p	pressure (variable)
p_P	air delivery pipe pressure, chap. 8
p_t	total gas pressure
p_{O_2}	pressure of oxygen in leaching system
$p_{O_2(std)}$	reference pressure for oxygen solubility, eqn 12.29
P_{H_2O}	water vapor pressure at the leaching system temperature
Py/Cp	molar ratio of pyrite to chalcopyrite in the ore
Δp_{DB}	air pressure drop passing laterally through the ore heap drain blanket; eqn 8.11
Δp_H	gas (air) pressure difference through heap or dump
Δp_{Or}	air delivery pipe orifice pressure drop, chap. 8
ΔP_{IP}	pressure difference between injection and production wells in a wellfield configuration

P_m, P_{Au}	metal (gold) price
PV	financial present value, see eqn 10.13 and chap. 9
$q_{t,r}$	*local* mass fraction of mineral within ore fragment at radius, r, that has been oxidized and dissolved at time, t, eqn 12.3
Q	volumetric flow rate, usually to/from wells; see chap. 11 and eqn 10.8
Q_g	gas flow rate; see eqn 8.13
Q_l	solution percolation volumetric flow rate, see eqn 7.6
Q_L	seepage flow rate through leach pad liner, eqn 4.3
r	radial position in the quasi-spherical ore fragment (rock)
r_0	external radius of the ore fragment (rock)
r_c	radius of unreacted core of the ore fragment; see Fig. 5.7
r_z	average radius of the reaction zone in the reaction zone mixed kinetics model; see eqn 12.17
r_{Py}	mineral (pyrite) grain radius in ore
r_{Py^*}	largest mineral (pyrite) grain radius in ore
Δr	thickness of concentric ore fragment shells in finite difference equations
r_w	radius of wellbore, and air injection pipe
r^*	radius of largest ore particle in a distribution of ore particle sizes
R	gas constant—used in thermochemical calculations; e.g. eqn 5.37
R_{ox}	pseudo-homogeneous mineral oxidation rate per unit volume of ore fragment (rock); see eqn 12.1
R_Q	volumetric rate of chemical heat generation in heap; see eqn 12.30
s_s	skin factor for a well with skin effect, eqn 11.10
s_w	drawdown at a well, $H - h_w$; also see eqn 11.6
S_0	oxygen solubility coefficient; see eqn 12.29
S_5	five spot well spacing, see Fig. 10.17
t	time
t_{avg}	average time, in eqn 7.13 for the tracer effluent (exit)

$t_{Py(100)}$	time to oxidize all (100%) of pyrite grains in ore when mineral kinetics is rate controlling, chap. 9
t_{100}	time for complete (100%) extraction from ore particles; see chap. 5
Δt	increment of time in finite difference equations
T	temperature
u_g	superficial gas (air) velocity (vertical) flowing through heap or dump
u_l	superficial solution velocity (vertical) percolating through heap or dump
v_g	volume fraction of heap between rocks occupied by air
v_s	volume fraction of heap occupied by solids; does not include open internal rock porosity
v_l	volume fraction of heap occupied by solution
$v_{l(0)}$	volume fraction of heap occupied by solution at draindown
V_{Cu}	molar volume of copper in rock, $MW_{Cu}/\rho_{ore}G$
W	solution percolation volumetric flow through heap, eqn 4.1 [also see Q_l]
W_0	total amount of accessible mineral that can be dissolved from a production well in eqn 10.6
W_p	width separating forced air injection pipes in eqn 8.14
Δx	distance over which a hydraulic gradient is expressed, eqn 7.7
ΔX_L	seepage barrier (liner) thickness, eqn 4.3
x	elevation index within heap or dump
x_f	distance between fan faces in fan pattern flooded leaching; see Fig. 11.8
x_0	half of the distance between wells, eqn 11.6
X	partial height of heap or dump at position x
X_D	biooxidation penetration distance into an ore heap based on oxygen diffusion in the stagnant gas within the heap; eqn 8.1
y	rock size index in a distribution of rock sizes, eqn 2.3
Y	total discrete sizes in the rock size distribution, eqn 2.3

δ	thickness of reaction zone (arbitrary) in reaction zone model, eqn 12.17
δ_1	surface tension of solution in fragmented ore, eqn 7.1
ε	microporosity, volume fraction of open micropores in the ore fragment
ω	Dixon number, a dimensionless factor defined by eqn 5.30 that characterizes lixiviant flow through an ore heap
κ_s	intrinsic sulfide mineral grain oxidation rate constant, eqn 9.3
K_s	intrinsic sulfide mineral grain oxidation rate constant combined with reactant concentration term, eqn 9.4
ρ_H	oxygen depleted air specific gravity at the top of an ore heap
ρ_{ore}	specific gravity of ore fragments (rocks)
ρ_m	specific gravity of reacting mineral grains; see eqn 12.22
ρ_o	specific gravity of air at standard temperature and pressure
ρ_1	specific gravity of the leaching solution
μ	fluid dynamic viscosity (solution or air), eqns 7.9 and 8.2
β	group defined by eqn 12.23
β'	group defined by eqn 12.27
γ	group defined by eqn 12.28
Ψ	dimensionless sphericity factor, eqn 12.18
τ	tortuosity (average) of open pore paths in the ore fragments
θ	dimensionless time, $= t/t_{avg}$, eqn 7.14
θ	solution/ore wetting contact angle, eqn 7.1
Θ	fan angle, radians, between wells, eqn 11.13

PREFACE

This text is derived from a course in solution mining taught to metallurgical and mining engineers at the University of Idaho. The course was initiated to provide an introduction to the rapidly expanding field of solution mining, particularly for the extraction of gold, silver and copper. The annual economic value of solution mined metals in the United States now exceeds that of metals extracted by underground mining. But solution mining is a relatively new field, at least at its present scale. It has not been part of the established academic curricula and has been taught only as an elective subject at a few institutions and in short courses given off campus.

Although several solution mining symposia proceedings have been published in recent years, primarily by the Society of Mining, Metallurgy and Exploration, an integrated text with problems suitable for undergraduates, practicing engineers and geologists was not available. Keeping in mind this broad audience spectrum, the book has been designed to require only a preliminary understanding of basic inorganic chemistry. Although mathematical modeling has made contributions to understanding solution mining phenomena, it is eschewed somewhat in the interest of a wider audience of practitioners, except for chapter 12.

I first became interested in solution mining in 1971 after spending a summer at Lawrence Livermore Laboratory (LLL) where a large pilot plant scale pressurized leaching experiment was underway to experimentally simulate copper extraction by flooded leaching at depth. Livermore was then pursuing peaceful uses of nuclear explosives, and investigating *in situ* rubblization of ore using nuclear explosives prior to leaching. Experimental results were explained by a computer simulation model published in 1972. Before that I had been stimulated by Prof. Milton Wadsworth, while a student at the University of Utah.

During summers of the early 1970s, I solution mined gold contained in low-grade ore heaps left from a previously operated Nevada gold mine. Since then, heap leaching gold ore has become a billion-dollar industry in Nevada.

From 1973 to 1978, as manager of hydrometallurgy at the Kennecott Research Center in Salt Lake City, I participated in major laboratory and mine site experiments to better understand copper leaching from mine waste dumps. This major source of domestic copper production involves many interrelated complex phenomena, and the Kennecott research during this period (subsequently released) was the most extensive yet conducted on this important aspect of metal solution mining.

Ore testing of precious metal deposits returned to the fore during the 1980s, along with studying the solution mining of borates, while I was directing the Anaconda Minerals' Tucson Research Center. Recently, I have been interested in biooxidative pretreatment of low-grade refractory gold ores in bacteria inoculated wet stockpiles, prior to cyanide heap leaching.

I am indebted to many colleagues over a span of more than twenty years for the experiences, ideas and friendships shared during our mutual interest in solution mining. Principal among these are Jay Agarwal, John Apps, Roshan Bhappu, Bob Braun, Al Bruynesteyn, Larry Cathles, Don Davidson, Jerry Fountain, Joe Harrington, Rudy Jacobson, Jonathan Jackson, Bill Larson, Art Lewis, Ed Malouf, Dave Milligan, Keith Prisbrey, Bruce Ream, Dave Reese, Ken Richards, Ron Roman, Joe Schlitt, John Sibert, Rush Spedden, Dirk Van Zyl, Milton Wadsworth and Rolf Wesely.

I am especially grateful to Joe Schlitt, Milt Wadsworth and the late Don Davidson for careful reviews of the first-edition manuscript and many useful suggestions. Rudy Jacobson, Milt Wadsworth and David Dixon were helpful with this new edition.

Text and illustration preparation and editing were accomplished with the able assistance of Carol McAleer, Allan Jokisaari, Judy Reisenauer and Pat Hautala, for which I am most grateful.

ONE

Introduction

SCOPE OF SOLUTION MINING

Solution mining is variously defined. In this text it includes all forms of extraction of materials from the earth by leaching and fluid recovery, both by *in situ* methods and heap leaching of excavated ore. The emphasis is on minerals, brines and other naturally occurring solutions, but the theory and practice developed for these materials can be extended to contaminants introduced into the earth by human activity. Several minerals are readily soluble in water, forming brines, and recovery of these fluids represent the earliest applications of solution mining. Extraction of solid minerals by an aqueous solution flowing through and **leaching** (dissolving) them from their host rock has become of increasing importance in solution mining. This includes both readily soluble **evaporite** minerals and metallic **ore** minerals requiring leaching reactions with acids or other chemical lixiviants and, often, oxidation of the mineral. In outlining the scope of solution mining in this introductory chapter, some words will be contextually defined as they are conventionally used in solution mining practice.

Solution mining is an interdisciplinary field involving geology, chemistry, hydrology, extractive metallurgy, mining engineering, process engineering and economics. A few of the factors to be considered in solution mining projects are: leaching chemistry, rock (gangue) chemistry, solution flow in the ore mass, air flow in the ore mass (percolation leaching), open void space in the ore mass available for flow, ore (rock) microporosity, transport within dead end micropores by ordinary chemical diffusion, metal/mineral recovery technology and operations and their affect on leaching extraction, environmental containment, solution losses, brine chemistry, solar evaporation pond engineering and reclamation.

The approach of this text is to include description, theory and practical aspects of current solution mining technology. Principles will be integrated with specific metal/mineral systems and practice, beginning with simple systems and proceeding toward increasing complexity. Understanding the important factors, their relationships with each other and their influence on the technological and economic outcome are not sacrificed to detail. Mathematical modeling of solution mining systems should aim at providing improved understanding of these complex phenomena rather than at numerical certitude.

At the present time water is the only solvent base of commercial interest, with the exception of molten sulfur recovered using hot water in the Frasch process, so this text's discussion will be focused on aqueous systems. Generally the principles are transferrable. *In situ* or "true" solution mining involves extraction of minerals from the undisturbed ore in place. Leaching the ore **rubblized** within the ore body is sometimes referred to as "modified *in situ*" solution mining. Excavation of the ore using conventional mining methods followed by **heap leaching** in prepared ore stockpiles on the surface and **dump leaching** of mine overburden (waste) are important applications of solution mining that will be covered in this text. The **extraction** operation (leaching, brine recovery, etc.) is coupled with a metal/mineral **separation** operation, nearly always on the surface and adjacent to the extraction operation, for separation of the dissolved metals or minerals from the recovered fluid solution or **leachate**. Large volumes of solution circulate between the extraction and separation operations. Both gravity flow and pumping are used to transport the leaching fluids. Leachates generated by the extraction operation are often referred to as "**pregnant solutions**" or pregnant liquors, while fluids being returned to the extraction operation are often referred to as "**barren solutions**" for obvious reasons. Sometimes solution is bled from the otherwise closed circulating solution to control the accumulation of undesirable impurities. This text will focus on extraction operations and will either omit or only peripherally include mineral and metal recovery from solutions. The science and engineering of metal separation technologies are described in hydrometallurgy texts including Habashi (1980) and Van Arsdale (1963).

Sufficient **permeability** in the ore mass to permit solution flow is an important, and often restricting, factor in solution mining. *In situ* leaching must rely upon open void space or natural fractures in the ore for solution flow paths. Sandstones and vuggy evaporites are often sufficiently open to provide good permeability. Evaporites contain large amounts of soluble mineral and therefore tend to become more open as extraction proceeds.

Many metallic minerals were deposited by geochemical processes that involved hydrothermal solutions flowing in natural rock fractures. If these fractures remain open, they can be a source for extraction by solution mining. However, host rocks for sulfide ore deposits typically contain fractures with a microporosity of only about 1–6% and very low permeabilities—often too low for commercial production rates in solution mining. Also, major fractures are a source of high flow channels that short-circuit the solution, not allowing it to sweep uniformly through most of the ore mass. This leads to low **sweep efficiency**, low yield of extracted metal/mineral, and often economic failure. Furthermore, the **uncertainty** in sweep efficiency causes uncertainty in economic forecasting, which prevents many solution mining projects from being implemented. Not only adequate permeability but fairly uniform permeability is required.

Success often requires a substantial increase in permeability by fragmenting the ore in place, **rubblizing**. Effective fragmentation nearly always requires a substantial increase in volume, 15–20%, to provide open space for flow channels. Planned deliberate caving, blasting and natural subsidence over underground excavations and abandoned mines are ready sources; the overlaying subsidence provides a large volume of broken ore. Leaching caved cappings over otherwise depleted underground mines is a common application of solution mining.

Both **flooded leaching** and **percolation leaching**, or trickle leaching, are employed in solution mining. Flooded leaching pertains to situations where the ore mass is saturated with the solution; usually only one fluid phase is present and the operation typically occurs below the water table, or the solution is otherwise contained. Flooded leaching has many flow characteristics in common with ground water flow. The mathematical treatment and numerical models describing flow problems in **hydrology** are often readily transferrable. Percolation leaching involves downward flow of unsaturated solution by gravity; two fluid phases, solution and air, are always present within the ore mass. Other distinguishing characteristics of percolation leaching and flooded leaching are presented in Table 1.1. Many aspects of solution mining are treated in the Mining Engineering Handbook (2nd edition, 1992).

Environmental containment of solution must always be maintained in a solution mining operation, and various approaches to achieve this will be discussed in the text. **Remediation of contaminated ground and water, including groundwater, is covered in the last chapter.**

Table 1.1   Characteristics of Percolation and Flooded Leaching.

PERCOLATION LEACHING

    – Solution trickles downward by gravity in an adequately permeable ore mass while dissolving the valuable mineral(s)

    – Solution must be uniformly spread over the top of the ore mass and collected at the bottom of the ore mass, over an impervious bottom layer to prevent further downward percolation

    – Adequate permeability is usually achieved by fragmentation from mining and/or by additional ore crushing. Excessively fine or clay-like ore with low permeability may be **agglomerated** to improve permeability

    – The solution is not saturated and air is present, which may or may not be important to the leaching chemistry; if oxygen is consumed the air may be partially oxygen depleted

    – Important applications are heap leaching gold, silver, copper, and uranium; copper mine waste dump leaching; leaching mine subsidence for copper and uranium

FLOODED LEACHING

    – Solution flows because of a pressure gradient, often from an injection well(s) to a production well(s), through an adequately permeable ore mass

    – Air is not normally present, although injection of air or oxygen as a chemical reactant bubbling upward through the otherwise solution saturated ore mass may occur; this is an example of two-phase flow

    – Leaching occurs only below a water table, either natural or artificially induced

    – The permeability must be nearly uniform to provide good sweep efficiency and extraction; the ore zone must be bound by ground of substantially lower permeability to contain solution flow

    – Important applications are brine production from wells, leaching uranium and copper (presently limited but of considerable   experimental development) below the water table

---

Solution mining to produce brines is applied to both buried and surface evaporites and is a major source for the production of numerous water soluble salts and minerals, including common salt (sodium chloride), potash, magnesium, lithium, trona (sodium carbonate), and boron minerals. Sea water, natural playas (salt lakes) and underground brines are all important sources of mineral bearing solutions. **Solar pond evaporation** is an important and relatively inexpensive method of concentrating brines before entering a mineral recovery or separation plant. Often several salts are separated and recovered from complex brine mixtures.

    Several industrially important metallic minerals are not soluble in water or natural brines but are soluble in aqueous chemical solutions, which are used to extract them in solution mining operations. Oxidized minerals of copper and uranium are soluble in acidic solutions at a sufficiently low pH—below their hydrolysis pH. Sulfuric acid is commonly used, primarily because of its low cost. Oxidized uranium is also soluble in ammonium carbonate solutions, and this lixiviant is used when the uranium ore contains limestone, dolomite or other major acid consuming gangue minerals.

Chemical oxidation of the economic minerals, or other minerals in the ore, is often required to render them soluble. Hydrogen peroxide is often used for this purpose in uranium leaching. Although oxygen is required to solubilize gold in cyanide leaching, the amount is small. For most heap leaching applications where ore is free of sulfides and organic material, oxygen dissolved in the leach solution is adequate. Leaching sulfide or carbonaceous ores is another matter. Large amounts of oxygen are required to oxidize sulfide minerals and for many ores and mine wastes air is the only affordable source of oxygen. Furthermore, the solubility of oxygen dissolved in aqueous solutions is insufficient for commercial production rates. Hence, these leaching systems must have both the solution phase and the air phase present and flowing through the ore mass being leached, a case of **two-phase flow**.

Because of oxygen's low solubility near room temperature, leaching of sulfides and other minerals requiring oxidation would be very slow except for the presence of ferric ions. Fortunately, iron is a ubiquitous element in the earth and most nonferrous ore mineral assemblages. Ferric ions are often present in the ore leaching solution as the primary oxidant. However, the regeneration of ferric ions (from the ferrous ions resulting from mineral oxidation) is kinetically very slow without the assistance of *Thiobacillus ferrooxidans*. These bacteria catalyze the oxidation of both ferrous ions and sulfide minerals. Thus, **biooxidation** of minerals is an important aspect of many solution mining applications. The role of bacteria and bacteria products such as enzymes has expanded into many aspects of solution mining, including environmental treatment of spent solutions, waste water, and hazardous residues as well as mineral oxidation. **Biohydrometallurgy** has recently evolved as a new field of research and industrial practice, which has been exhaustively reviewed in a text by Rossi (1990) that includes over 1,600 references.

IMPORTANCE OF SOLUTION MINING

When technically feasible, the specific treatment cost of solution mining can often be substantially lower than any other approach. Consequently, as lower grade ores are being mined, solution mining is often the only available

Table 1.2 Importance of Solution Mining for USA Selected Metal/Minerals.

Metal or Mineral	Approximate Primary Production
Gold	35%
Silver	25%
Copper	30%
Uranium	75%
Common Salt	50%
Potash	20%
Trona	20%
Boron	20%
Magnesium	85%
Sulfur	35%

route and its use is rapidly expanding. Solution mining ore in heaps and leaching mine waste dumps are well matched to excavation (mining) of bulk ore deposits using large-scale surface mining equipment, which has evolved during the last few decades to provide remarkably low specific mining costs, sometimes less than one US dollar per tonne of ore (1 tonne = 1 metric ton = 1 Mg).

Solution mining is already a major source of several metals and minerals. The author's estimates of the percentage of the primary production attributed to solution mining in the United States are shown in Table 1.2. These estimates are in part based on the U.S. Bureau of Mines Mineral Commodity Summaries. In 1990, the added value of solution mining operations in the western United States was approaching two billion dollars. This figure does not include the value or equivalent cost of associated conventional mining operations to provide ore for leaching in heaps and mine dumps, nor the value of subsequent metal separation operations such as solvent extraction, electrowinning and metal refining. The magnitude of solution mining in the western United States is now easily greater than that of underground hard rock mining of non-fuel minerals, and it is expanding rapidly. Many low grade gold mines, opened in the last decade and operating profitably, would not be possible without the low-cost leaching of ore in heaps.

SHARPENING OUR INTUITIVE UNDERSTANDING OF LEACHING

Let us begin with a simple leaching system and a few experiments the results of which we can easily comprehend, and even perform without benefit of a laboratory. This might be called "the kitchen series." Consider an event that we are all familiar with—dissolution of sugar in a cup of coffee; or to be more observable, dissolution of sugar in a cup of water. The system is simple because:

— The solubility of sugar in water is very high and unlimited at least with respect to the usual teaspoon of sugar in a cup

— No chemical reaction with another solute or component in water is required to obtain dissolution

— There are no interfering solids, at least yet, assuming we use nominally pure sugar.

For sugar in each of the following four forms, rank the required time (1 through 4 with 1 the least time and fastest rate) to completely dissolve one gram of sugar in a cup of water at room temperature:

powdered sugar   _____

rock candy   _____

granular sugar   _____

cubed granular sugar _____

This may be obvious to you, but if not, try the experiment. What will be the trend in results if we use boiling water for each of the four sugars? What did we learn from this—or did we already know it? Mineral particle size is important in leaching, with smaller mineral grains leaching faster than coarser mineral grains; and heat usually accelerates the rate of leaching.

Now consider two series of experiments, each with powdered sugar, granular sugar and 1 mm rock candy, sized by sieving. In the first series, three 20 g spheres of plaster of paris are formed; each contain 0.20 g (1 wt pct) of one of the sugars uniformly mixed with dry plaster prior to adding a little hardening water and forming the sphere.

The second series of experiments are made by forming three identically sized spheres using coarse sand with a very small amount of glue in a volatile solvent as a binder, just enough to hold the sand grains together.

After the spheres have dried and hardened overnight, obtain six large cups or beakers, place one of the spheres in each and fill it with water. The question is which spheres will permit faster sugar extraction? A taste test of each mug at intervals of time will allow you to rank each of the six spheres with respect to the rate of sugar extraction using only the following allowed answers: too slow to taste, slow, slowest, fast, faster, fastest.

PLASTER   **SAND**

plaster of paris with powdered sugar   ————

plaster of paris with granular sugar   ————

plaster of paris with rock candy   ————

sand sphere with powdered sugar   ————

sand sphere with granular sugar   ————

sand sphere with rock candy   ————

Hint: the microporosity of the sphere composed of cemented sand grains is higher than the microporosity of the plaster of paris.

As shown in Fig. 1.1, **fractional extraction** increases with leaching time but the **rate** of extraction declines as the solute has farther to travel to escape the plaster or sand spheres (and the rock matrix in ore leaching), and the concentration **gradient**, driving diffusion, decreases.

The fractional extraction rate is expected to decrease as the sphere size increases.

Next, imagine the experiment sketched in Fig. 1.2. Actually this is a comparison of the results from two similar experiments but with a distinctive difference. Identical spheres of powdered sugar mixed into coarse sand are used in each experiment. In both experiments the spheres are mounted on a pedestal inside a vertical glass tube. In the first experiment the tube is flooded with water that flows past the sphere at a velocity of 0.1 mm/s (call it a superficial velocity for the process engineers). In the second experiment water drips from an orifice immediately above the sphere onto the top of the sphere and then runs off, simulating a percolation leaching experiment. The volumetric water flow rates are identical in each case. Which experiment, if either, will yield the faster sugar extraction? The correct answer is: because water wets both spheres there will be little or no difference. Percolation leaching may be a little faster for sufficiently small spheres because of the higher liquid film velocity adjacent to the sphere's surface.

Next, repeat the prior experiment comparing flooded leaching and percolation leaching except thoroughly inundate the surface of each sand

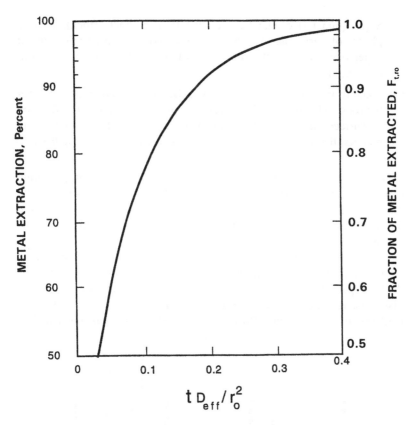

Figure 1.1. Fractional extraction of completely dissolved solute from a porous sphere (Crank, 1956).

sphere with hair spray **before** beginning the comparative leaching experiments. That's right, common hair spray. According to the can in my wife's bathroom, hair spray contains organic solvents, methacrylates, isostearic hydrolyzed animal protein and other polymers—in other words, greasy stuff. If you use enough hair spray it should prevent any leaching of the sugar.

What are the factors that determine the rates of sugar extraction in these various cases? Consideration of these experiments suggests the following:

(1) The sugar (mineral) must be soluble in the water (solution).

(2) The rate of dissolution may depend on the sugar (mineral) grain size.

(3) Sugar (mineral) cannot be extracted from a sphere (rock), such as plaster of paris, that has no open microporosity to permit entry of water (solution) into the sphere (rock) and access to the sugar (mineral) grain.

(4) Sugar (mineral) cannot be extracted if the water (solution) does not **wet** the rock and enter the micropores by capillary action; this is a function of surface energy and viscosity; hair spray coated the sphere with material that prevents water from wetting and entering the capillaries.

Other factors affecting extraction but not discovered in the preceding experiments are: (5) sphere (rock) size, (6) ability of solutions to flow by the sphere (rock), sweep away dissolved mineral and avoid saturation at the sphere (rock) surface, and (7) the total amount of sugar (mineral) in the sphere (rock).

Consider the simplified case of a small amount of sugar **disseminated** uniformly within a porous sphere of cemented sand grains. As water enters the pore space, the amount of sugar present is so limited that all of it quickly dissolves. Post examination of the sectioned wet sphere would show no evidence of sugar grains. How does the dissolved sugar, contained in

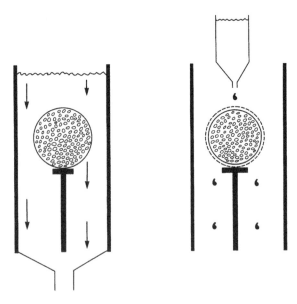

Figure 1.2. Sugar extraction from porous spheres by flooding (left) and percolation (right) leaching.

water-filled micropores within the sphere, exit the sphere? It occurs by ordinary chemical diffusion and **not** by flow. The solution inside the rock is stagnant. Diffusion is driven by a sugar solution concentration that is higher inside the sphere (rock) than at its surface. Flow only becomes an important transport process once the dissolved material is out of the sphere (rock); then, soluble sugar is swept away by the solution flowing around the sphere (rock). In the case of percolation leaching this is a slowly flowing film on the surface of the sphere (rock).

MODELING EXTRACTION RATES CONTROLLED BY CHEMICAL DIFFUSION

For most mineral leaching systems, fragmented rocks are roughly equidimensional and can be adequately approximated as spheres. Obviously this would not apply to asbestos or other acicular mineral morphologies, but these are rarely encountered in commercial solution mining systems.

The spherical approximation simplifies a mathematical description of transport processes within the rock, using one-dimensional polar coordinates. Fracture patterns in an ore body useful for *in situ* leaching often do not have a preferred orientation and can be similarly approximated. However, the ore mass will have a wide **distribution** of rock sizes from giant boulders (in the case of run-of-mine ore and mine subsidence) to very fine particles. Crushing prior to heap leaching eliminates the boulders and leaves a smaller maximum size, typically about 20 mm, but the rock size distribution will remain wide, with the fractional mineral extraction from the small particles occurring much more rapidly. This situation is usually modeled by dividing the ore mass into a histogram of several (often a dozen) discrete rock sizes, with each size described and simulated by a separate numerical subroutine.

Conventional metallurgical and mineral extraction processes are normally conducted using ground ore to accelerate reactions within process vessels of limited size with reaction times measured in seconds to minutes, and occasionally a few hours. Solution mining processes usually occur over much longer periods. Rather than a few tonnes of material in the reactor, many thousands and even millions of tonnes are typically involved. With solution mining, the mineral dissolution process (commonly a heterogeneous chemical reaction between the solution and mineral grain) and the internal rock pore diffusion process are usually slower than transport processes outside individual rocks. Therefore, these two slow-step processes tend to govern the overall rate of extraction.

Consider our final sugar example: a simplified case of a **small amount** of powdered sugar disseminated in a porous sand sphere that is submerged in water. As water enters the micropores of the sand sphere, the small amount of very fine sugar quickly dissolves. Because a limited amount of sugar was used, all of it dissolves without exceeding the solubility limit of the water contained within the rock micropores (pore water); this is our definition of "small amount." Under these restrictive circumstances, the extraction rate is dependent only on diffusion of the sugar solute through the tortuous pore paths within the sphere.

Thus, the leaching extraction model consists of the diffusion equation for a porous sphere with c the local concentration of solute in the pore fluid:

$$\varepsilon \frac{\partial c}{\partial t} = D_{eff} \left(\frac{\partial^2 c}{\partial r^2} + \frac{2}{r} \frac{\partial c}{\partial r} \right). \tag{1.1}$$

The diffusivity, D, will be that for the sugar solute in water. Since diffusivities for dissolved species in water do not vary much, a reasonable approximation at room temperature is $D = 1.75 \times 10^{-3} \, mm^2/s$ ($1.5 \, cm^2/day$). The diffusivity of the sphere must be modified by the fractional **microporosity**, ε, of the sphere (internal fractional volume of water filled open micropores), and by a factor, τ, to account for the **tortuosity** of the micropores and the resulting greater diffusion path length. Thus, an **effective diffusivity** results where

$$D_{eff} \equiv \frac{D\varepsilon}{\tau}.$$

The rock interporosity is sometimes referred to in the hydrogeological literature as secondary porosity to distinguish it from the intraporosity between rocks in a clastic sediment, which is referred to as the primary porosity. An ore heap can be characterized as a clastic sediment. In this text, with respect to ore heaps, the word "microporosity" is restricted to the interporosity of the rocks, while the intraporosity is termed the "void space," or, on occasion, the macroporosity.

Comparison with experimental and calculated results for diffusion in porous media usually results in a tortuosity factor of $\tau = 2$. Note that one unit of time used to express the effective diffusivity is days. Solution mining is a slow process measured in weeks, months and sometimes years. Periods less than a day are hardly relevant. With standard values for the diffusivity and tortuosity (used throughout this text unless otherwise stated) the effective diffusivity is:

$$D_{eff} = 8.7 \times 10^{-4}(\varepsilon) \quad [\text{mm}^2/\text{s}],$$

$$D_{eff} = 0.75 \times 10^{-4}(\varepsilon) \quad [\text{cm}^2/\text{d}].$$

For initial and boundary conditions of a negligible sugar concentration outside the sphere and a uniform initial concentration of dissolved sugar within micropores of the sphere, c_i, the diffusion equation has been solved (Crank, 1956) to yield the following expression:

$$F_{t,r_0} = 1 - \frac{6}{\pi^2} \sum_{n=1}^{\infty} \frac{1}{n^2} \exp\left(\frac{-D_{eff}\, n^2 \pi^2 t}{r_0^2}\right), \tag{1.2}$$

where F_{t,r_0} is the fractional extraction of sugar in time t for a porous sand sphere of radius r_0. This infinite series can be adequately approximated as a finite series for computer solution, since the terms in the series quickly become negligible as the index, n, increases. **Figure 1.1 is a dimensionless plot of the relationships expressed by eqn 1.2.**

EXAMPLE PROBLEM

Uniformly sized porous spherical particles containing sugar were leached over a period of one full day, 24 h and 60% of the sugar was extracted. If the leaching time is increased to 5 days, predict the percent of sugar that will be extracted using Fig. 1.1 ... and without using eqn 1.2.

ANSWER
(1) Find the value of tD_{eff}/r_0^2 at 60% extraction using Fig. 1.1 and a straight edge or ruler:

$$tD_{eff}/r_0^2 = 0.04.$$

(2) If we increase the time by a factor of five, we must increase tD_{eff}/r_0^2, by a factor of five, hence:

$$tD_{eff}/r_0^2 = 0.20.$$

(3) Again using Fig. 1.1 and a straight edge, find the percent extraction:

$$\text{Answer} = 92\%$$

Note that several parameters needed to solve the problem with eqn 1.2 were not given (rock microporosity and rock radius), yet with Fig. 1.1, you can solve the practical problem. While you don't know the values of microporosity and radius, you can, with the given data coupled with Fig. 1.1 defines the value of D_{eff}/r_0^2. Whenever you have a set of conditions definable by Fig. 1.1, you can change a variable, such as time in this problem, and find the complimentary variable. You can change more than one variable if you know the initial values of those variables. For example, you could change both sphere microporosity and sphere radius to new values and compute its effect.

Continuing with this problem, what is the required time to obtain 60% extraction if the sphere radius is tripled. Answer: Because of the radius squared effect D_{eff}/r_0^2 will be $1/(3)^2$ times its former value (i.e., one-ninth). Therefore the time to obtain the *same* percent extraction must increase from one day to nine days.

The variables, time, percent extraction and sphere size are easy to measure in a laboratory experiment. Microporosity, and therefore D_{eff} are more difficult, but not impossible. Normally in doing laboratory work it is most useful to scale from one set of conditions to another using Fig. 1.1.

Recognize that the fraction reacted, F_t, is a function of $1/r_0^2$. Hence, doubling the rock radius, r_0, increases the time to obtain the same fraction extracted by four.

Figure 1.1 is a **dimensionless parametric process design curve** for a fairly simple leaching system. Dimensionless parametric design curves for more complex ore leaching systems will be introduced later in this text.

If desired, concentration profiles of sugar within the sphere can be computed by solving the diffusion equation for the initial and boundary conditions (Crank, 1956),

$$\frac{c_i - c}{c_i} = 1 + \frac{2r_0}{\pi r} \sum_{n=1}^{\infty} \left(\frac{(-1)^n}{n} \sin\left(\frac{n\pi r}{r_0} \right) \right) \exp - \left(\frac{D_{eff} n^2 \pi^2 t}{r_0^2} \right). \quad (1.3)$$

Figure 1.3 schematically shows the sugar concentration within the porous sphere under four conditions: (a) before the sugar containing sphere is submerged into the water (no dissolution of sugar), (b) immediately after submergence and water entry into the micropores causing "instantaneous" dissolution [*time* $= t_0$], (c) a later time interval t^+, and (d) a still later time interval t^{++}.

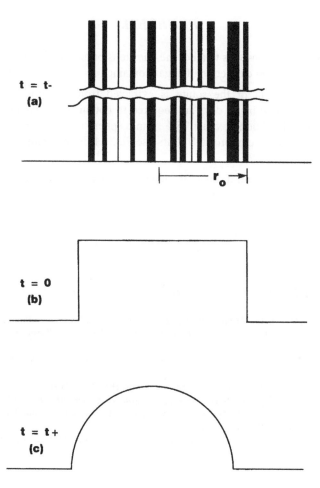

Figure 1.3. Concentration profiles (schematic) across a porous sphere at four time intervals, before and during extraction of a solute.

PROBLEMS

1. Provide your answers to all of the questions posed in this chapter on the dissolution and extraction of sugar contained in spherical solid bodies.

2. Ninety percent of the solute is extracted from a porous sphere in 10 days. How many days are required to obtain only 75% extraction? What is the time to extract 75% from a sphere of the same size with a 25% greater microporosity? Hint: Use Fig. 1.1 directly rather than cumbersome eqn 1.2.

3. Estimate the **initial** internal sphere microporosity in the preceding problem, with 10 days to extract 90%, if the sphere diameter is one inch (2.54 cm).

4. Estimate the time to obtain 98% extraction of sugar for a sphere with the same microporosity if its diameter is increased to four inches (10.16 cm).

REFERENCES AND SUGGESTED FURTHER READING

Crank, J. (1956). *The Mathematics of Diffusion*, Oxford Press, London, Chap. 6.
Habashi, F. (1980). *Extractive Metallurgy, Vol 2, Hydrometallurgy*, Gordon and Breach, New York.
Mineral Information Office, U.S. Bureau of Mines, *Mineral Commodity Summaries, 1990*.
Rossi, G. (1990). *Biohydrometallurgy*, McGraw-Hill, New York.
Schlitt, W.J. (1992). *SME Mining Engineering Handbook*, 2nd Edition, Hartman, H.L. (Ed.), Society for Mining, Metallurgy and Exploration, Inc., Littleton, CO, Section 15.
Van Arsdale, G.D. (1963). *Hydrometallurgy of Base Metals*, McGraw-Hill, New York.

TWO

Heap Leaching
Gold (Silver) Ore
—Theory

Enough of sugar; does an important mineral leaching extraction system correspond to the simple example of the previous chapter? Yes; and it's a more exciting commodity—**GOLD**.

DISSEMINATED GOLD ORE DEPOSITS

Disseminated gold deposits, sometimes referred to as Carlin-type deposits, are the basis for most of the gold mining industry in the United States. When oxidized by nature, these ore deposits are readily heap leached; the gold is present as free gold particles, one micron (10^{-6} m) or smaller, that are generally not visible except by scanning electron microscopy. The gold, present in concentrations often less than 3 ppm, usually is hosted in sedimentary rocks. Rock microporosity is usually low but open, and with the exception of some highly silicified rocks, the gold is accessible through micropores and microfractures. However, there are jasperoids of acceptable gold grade that cannot be mined and heap leached because of their low open microporosity. This is caused by a high degree silicification that has occurred after or concurrent with gold deposition and blocks the micropores. These ore deposits are usually associated with ancient hot springs.

Much of the disseminated gold is believed to have been originally deposited as blebs occluded within pyrite or arsenopyrite grains and, therefore, not originally accessible to dissolution by cyanide solutions. However, many of the shallow ores are now above the water table and were geochemically oxidized, after metal deposition, to alter the sulfide minerals and render the gold leachable. The unoxidized ores containing sulfides may also contain pyrobitumen, a residue of thermally mature petroleum, or other organic material, indicating deposition under reducing environments. These unoxidized ores are almost invariably refractory to cyanide leaching.

Most of these disseminated sedimentary deposits contain silver, which is sometimes the major economic metal. Heap leaching technology for silver is essentially the same as for gold.

Heap leaching is not useful for all gold ore deposits. Gold particles are dissolved very slowly and heap leaching does not work well on placer deposits because the gold grains are too large. Sulfide and arsenides occlude the gold particles so that cyanide cannot reach them. Organic material in ore can be **a sink for** adsorbing the gold cyanide complex. Consequently, in these carbonaceous ores, the gold is solubilized but not removed from the ore mass. Gold present as tellurides is less common but very difficult to leach with cyanide solutions. High cyanide losses and solution fouling can occur when significant amounts of copper, arsenic, iron and other **cyanicides** are present.

All ores should be evaluated for cyanide gold extraction and cyanide consumption in bottle roll or agitation leaching tests. If these tests are successful, follow them with column testing to simulate heap leaching behavior before proceeding with a new heap leaching project. Nevertheless, the following theoretical discussion is a useful adjunct to ore testing and leaching process understanding.

GOLD LEACHING CHEMISTRY

Gold particles are oxidized and solubilized in the presence of basic cyanide solutions:

$$4Au + 8CN^- + O_2 + 2H_2O = 4Au(CN)_2^- + 4OH^-. \qquad (2.1)$$

This is an electrochemical oxidation reaction that proceeds only because of the great stability of the gold cyanide complex. Oxygen is required for gold dissolution and supplied by air sparging in the leaching of milled ore.

But, the air present in percolation leached heaps is usually adequate because of the small amount of gold present and the limited need for oxygen. Unlike copper sulfide mineral leaching (discussed in a later chapter), the oxygen dissolved in the initial cyanide solution wetting the ore and penetrating rock micropores at saturation is usually sufficient to dissolve all of the gold in the rock. Computations by Wadsworth (1996) based on only the dissolved oxygen at 7 ppm in the pore solution for a Carlin type ore with a measured internal microporosity of 0.036 (3.6 vol pct) indicate a gold dissolution capacity of about 0.07 opt (2 g/tonne); see Fig. 2.1. Furthermore, residual oxygen in the air within a heap, assuming 20 vol pct air void space after solution soaking, is at least two orders of magnitude greater than the amount of oxygen required to dissolve all of the gold typically present in heap leached gold ores.

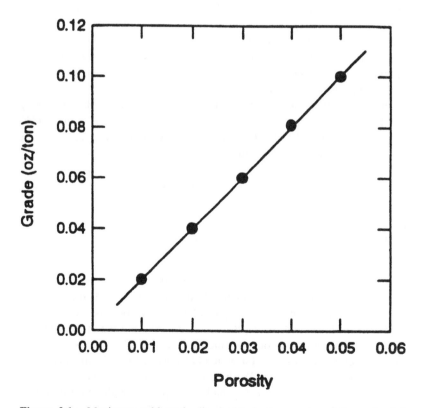

Figure 2.1. Maximum gold grade dissolvable by oxygen in solution saturated micropores of a gold ore (Wadsworth, 1996).

However, the presence of small amounts of sulfide minerals and organic material can compete with gold for the oxygen in some ores. Consequently, on occasion low residual levels of dissolved oxygen in the pregnant liquor draining from the heap will be encountered, and the lower oxygen concentration in the leach solution will retard the rate of gold extraction from the ore heap. When oxygen must diffuse into the micropores from outside the ore rocks, there is a small retardation on the gold cyanide complex diffusion out of the rock, which is normally the rate limiting step and described in the next section.

Wadsworth (1996) also examined limitations on cyanide diffusion into the solution-filled rock micropores for a case where the micropores of a wet ore are initially filled with water without cyanide and later exposed to the cyanide solution when heap leaching begins. Because concentrations of cyanide in the leaching solution, typically 50–200 ppm, are much higher than the oxygen saturation limit of 7 ppm, this process is much faster and would account for gold dissolution in hours rather than many days. In a sequence of transport and chemical reaction steps the slowest step will be the "bottleneck" and is the step that limits the overall rate. Because cyanide diffusive transport is a faster process, due to its much higher concentration, it is not rate limiting.

Further details of gold hydrometallurgy are covered by Burkin (1966) and Dorr (1950). Practice improvements are covered by Kudryk (1984) and Osseo-Asare and Miller (1982). The chemistry of silver leaching with cyanide is similar to that of gold:

$$4Ag + 8CN^- + O_2 + 2H_2O = 4Ag(CN)_2^- + 4OH^-. \tag{2.2}$$

MODELING DISSEMINATED GOLD ORE HEAP LEACHING EXTRACTIONS

Fortunately, water is the wetting fluid for nearly all rock minerals. During heap leaching, cyanide solutions trickle down over the wetted rocks with solution entering the micropores by capillary action. Because the gold particles are very fine and sparse, the accessible gold located in open micropores and fissures of the rocks is quickly dissolved. Once the soluble gold diffuses out of the rock, it is fairly quickly washed from shallow ore heaps that have good permeability. Washing ore heaps will be treated more thoroughly in a later chapter, but for shallow heaps with high uniform permeability this process is mostly completed in a very few days. Thus, fractional extraction, F_t, is reasonably approximated by the diffusion component of the process, eqn 1.2 (Bartlett, 1974). However, it is worth emphasizing that this is an estimate of the **fraction** extracted based on only the total extractable metal, which is the free gold accessible to the leaching solution in micropores of the rocks. Rarely is all of the precious metal contained in ore rocks extractable. Often gold has a higher extractability (percentage) than does silver in the same ore because, in part, silver more easily forms refractory mineral compounds.

Assuming a negligible concentration outside the rock because of efficient washing, the estimated soluble gold concentration profiles in a 20 mm diameter rock, calculated from eqn 1.3, are shown in Fig. 2.2 for various times after leaching has started. Calculations for this specific example were based on an assumed rock microporosity of five percent, $\varepsilon = 0.05$. Concentration profiles will be lower, at the same leaching time interval, for smaller rocks.

The time-dependent fractions extracted for 20 mm through 200 mm diameter rocks, each with five percent microporosity, are compared in Fig. 2.3, based on eqn 1.2. Rocks larger than 50 mm show insufficient extraction in the time covered by this figure, four weeks. Uncrushed ore from an open pit mine will usually contain a significant fraction of rocks greater than 200 mm, and a plot of fractional extraction versus log time, shown in Fig. 2.4, amply demonstrates that much longer times are needed to complete extraction from the bigger rocks.

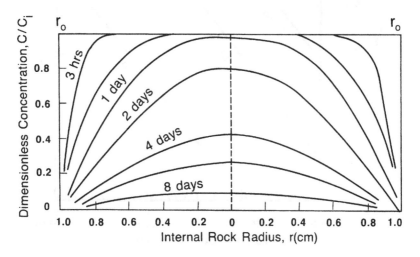

Figure 2.2. Soluble gold concentration profiles in a 20 mm diameter rock with 5% microporosity at times shown after beginning of heap leaching; eqn 1.3.

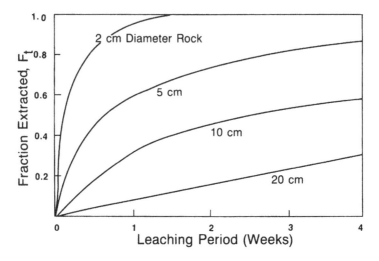

Figure 2.3. Time dependent fraction extracted for heap leaching gold from monosize rocks with 5% microporosity and the rock diameters shown; eqn 1.2.

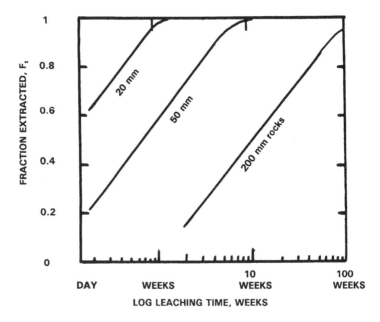

Figure 2.4. Fractional extraction versus log time for selected rock diameters; eqn 1.2.

Figures 2.2 and 2.3 indicate that for reasonably porous ore crushed to $-20\,mm$, most of the gold will be extracted in one or two weeks of heap leaching.

A lowered microporosity can increase the required time for leaching, even when all of the gold is **open pore accessible**. For most mineral deposits, as the rock microporosity declines increasing amounts of precious metal are locked away and ultimate extractions also decline. Reduced microporosity adversely affects fractional extraction. Results for 20 mm rocks are shown in Fig. 2.5.

Dividing the crushed ore size distribution into a few discrete screen sizes, each weighted by its mass fraction of the total, and with extraction calculations using eqn 1.2 performed on each group yields the time dependent fractional extraction, F_t, for the ore heap:

$$F_t = \sum_{y=1}^{Y} F_{t,y}(N_y), \tag{2.3}$$

$$1 = \sum_{y=1}^{Y} N_y, \tag{2.4}$$

where N_y is the mass fraction of rock group with size r_0 indexed to y.

The fractional extraction as a function of leaching days (time from first appearance of leach solutions draining from the heap) was estimated for

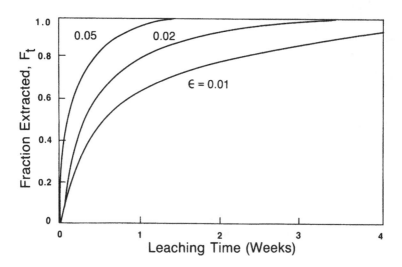

Figure 2.5. Effect of rock microporosity on estimated fractional gold extraction from 20 mm diameter ore rocks, eqn 1.2.

a **distribution** of rock sizes, using the preceding equations, eqn 1.2, and the largest rock size in each size range. The calculated results for a specific case are shown in Fig. 2.6. Extraction is faster for the size distributed rocks than for 20 mm monosize rocks because of the smaller rocks that are present. This conforms with gold ore heap leaching observations, at least when the heaps are not too high. In practice, a few days to soak the cyanide solution into the ore mass (while simultaneously dissolving most of the gold in place), a week or two for diffusion, and a final few days for washing soluble gold through the low permeability areas of the heap

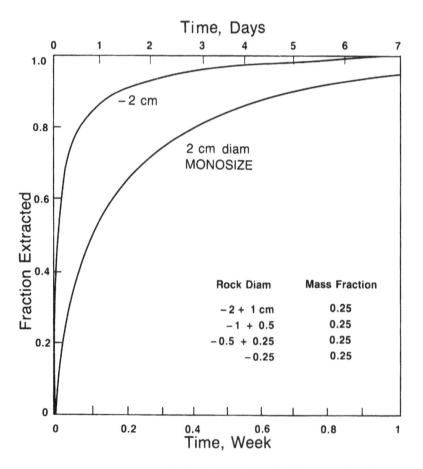

Figure 2.6. Time dependent fractional extraction of gold by heap leaching ore with 5% microporosity and the indicated − 20 mm rock size distribution.

should do nicely for shallow ore heaps. When crushed ore is emplaced, heap leached and removed from a reusable impervious leaching pad, the duration (forecast by Fig. 2.4), shows good agreement between this simple leaching model and heap leaching practice at gold and silver mines. Leaching periods of one week to several weeks are usually used with crushed ore, including crushed ore that has been agglomerated. Tighter ores (low microporosity) require longer periods and, as mentioned, some highly silicified jasperoids cannot be heap leached.

Run-of-mine ore, with rock fragment sizes much greater than 20 mm, will require much longer periods. When gold heap leaching run-of-mine ore is practiced, the ore is usually permanently stacked, with a series of layers, sometimes referred to as "lifts." Each lift is leached before the next **lift** is emplaced. Extraction from the larger rocks in buried lifts will continue as the leaching solution trickles down over many months and years.

ESTIMATING FRACTIONAL EXTRACTION OF DISSOLVED SOLUTE FROM A DISTRIBUTION OF ORE ROCK PARTICLE SIZES USING A DIMENSIONLESS EXTRACTION CURVE

There have been several studies of the particle size distribution resulting from crushing ores and other brittle solids. It has been observed that a logarithmic plot of the cumulative fraction of material finer than a given size versus the corresponding size is usually a straight line with a slope, m, of one, or slightly less. Different, but similar equations have been used to express this relationship, including the well established Gates–Gaudin–Schuhmann (GGS) equation (Schuhmann, 1960):

$$Y(r_i) = (r_i/r_*)^m \tag{2.5}$$

with: $0.7 < m \leqslant 1.0$, where $Y(r_i)$ is the cumulative fraction finer than size r_i and r_* is the radius of the largest rock in the broken ore.

The GGS equation has been coupled with the diffusion equation (eqn 1.1) to develop a numerical relationship for a distribution of particle sizes that is analogous to the dimensionless analytical relationship for monosize particles that was plotted in Fig. 1.1 (Bartlett, 1971). This leads to a relationship in which F_t depends on $tD_{eff}/(r_*)^2$. However a band in the plot of F_t occurs because of the variation in the Gates–Gaudin–Schuhmann breaking function, m, in eqn 2.5, with the boundaries of the band corresponding to the extremes in the breaking function, viz. $m = 1$ and $m = 0.7$. While the value of m can be determined from a log plot of the sieve analysis data, an additional operational problem and error may arise in estimating the largest particle radius, r_*, by extrapolation from a sieve analysis of the ore sample. It is often easier and more precise to use a sieve analysis to accurately determine a particle diameter for which the undersize mass of particles is equal to a percentage of the total ore sample mass, but less than 100%. Commonly, the ore sieve analysis is expressed this way using a d_{80}, which is the rock diameter at which the undersize particles comprise 80% of the total sample mass. With respect to the Gates–Gaudin–Schuhmann distribution when the breaking function is $m = 1$, then from eqn 2.5, $d_{80}/2 = 0.8r_*$. However, when $m = 0.7$ using eqn 2.5, $d_{80}/2 = 0.727r_*$, which is only about 91% of the d_{80} at $m = 1$. This results because there are proportionately more fine particles and the d_{80} shifts to a smaller size as the breaking function shifts from $m = 1$ to $m = 0.7$, while both are at the same maximum particle size.

Figure 2.7 is a dimensionless plot of the percent extraction of metal, which is $100F_t$ versus $tD_{eff}/(r_*)^2$ when $m = 1$. Clearly, at $m = 1$, $d_{80}/1.6$ can be substituted for r_*:

$$r_* \equiv d_{80}/1.6. \qquad (2.6)$$

Fortuitously, and most important, when $d_{80}/1.6$ is *defined* as r_* at other values of the breakage function within the range $0.7 < m \leqslant 1$, and compared with the correct extraction curves computed (Bartlett, 1971) there is a fairly good correlation. It is adequate to the imprecision of sampling ore and characterizing all of the ore, including r_*, from the sieve analysis obtained from the sample.

The definition of eqn 2.6 provides a convenient and practical method of analyzing the extent of metal extraction over time, using the single design curve in Fig. 2.7, with either r_* or d_{80} obtained from an ore sample sieve analysis characterization. Be cautioned that selecting either greater or lower characterizing ore particle diameters than d_{80} leads to distortion from the correct results at $m < 1$, which becomes significant as the selected characterizing size variation from d_{80} increases.

Figure 2.7 is directly useful in estimating fractional extraction, F_t, as a function of leaching time for any value of d_{80} and $0.7 < m \leqslant 1.0$, if the effective diffusivity is known. However, an accurate value of the internal microporosity, ε, is needed for D_{eff}, and this is difficult to obtain. While internal rock microporosity measurements can be made on small samples, it is difficult to determine if they are representative without making a very large number of expensive measurements. This is essentially a rock internal microporosity sampling problem.

Nevertheless, Fig. 2.7 can still be very useful, because it can first be used to determine an empirical value of D_{eff}, for example from the results

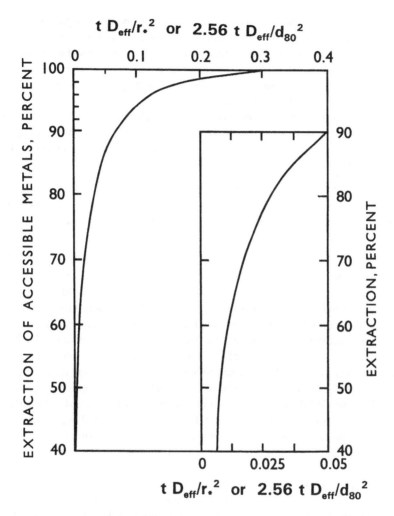

Figure 2.7. Fractional extraction design curve for completely dissolved solute from a multisize aggregation of rocks, as a function of effective diffusivity, D_{eff}, and the d_{80} or r_* in the rock size distribution.

of column ore leaching tests. This is done by comparing an experimental fractional extraction obtained at the corresponding leaching time. Ideally, several experiments will be performed to obtain a best value of D_{eff} by averaging the results:

$$D_{\text{eff}} = [\text{Abs}(F_t)] \, [(d_{80}/1.6)^2/t_{\text{L}}], \qquad (2.7)$$

where Abs(F_t) is the abscissa value obtained from the design curve of Fig. 2.7 that corresponds to the fraction extracted, F_t, at leaching time, t_L.

After a best value of D_{eff} has been determined, it can be used to estimate the time required to obtain a desired fractional metal extraction, again using Fig. 2.7. Conversely, for a specified leaching time, the percent of metal extracted can be estimated.

Furthermore, with a reliable value of D_{eff} determined, a change in the ore size distribution, e.g., from more or less crushing to a new value of d_{80}, can also be accommodated with these relationships to compute its leaching extractions with time.

Caution: The dimensionless group in the abscissa (*X*-axis) of Fig. 2.7, $tD_{eff}/(d_{80}/1.6)^2$, must be in consistent units. For example if D_{eff} is given in m^2/s, then the characterizing rock size, d_{80}, must be in meters and the time must be in seconds.

There are two principal problems associated with the use of Fig. 2.7 to determine the leaching rate characteristics of gold and silver ore, or any other ore where transport by diffusion after rapid dissolution of *all* of the mineral occurs governs the rate of mineral extraction. First, some of the mineral may be locked in plugged micropores and unaccessible to the leaching solution at the planned crush size; the gold is not adequately liberated. Second, a finite amount of time is required for the leaching solution to fully penetrate the internal micropores of the rocks and dissolve the gold, causing an induction period or delay before soluble gold appears in the leachate. While the rate of gold dissolution of micron sized gold particles is relatively fast, hours to several days will be required depending on the gold particle size. This can extend to much longer times, including weeks and months if nuggets are present. When gold particles larger than a few microns are in the ore they are tantamount to not fully accessible gold within the time of usual heap leaching cycles.

Figure 2.8 displays extraction versus leaching time for a large gold ore sample crushed to −2.5 cm (−1 in), thoroughly blended and divided into fifteen equal portions each of which was used in a cyanide column leaching experiment. The gold in this hypothetical ore is not visible under a microscope. Five leaching periods were used: 4, 8, 9, 12 and 17 days, with triplicate column experiments run at each period. The range of experimental results at each time period is shown in the graph.

Clearly, these results do not match an overlay of Fig. 2.7. However, if Fig. 2.8 is moved to the right and down, a very good fit results, as is shown in Fig. 2.9. This corrected fit can be explained by the following assumptions: First a two day induction period is required for lixiviant

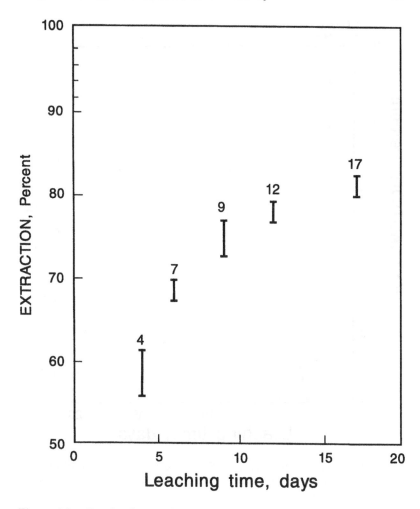

Figure 2.8. Fractional extraction of gold from triplicate cyanide column ore leaching experiments carried out for five time periods.

penetration into the ore and gold particle dissolution, which is reasonable, and second, apparently only about 82% of the gold contained in the ore is **accessible** to the lixiviant after crushing to -2.5 cm. The remaining gold is presumably locked in micropores and fissures that are sealed. With this corrected fit it is possible to make an estimate of the average internal microporosity in the ore and its effective diffusivity.

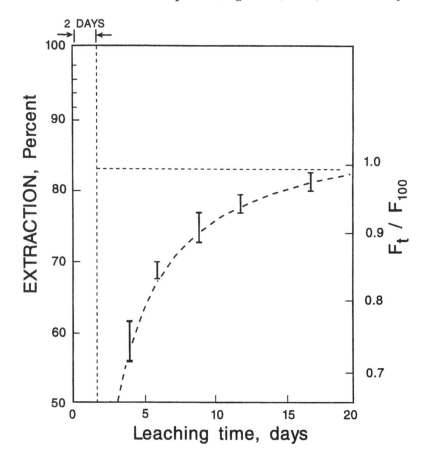

Figure 2.9. Matching the data from Fig. 2.8 to the design curve of Fig. 2.7, using a two-day induction period and limiting the accessible mineral to 82% of the total mineral present in the ore.

These data also suggest a further investigation toward improving leaching performance for this ore. For example, will crushing finer than $-2.5\,\text{cm}$ raise the ultimate possible extraction above 82%?

VALUABLE MINERAL ACCESSIBILITY TO THE LIXIVIANT AND EXTRACTION

The importance of valuable mineral accessibility to the leach solution cannot be overemphasized. The accessibility, or availability, of the mineral

Table 2.1 Silver Extraction by Column
Leaching at Various Crush Sizes for a
Bolivian Silver Ore Stockpile.

Ore Size	Extraction
−75 mm	15%
−25 mm	44%
−6 mm	64%
−2.36 mm (8 Mesh)	66%
−600 μm	84%
−150 μm	87%

being extracted from the rock to the leaching solution is similar to the concept of liberation of minerals prior to separation (concentration) in mineral
processing operations such as flotation. Mineral accessibility often
increases as rock sizes decrease. An extreme example of this is shown in
Table 2.1, which is a summary of column leaching experiments on a
Bolivian silver ore stockpile after crushing to various top sizes. These test
results dictated grinding the ore to 600 μm (28 mesh) followed by ore
agglomeration before heap leaching. Even then, the late stage leaching rate
was very slow and a long leach cycle of 120 days was selected for this
heap leaching operation.

Other ore characteristics that often affect valuable mineral accessibility
include ore grade, host rock, and the other minerals present in the ore mass.

ROCK PARTICLE LEACHING SIZE
VERSUS SIEVE PASSING SIZE

In laboratory ore testing practice, aggregates of monosize particles don't
exist. The closest approximation is an aggregate of particles passing one
sieve but not passing the next smaller size sieve. The particles will vary in
size between the two sieve aperture sizes and it is possible to choose the
mean of these to represent the "monosize" particle aggregation. However,
leaching transport rates depend on the rock radius squared. Hence for purposes of solution mining it is advisable to use an average of the apertures
squared for the squared average aperture,

$$d^2_{a(avg)} = \frac{1}{2}\,(d^2_{a(H)} + d^2_{a(L)}). \tag{2.8}$$

For the standard sieve sizes, where the next larger sieve aperture is $2^{1/2}$ greater, this leads to

$$d_{a(avg)} = 0.866d_{a(H)}, \tag{2.9}$$

which is slightly larger than the simple mean of the two separating sieve apertures. Computed results of these values are tabulated, with standard sieve sizes, in Appendix Table A-2. This approach can also be used for crushed rock that has been passed through a screen or grizzly. The use of d_{80} in conjunction with the dimensionless design curves of Fig. 2.7 is preferred because it automatically accounts for variations in the rock breaking function. However, it is often difficult to determine a d_{80} for coarse crushed ore and run-of-mine ore. When d_{80} is not available, but the ore has passed a grizzly or screen so that $d_{a(avg)}$ can be computed from the screen size using eqn 2.9, then the following relation is recommended as the best available estimate for use with Fig. 2.7:

$$d_{80} = 0.8d_{a(avg)} \tag{2.10}$$

and

$$d_{80} = 0.693d_{a(H)}. \tag{2.11}$$

Furthermore, if an estimate of r_* is needed, it can be obtained from eqns 2.6 and 2.11 yielding,

$$r_* = 0.433d_{a(H)}. \tag{2.12}$$

EXAMPLE PROBLEM I

Leaching tests of a gold ore crushed to -10 mesh showed that the maximum extraction over an extended period was 90%. Multiplicate column tests of the ore crushed and screened to pass 1-1/2 in yielded an average extraction of 63% in 7 days. Using Fig. 2.7, estimate the leaching time required for this crushed and screened ore to obtain 85% extraction?

ANSWER
(1) Because only 90% of the gold is extractable (10% is locked), the extraction of *accessible* gold in 7 days is:

$$63\%/0.90 = 70\%$$

and at 85% total extraction the extraction of accessible gold is:

$$85\%/0.9 = 94.5\%$$

(2) Using Fig. 2.7, the value of $tD_{eff}/(1.6/d_{80})^2$ at 70% extraction is 0.016.
(3) Using Fig. 2.7 the value of $tD_{eff}/(1.6/d_{80})^2$ at 94.5% extraction is 0.125.
(4) Because the values of D_{eff} and d_{80} have not changed, the leaching time must increase from 7 days, and do so in proportion to the increase in the value of $tD_{eff}/(1.6/d_{80})^2$; hence:

$$t = 7 \text{ days } (0.125/0.016) = 54.7 \text{ days.}$$

EXAMPLE PROBLEM II

For the given data of the preceding problem, estimate the value of the effective diffusivity and microporosity of this ore.

ANSWER

(1) First, we must determine r_*, and since we don't know d_{80}, we will estimate r_* from eqn 2.12:

$$r_* = \frac{1}{2} d_{a(avg)} = 0.433 \, d_{a(H)} = 0.433(1 \text{-} 1/2 \text{ in}) = 0.65 \text{ in}$$

$$r_* = 16.5 \text{ mm} = 1.65 \text{ cm}$$

$$r_*^2 = 2.72 \text{ cm}^2$$

(2) for the given data ($t = 7$ d and $F_t = 0.70$) and this value of r_*,

$$r_*^2/t = 2.72/7 \text{ cm}^2/d = 0.389 \text{ cm}^2/d$$

and from the previous problem, $tD_{eff}/(r_*)^2$ is 0.016 then D_{eff} is:

$$D_{eff} = [tD_{eff}/r_*^2] \, (r_*^2/t) = 0.016 \times 0.389$$

$$D_{eff} = 6.2 \times 10^{-3} \text{ cm}^2/d$$

However, for aqueous solutions,

$$D_{eff} = 0.75(\varepsilon) \text{ cm}^2/d$$

Hence, $\varepsilon = 6.2 \times 10^{-3}/0.75$

$$\varepsilon = 0.0083$$

$$\varepsilon = 0.83 \text{ percent.}$$

MEASURING INTERNAL ROCK MICROPOROSITY, ε

Internal rock microporosity is measured by comparing the mass of a dried rock sample with the same sample first dried and then saturated with an injected liquid. Drying is conducted at a modest temperature to prevent removal of chemically bound water, e.g. 105°C for 48 hr or more. If total microporosity is sought then pressure injection of mercury using a mercury porosimeter can be used. Pressure forces the mercury into all of the open micropores regardless of pore size. For purposes of heap leaching it is easier and more valid to soak a previously dried sample in water for an extended period of time, typically at least 48 hr. Micropores not filled by this procedure are generally too small to be involved in the leaching process. The volume of the rock sample, including its internal microporosity, is determined from the volume of water displaced upon submergence of the blot-dried but soaked sample. The water soaked sample is blot-dried and immediately weighed. Its mass is compared with the dry sample mass to obtain the mass and thereby the volume of internal water contained in the sample. Comparing the two volumes yields the fractional volumetric microporosity, ε.

PROBLEMS

1. Estimate and plot on graph paper Fraction extracted, F_t, versus weeks of leaching time for solubilized gold in a disseminated ore with 0.03 microporosity for each of the following monosize ore fragments:
   (a) 2 mm diam. rock,
   (b) 20 mm diam. rock, and
   (c) 80 mm diam. rock.

2. On/off heap leaching of crushed ore with a reusable pad at the New Ophir Mine is providing an average gold extraction that is 80% of the accessible gold in the ore. Using Fig. 2.7, estimate the extraction of accessible gold that will be obtained if the leaching time is *doubled*.

3. For an ore with a d_{80} of 2.10 in, recommend a leaching time to obtain 90% extraction. The internal microporosity is unknown but not expected to be less than 0.005 (0.5%).

4. Using the data presented in Figs. 2.8 and 2.9, estimate the average value of the internal rock microporosity in this ore, which was repeatedly crushed and sieved until all of it passed a one inch *nominal* sieve.

5. An on/off gold heap leaching operation using two crushing stages to produce rock that passes through a two inch screen (nominal) obtains 55% extraction of the accessible gold in the ore. (a) Using Fig. 2.7 and Table A-2, predict the percent extraction of accessible gold that will be obtained if a third crushing stage and larger screen decks are installed

to produce ore passing through a 1 in (nominal) screen. (b) As a safety factor recalculate the estimated percent extraction of accessible gold if the largest rock size is equal to the size of the one inch screen aperture.

6. Estimate the ore specific gravity and microporosity from the following experimental data. The crushed ore was wet screened to remove −10 mesh fines and dried. Then, a dry sample weighing 213.0 g was submerged into a large graduated cylinder containing water, with a corresponding increase in the water level of 81.0 ml after all air bubbles had been released. After 72 hr, the contents of the graduated cylinder were washed out, filtered with a large buchner funnel and allowed to blot-dry by sucking air through the filter cake. The filter cake weight, including the filter paper weighing 1.0 g, was 216.7 g.

REFERENCES AND SUGGESTED FURTHER READING

Bartlett, R.W. (1971). Pore diffusion-limited metallurgical extraction from ground ore particles. *Met. Trans.* 2, pp. 2999–3006.

Bartlett, R.W. (1974). Application of diffusion models in estimating heap leach extraction. *SME Preprint 74-AS-351*, SME Fall Meeting.

Burkin, A.R. (1966). *The Chemistry of Hydrometallurgical Processes*, D. Van Nostrand, Princeton, NJ.

Dorr, J.V.N. (1950). *Cyanidation and Concentration of Gold and Silver Ores*, McGraw-Hill, New York.

Kudryk, V. et al. (Eds.) (1984). *Precious Metals: Mining, Extraction and Processing*, The Metallurgical Society of AIME, Warrendale, PA.

Osseo-Asare, K. and Miller, J.D. (Eds) (1982). *Hydrometallurgy Research, Development and Plant Practice*, The Metallurgical Society of AIME, Warrendale, PA.

Schuhmann, R. (1960). *Trans. SME-AIME*, 217, pp. 22–25.

Wadsworth, M.E. (1996). Advances in gold and silver leaching practice—chemical and physical factors. *Proceedings of XIX Int'l Mineral Processing Congress*, Vol. 2. SME, Littleton, CO, pp. 3–8.

THREE

Heap Leaching Practice (With Applications to Gold and Silver Ores)

While the applications presented in this chapter are on gold and silver ores, most of these heap leaching practices are also directly relevant to the heap leaching of copper and other base metal ores. Information presented in this chapter will not be repeated in subsequent chapters on copper ore heap leaching.

OVERVIEW

The primary advantages of heap leaching are low capital cost and low operating cost when compared with the available alternative, agitation cyanide leaching of milled ore. Many mines are equipped with heap leaching for the low grade ore and milling-agitation leaching for the high grade ore. In many instances heap leaching was installed later to treat sub-economic rock or mine overburden. The disadvantages of heap leaching are less certain forecasting of metal recovery and in most instances lower recovery. In a few instances higher recoveries have been claimed for heap leaching than agitation leaching because of the much longer time required for heap leaching.

Much of this chapter and several figures were derived from Van Zyl et al. (1988). A general layout of a heap leaching operation is shown in

Fig. 3.1. Typically, the ore heap is placed on a moderately sloping (2–6 degrees) impervious plane, called the **leaching pad**, to facilitate lateral drainage. Pregnant solution is recovered by gravity and barren solution is returned by pumping. Large surge capacities are required in the pregnant and barren solution ponds to handle potential solution volume imbalances caused by many factors including unusual weather, plant breakdowns, etc. Unlike a conventional processing plant, the heap cannot

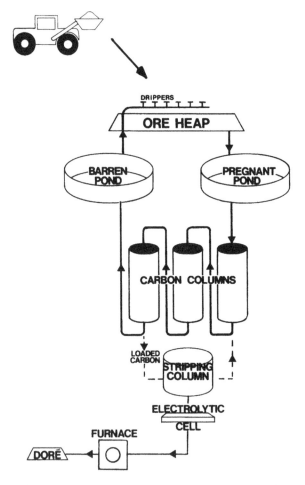

Figure 3.1.   General layout of a heap leaching operation.

be turned off. Drainage will continue to occur for several days after solution application ceases.

Make-up chemical reagents, added to the barren solution being returned to the ore heap, include water to supplant evaporation losses in dry climates, sodium cyanide, and usually caustic (NaOH) to provide protective alkalinity in a pH range between 10 and 11. Caustic is often preferred over less expensive lime in order to reduce calcium carbonate scaling in the pipelines. Small amounts of proprietary scaling inhibitors are also continuously injected into the returning barren solution. Consumption of caustic is often reduced by adding less expensive lime, in a small fixed amount, to each truck while hauling ore to the heap. Dumping onto the heap provides crude, but adequate mixing and the excess calcium is retained in the heap rather than depositing in pipelines.

Metal recovery occurs either by (1) adsorption on hard-shell activated carbon or (2) zinc precipitation, after purging excess dissolved oxygen, using the Merrill–Crowe process. Both zinc precipitates and metal stripped from the activated carbon are smelted to a doré bullion. Because the metal loading capacity of activated carbon is limited, the economic value of silver loaded carbon is quite modest. For this reason, zinc precipitation is usually selected when an economically significant amount of silver is being extracted from the ore.

HEAP CONSTRUCTION

Heaps are constructed by (1) dumping from haul trucks, (2) stacking with a front-end (wheel) loader, and (3) mechanical stacking using conveyor belts. Two truck dumping methods, (1) over end dumping and (2) stacking, or plug dumping, are illustrated in Fig. 3.2.

The major problem with heap construction using haul trucks is the compaction and extreme loss of permeability in the upper few feet of the heap caused by truck wheel pressure. This is less serious with hard ore, but soft ore will crush and produce a virtually impervious top layer. Because dozers and other track vehicles produce very low ground pressure, they are used to level and rip the surface of the heap, as deeply as possible, after construction and prior to leaching. This breaks up the compaction zone and improves permeability.

Front-end loaders travel on the leach pad surface to stack the ore in a heap. Consequently, the loader does not cause compaction on top of the heap. Depending on their size, front-end loaders can construct heaps from 3 to 5.5 m high.

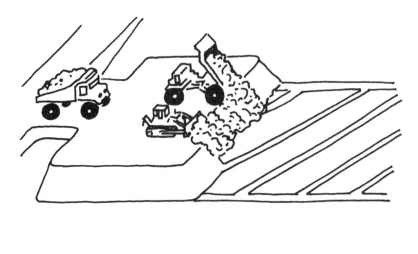

Figure 3.2.   Two variations in heap construction using haul trucks.

Leach pads are reused at many mines. With this **on/off** method, pre-pared ore is (1) placed on the pad, (2) leached for a scheduled period, (3) washed, and (4) removed from the pad and hauled to a separate waste pile for permanent disposal. A cyanide destruction step may occur prior to, or in conjunction with, final disposal. Providing for continuous operation requires several heaps with each step in the operation occurring at one or more of the heaps.

Heap leaching with a sequence of successive ore layers, or **lifts**, on a dedicated pad is usually practiced with run-of-mine ore. Usually, the leached ore is never removed, so the location of the leaching site and the leach pad design must be environmentally adequate for permanent disposal of the leached ore residue. Large boulders buried deep in the heap can continue to slowly leach over many months and even years. This method is often selected at mines with steep terrain where sufficiently level ground is not available for the reusable leach pad method. The use of sequential lifts is often associated with leaching in a canyon or small valley, and is sometimes referred to as "dump leaching" because it follows the practice at many copper mines of leaching mine overburden dumped in nearby canyons. It is also referred to as the "**Valley Fill**" method of heap leaching. Successive lift leaching adjacent to a steep slope, with a retaining structure to prevent downslope failures and collect solution drainage, is shown in Fig. 3.3.

Construction of ore heaps with conveyor belts is often selected for crushed ore (Bernard, 1993). They are the most gentle way of handling agglomerated ore and usually have lower operating costs than truck haulage systems. Conveyor belt stacking is divided into portable systems and mechanical systems. A series of several inclined portable conveyor belts, typically 20–30 m in length, that roll under each other are often used for portable systems. They are called "grasshoppers," because, viewed from their side at a great distance, each conveyor looks like a grasshopper.

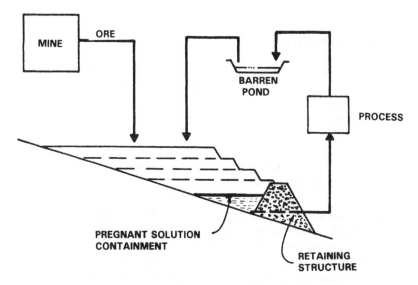

Figure 3.3. Multiple lift heap leaching in steep terrain (valley fill method).

A radial belt stacker, located at the end of the last telescoping grasshopper belt, deposits a conical pile of ore at its perimeter. Stacking begins at the far end of a heap or pad with the grasshoppers extended to their maximum distance. As the ore is placed the grasshopper belts gradually retreat under each other. A series of grasshopper belts is also very useful for mixing solid, agglomerating materials introduced at the first grasshopper and mixed with each drop to the next grasshopper conveyor belt. Wetting the ore with leaching solution introduced at one or more drop points is also easily accomplished and useful for ore agglomeration.

Custom designed mechanical stacker systems are being used with heap leaching of crushed ore at both gold and copper mines. They are usually installed at large operations with on/off pads, although they have also been used with multilift heaps. A movable bridge conveyor belt perpendicular to a central fixed conveyor belt is often used. The bridge conveyor belt is periodically moved laterally, either with crawler tracks or on rails and it has a movable discharge so that ore can be discharged at any point along its length.

Another mechanical stacker system design consists of a conveyor belt bridge spanning the distance between a center post and an end crawler. The bridge slowly walks in a circle rotating about the center post with the conveyor belt discharging ore at all points along the bridge and laying down a circular shaped ore heap (see Fig. 3.4).

ORE PREPARATION

Ore preparation may include ore blending for grade control or other characteristics, crushing and agglomeration. All or none of these operations may be practiced.

Ore Blending is often accomplished by depositing ore from the mine into a few adjacent working stockpiles, followed by reclaiming ore (with a front-end loader) from more than one ore stockpile. A front-end loader is also useful in starting, stopping and controlling the feed rate of ore introduced to a crushing or agglomeration plant.

Crushing converts rocks to a size small enough for extraction to be completed during the leaching cycle—a close approximation to the maximum extraction that can be obtained with unlimited leaching time. Finer crushing shortens the required leaching time but has two disadvantages: (1) increased operating cost and (2) increased production of fines, which lowers ore heap permeability and may prevent uniform solution percolation through the heap.

Figure 3.4. Ore stacking at a Nevada gold mine (courtesy RA Hanson Company, Inc.).

Crushing to about −20 mm or −25 mm is a good compromise for an on/off system. This can be achieved with two crushing stages: a primary crusher such as a jaw crusher sized to the largest mine boulders followed by a secondary crusher, typically a cone crusher. Closed circuit screening and a third crushing stage may be used.

A two-stage crushing operation with closed circuit screening is illustrated in Fig. 3.5. This configuration consists of a receiving bin, feeder, primary jaw crusher, vibrating screen and a secondary shorthead cone crusher. With a 1.7 m (5.5 ft) shorthead cone crusher, this plant is capable of crushing 2,000 tonne (2,000 Mg) in one 8 hr shift. It is fairly typical of the crushing plants at western USA gold heap leaching operations. Oversizing the crushing plant to permit a one-shift operation is usually cost effective because of the labor savings. Crushing to a −10 mm, or somewhat less, can be accomplished with tertiary crushing and is occasionally practiced. Lime, for protective alkalinity, is often introduced to the ore stream on a conveyor belt during the crushing operation.

Examining the proportionate times for leaching 90% of accessible metal from ore crushed to different sizes leads to the relationships shown in Table 3.1, which are governed by the rocks squared radius effect. Observe that the knee of the extraction curve in Fig. 2.7 occurs at about 90% extraction of the accessible metal so that greater extractions than 90% require proportionally much greater leaching times.

Use of a large primary crusher will usually provide little advantage for valley-fill leaching or for multiple ore lifts, provided the ore is properly fragmented by blasting in the mine and leaching continues for many years. However, a primary crushing step is necessary if conveyor belt ore haulage is used.

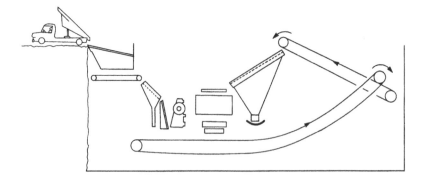

Figure 3.5. Two-stage crushing with closed circuit screening.

Table 3.1 Approximate Leaching Time for 90% Extraction of the Available Precious Metal as a Function of Crushing Size.

Nominal Crusher Setting	Approximate Leaching Time
8 in	2 years
6 in	1 year
2 in	2 months
1 in	2 weeks
3/4 in	1 week
1/2 in	3 days

Single stage crushing is inadequate for an on/off heap leaching system using a reusable pad because the time to obtain 90% extraction of the available metal is unacceptably long, one to two years. Two stages of crushing are necessary to reduce the leaching cycle time to a desirable limit, and three stages of crushing prior to on/off heap leaching are often used at large operations.

Agglomeration provides improved permeability for uniform percolation through the heap and efficient metal extraction. Agglomeration is used with mill tailings that are being reprocessed and with ores that contain too many fine particles (−150 mesh) or too much clay to provide good solution percolation without agglomeration.

Segregation will occur during ore stacking in heaps as larger rocks roll to the bottom while fines predominate in the upper part of the pile. This may cause variations in permeability with depth and non-uniform percolation through the ore.

The U.S. Bureau of Mines at Reno, Heinen et al. (1979), pioneered agglomeration in conjunction with gold heap leaching in the mid-1970s and developed the common binders (lime, portland cement, and their mixtures) that are being used today. The first commercial applications were in 1980. The Bureau of Mines determined the three critical parameters for producing stable, permeable agglomerates with adequate handling strength: (1) amount of binder, (2) moisture content and (3) curing period to strengthen the agglomerates. These parameters powerfully affect flow rates in agglomerated ore heaps, as shown by the examples from Li and Plouf (1987) in Fig. 3.6. Maximum percolation flow rates, above which flooding occurs, and flooded flow rates are plotted versus the magnitude of each of the three parameters.

The agglomeration goal is to provide the least-cost satisfactory agglomerate. The binder, varying between 2 and 9 kg per tonne of ore, is the major cost element. Optimum conditions are determined by laboratory ore

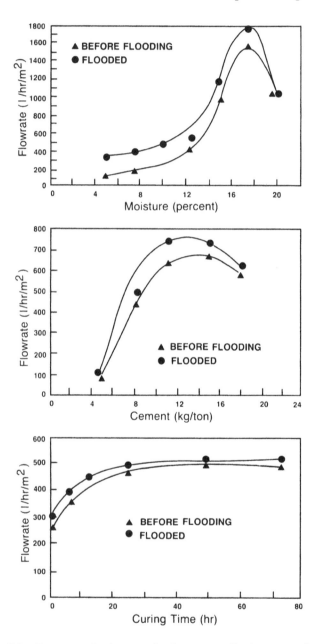

Figure 3.6. Parameters having a major impact on effective ore agglomeration (Li and Plouf, 1987).

testing followed by adjustments in actual operations. Usually the presence of more fines in the ore will require more binder. The resulting agglomerates are far from the uniform hard pellets produced in the iron ore industry on a balling pan and indurated by firing. Agglomerates that are easily broken with finger pressure can be adequate for heap leaching.

Ore, binder and the correct amount of water must be thoroughly mixed before curing. Continuous agglomerating equipment includes (1) rotary (tumbling) drums and (2) a series of several short conveyor belts, with mixing occurring by rolling on the belt at each conveyor drop point. Binders are usually added as dry solids at the feed end of the agglomerator while water is sprayed into the ore. Insufficient water yields a dry mix while excessive water causes a muddy effect, both with inadequate agglomerate strength.

Both lime and cement binders are alkaline materials that are added in sufficient quantity to provide protective alkalinity. A strong sodium cyanide solution is often injected into the agglomerator. Consequently, gold dissolution may begin during curing and be virtually completed by the time the agglomerated ore is stacked into a heap for leaching. When this occurs, diffusion–extraction and washing the heap completes the leaching process.

Often old tailings contain sufficient gold or silver to be profitably recycled by heap leaching. The tailings are usually crusty and must be mildly pulverized before agglomeration. For this purpose, an impact breaker is ideal. Tailings invariably require agglomeration with a binder prior to heap leaching, typically in a rotating drum, as shown in the process flowsheet of Fig. 3.7.

The particle size distributions obtained by sieving a −25 mm crushed ore before and after it was agglomerated are shown in Fig. 3.8, Van Zyl (1987). For this particular sample, −200 mesh fines, originally at 40 wt pct, were nearly eliminated by agglomeration. Ore heap permeability can often be considerably increased by agglomeration, sometimes by a factor of 1,000, or more.

Polymeric ore agglomeration binders have also been developed (Gross and Trominger, 1990). These are water soluble long chain organic polymers, and they are usually used in conjunction with cement and lime binders to improve agglomerate strength and reduce the amount of fines remaining after agglomeration. The polymeric binders are incorporated into the water or leaching solution added to the ore during agglomeration. Typical dosages are about 100 g of polymer per tonne of ore.

Ore fragmentation during mining can accelerate the rate of percolation leaching and often eliminate the cost of crushing the ore, which is

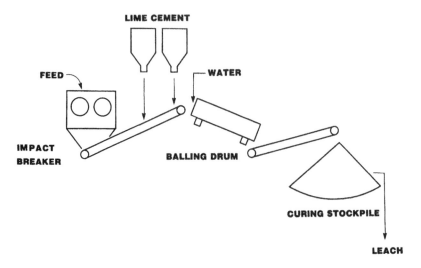

Figure 3.7. Agglomeration in a balling drum prior to heap leaching.

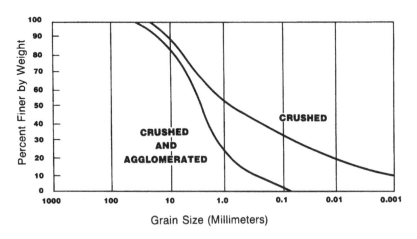

Figure 3.8. Screen size distribution of $-25\,mm$ crushed ore before and after agglomeration (Van Zyl, 1987).

approximately $0.30 per tonne. Ores that are heap leached in place by adding successive lifts, rather than being removed after the leaching period, are often not crushed. This is true for many gold (silver) ores and almost invariably true for copper ores and mine waste that are heap or dump leached.

The presence and the inter-relationship of pre-existing fractures in the rock are the primary determinants for obtaining adequate fragmentation during mining. Some ores and rock matrices are easily fragmented by explosive blasting and others break into large blocks of rock. The drilling and blasting practices, including blast hole patterns and selection of explosive are secondary factors. Explosive blasting normally pulverizes rock only out to one or two charge (drill hole) radii. Some rock rupture can occur out to five to fifteen charge radii. Beyond that, opening of pre-existing fractures accounts for breakage. A closer spaced pattern of drill holes may improve fragmentation, but blast hole spacing is usually greater than fifteen radii and closer spacing of drill holes coupled with greater amounts of explosive may not be cost-effective. Each ore must be evaluated to determine the fragmentation likely to result from drilling and blasting practice. Prediction methods are being developed (Kim et al., 1989).

CHOOSING BETWEEN REUSABLE PAD AND VALLEY-FILL HEAP LEACHING

Because the valley-fill method, with successive leached ore lifts, avoids crushing the ore and removing the spent ore it is less expensive than the on–off heap leaching method. Valley-fill has three disadvantages. First there is a longer delay in extracting and recovering metal for sale. This revenue delay has an imputed cost through the time value of money. Second, there is a large inventory of pregnant solution and dissolved metal stored in the heap that is not recovered until mining and ore stacking ceases and final washing begins. A third possible disadvantage is a lower metal accessibility and eventual lower ultimate extraction from the larger rocks in the ore mass. This latter effect, if it exists at all, is very difficult to quantify and it is offset at least somewhat because of the continued leaching of the ore, typically for many years.

There is a third, intermediate method practiced at several mines and that is to crush the ore and stack it in successive lifts without ever removing the spent ore. This method is well-suited to mechanical stacking systems, and an example is shown in Fig. 3.4. This method avoids the long delay with large rocks, but there is still a considerable inventory of pregnant liquor stored in the ever increasing volume of spent ore.

The following comparison is made between the reusable pad or on–off method of an ore crushed to minus 19 mm ($-3/4$ in) and the valley-fill method for run-of-mine ore at two assumed cases. Calculations were made using Fig. 2.7 and an assumed rock microporosity of 0.03. In the first valley-fill case, it is assumed that there is no loss of metal accessibility to the leaching solution with increasing rock size and the largest rock diameter in the ore is 200 mm. In the second case, ten percent of the ore mass consists of rocks larger than 200 mm, which are assumed to be essentially unleached because of their large size. Results over the first 12 months are plotted in Fig. 3.9. Both curves of the run-of-mine ore are nearly flat after two years, so that very little additional metal is produced. At this time scale the extraction delay for the minus 19 mm (−3/4 in) ore is negligible and does not appear on Fig. 3.9.

Analyzing these results show that 5–10% of the accessible gold in the very large rocks will not be recovered and that the average revenue delay for the gold that is recovered is about 1/3 to 1/2 of a year. At typical discount rates, this delay is financially equivalent to an *additional* loss of three to five percent of the gold.

Thus far, the soluble gold inventory in the ore heap has not been considered. Most of this gold will likely be recovered by extensive rinsing during reclamation after mine closure. Nevertheless, if the active life of the heap is several years, the present value of the rinsed gold at the beginning of the mine will be negligible and of little value in computing the economic

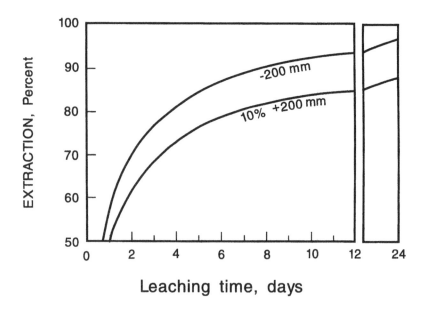

Figure 3.9. Gold extraction versus leaching time for run-of-mine ore, $\varepsilon = 0.03$.

feasibility of the heap leaching project. With multilift or valley fill leaching the fresh ore generating soluble gold is on the top of the heap. Consequently, the average solution grade throughout the heap during active leaching will likely be equal to, or slightly greater than, the pregnant liquor grade leaving the heap, and a significant amount of soluble metal will be held in the heap. Typically, it is about 5% of all of the metal extracted during active heap leaching, before reclamation at gold mines.

With these kinds of approximate comparisons, coupled with operating and capital costs for the two methods, it is possible to make a rational choice of leaching method. Higher grade ores will lead to crushing, faster recoveries and early rinsing using a reusable pad. Lower grade ores will likely benefit from the lower costs of the valley-fill method.

EXAMPLE PROBLEM I

In the following multilift gold heap leaching operation, what is the percentage of the total gold contained in the ore that is dissolved but held in the solution inventory within the heap? Twenty percent of the surface area of the heap is under active leaching at all times with the underlying ore containing 15 wt pct solution, while the remaining ore being rested is drained and contains 10 wt pct solution. The average pregnant liquor and barren liquor grades are 0.35 and 0.05 g/tonne of solution, respectively, and the average ore grade is 1.0 g/tonne of (dry) ore.

ANSWER
Since the rich solution is at the top of the heap with the fresh ore, the solution concentration within all of the ore heap is at least equal to the pregnant liquor solution (PLS) grade of 0.35 g/tonne. Therefore

PLS fractional mass $= 0.2(0.15) + 0.8(0.10) = 0.11$ [11 wt pct]

Ore fractional mass $= 1.0 - 0.11 = 0.89$ [89 wt pct]

Fractional total gold in PLS $= (0.11) \times (0.35\, \text{g/tonne PLS})$

Fractional gold in ore feed $= (0.89) \times (1.0\, \text{g/tonne Ore})$

Fraction of total gold in heap solution inventory:

$$(0.11/0.89) \times (0.35/1.0) = 0.0433$$

Percentage of total gold in heap solution inventory $= 4.33\, wt\, pct$

EXAMPLE PROBLEM II

If 66% of the gold is being extracted and recovered in the preceding heap leach, how much additional gold is potentially extractable when rinsing the heap during reclamation at the completion of the mine life?

ANSWER
Additional potential expressed as a percentage of the gold previously extracted during the active mine life:

$$0.433(100/66) = 6.55 \, wt \, pct \text{ of previous extracted gold}$$

SOLUTION DISTRIBUTION

Leaching solutions are distributed on the ore heaps using technology and materials originally developed for agricultural irrigation. Plastic pipe, either polyvinylchloride (PVC) or high density polyethylene (HDPE), is relatively inexpensive, lightweight, easy to install and modify, and noncorroding. Zinc oxide and other additives adequately retard ultraviolet degradation by sunlight in PVC. Cemented and threaded couplings are used for permanent joints and a locking ring arrangement with rubber gaskets is used for temporary connections that can be quickly assembled or disassembled. Plastic pipe is available in the standard sizes, couplings, and pressure ratings traditionally used for steel pipe.

Final solution distribution onto the heap can occur using a variety of spreading devices that include (1) **rotating impact sprinklers**, (2) **wobbler sprinklers**, (3) Bagdad **wigglers**, and more recently (4) **pressure drip emitters**. These are illustrated in Figs. 3.10 and 3.11. These sprinklers and drip emitters are also constructed of plastic to prevent corrosion and minimize cost.

Impact sprinklers, seen on golf courses, are usually arrayed in an extended "five spot" pattern with separations of 7–10 m so that there is some overlap of the spray circles as illustrated in the bottom sketch of Fig. 3.12. Providing impact sprinkler coverage over the designed circular area requires a discharge pressure between about 170 and 520 kPa (25 and 75 psig). Solution application rates vary between 2 and 3 cm^3/s m^2 (0.003 and 0.005 gpm ft^2) and cannot be turned down below these rates. This is equivalent to a superficial percolation velocity of 2×10^{-3}–3×10^{-3} mm/s.

Coverage within the spray circle is fairly uniform but evaporation losses are high in hot dry climates, up to 35%. For complete coverage of the ground surface with minimum overlap, the sprinkler heads should be

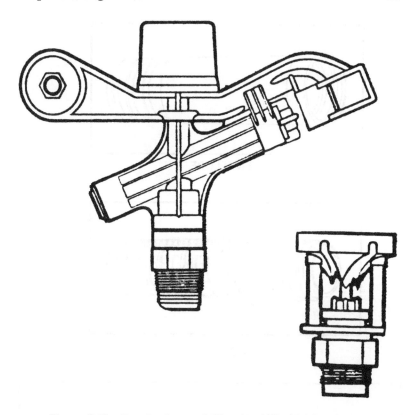

Figure 3.10. Rotating impact (left) and wobbler (right) sprinklers.

spaced at 1.73 times the sprinkler radius. Because of the slight overlap of sprinkler circles, this will yield an average solution application rate that is 1.2 times the individual sprinkler rate. Roman and Poruk (1996) provide details on engineering an irrigation system for heap leaching.

Wobblers cover a smaller area per unit than do impact sprinklers and they provide a larger solution droplet size, which on impact with the surface may cause decrepitation of weak ore agglomerates. Drip tubes have also been used, by this author at a small gold leaching operation, and by others. Drip tubes are 1–4 m in length and consist of smallbore flexible plastic tubing lying on the heap surface. While a drip tube is a point source of solution, its terminus can be periodically moved to a new location within a circle radius equal to the drip tube length.

Drip emitters require closer spacing of the terminal pipelines but they reduce evaporation losses. At each emitter, a small discharge of water

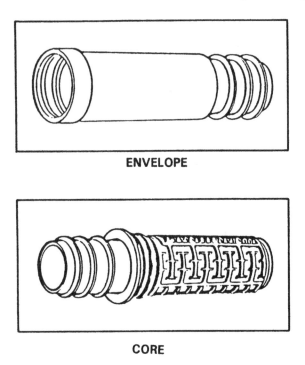

ENVELOPE

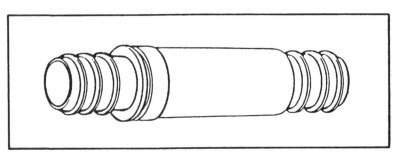

CORE

Figure 3.11. Construction of a pressure drip emitter showing the core and envelope.

flows through a tortuous path of grooves between the core and envelope of the emitter (see Fig. 3.11) while losing pressure. The drip flow depends on line pressure, which is usually less than that of sprinklers, typically 35–140 kPa (5–20 psig). The ends of each drip emitter have corrugated male connectors for slipping into the distribution tubing.

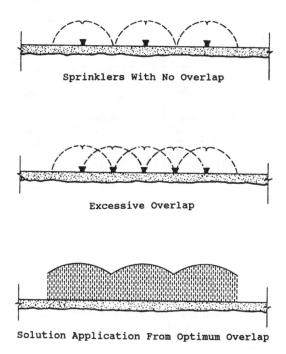

Sprinklers With No Overlap

Excessive Overlap

Solution Application From Optimum Overlap

Figure 3.12. Non-uniform solution application with impact sprinklers.

A typical drip emitter distribution system is shown in Fig. 3.13. The drip emitter lines are flexible, lightweight, and easily movable 13 mm (1/2 in) polyethylene tubes. Carbon and antioxidants are added to the polymer formulation for protection against embrittlement caused by solar radiation. These tubes are connected to a PVC or HDPE manifold, usually from 75 to 150 mm (3–6 in) diameter. On most ore heaps, emitters and lines are spaced so that each emitter covers an area of about one square meter. Parallel drip emitter lines on the surface of an ore heap are shown in Fig. 3.14.

Normally, sprinkler heap leaching of gold and silver cannot occur in extremely cold winter climates because the solutions freeze. However, a shallow ore cover over the distribution lines and drip emitters, illustrated in Fig. 3.15, permits operation through winter in severe climates, which can be a distinct operating advantage. However, in northern Nevada, where temperatures drop well below freezing, drip emitters are also used in winter without an earth cover on multilift heaps, provided the leaching solution enters a distribution line from both ends. With a multilift heap there is a very large reservoir of stored leaching solution in the heap and,

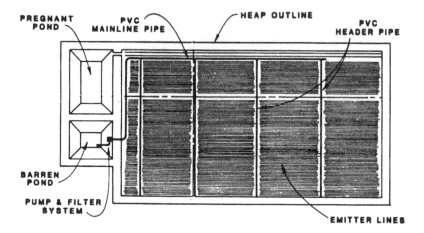

Figure 3.13. Typical solution distribution plan using drip emitters.

consequently, the average leaching solution temperature is well above the freezing point throughout the year. Freezing only occurs when stagnant solution remains in the distribution pipe for an extended time, allowing it to cool to the freezing point. Freezing begins at the far end of a distribution line, where the pipe flow velocity is negligible. Injecting solution from both ends of the distribution pipe tends to prevent flow stagnation, which otherwise would allow freezing to occur.

Drip emitters permit a much wider range of solution application rates than the other sprinklers, which is an important advantage. Unusually low application rates can be obtained by simply throttling back the distribution line flow rate, which lowers the pressure drop across each drip emitter. This can be particularly useful toward the end of a leach cycle when the gold extraction rate from the heap has significantly declined. Decreasing the solution flow rate offsets the decline in pregnant solution grade that would otherwise occur.

The Coeur–Rochester, Nevada silver–gold heap leaching operation (Anderson, 1989) uses drip emitters and begins each new heap at a high flow rate to obtain rapid saturation of the ore. Later in the heap life the application rate is decreased to $0.7\,cm^3/s\,m^2$ ($0.0015\,gpm\,ft^2$). With numerous heaps in various stages of leaching completion at the mine, the total mine solution flow rate is nearly constant. Adoption of this solution application strategy reduced total solution flow at the mine by 40% while maintaining the same ore processing and metal production rates.

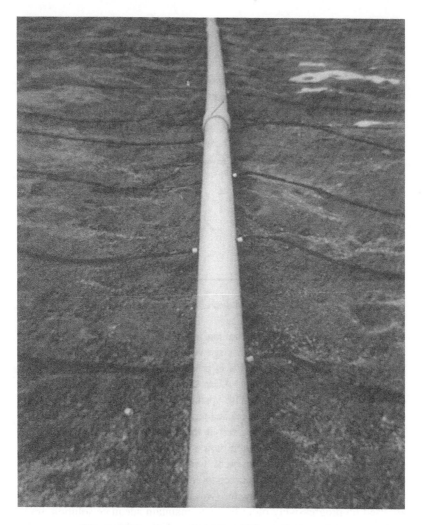

Figure 3.14. Drip emitters installed on a heap surface.

A variety of patterns for solution distribution are used at different mines. Pipelines on the heaps must be removed before reloading the pad with fresh ore or establishing a new lift. Because of its light weight, plastic pipe in diameters up to 100 mm can be manually moved with ease. Light weight, combined with corrosion resistance, low cost, and easy disconnects, makes plastic pipe ideal for ore heap leaching.

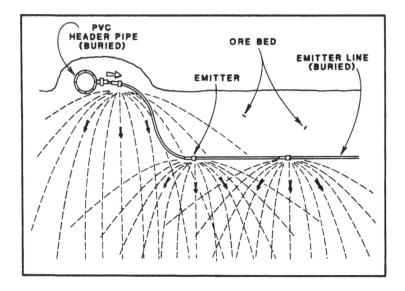

Figure 3.15.   Shallow buried drip emitters to prevent winter freezing.

The Zortmann–Landusky, Montana mine uses large multiple lift heaps containing run-of-mine ore with large vertical steel pipes passing through the entire ore mass. Each new lift is preceded by a six-meter extension to these permanent steel headers. After the new lift is emplaced, 50 mm diameter PVC portable pipes with sprinklers are spread on the top of the ore lift and connected to the steel headers.

Because run-of-mine ore is used at the Zortmann–Landusky mine, full metal recovery requires a leaching period up to 5 years in order to complete extraction. This result is consistent with estimates from the diffusion–extraction model of the previous chapter; for example, see Fig. 2.4. After the last lift is emplaced, rest cycles (without leaching) of increasing length are used on progressively older heaps to maintain adequate solution grade as the heap production declines with age. Old multilift heaps may die slowly, but they do die.

SOLUTION PONDS

Large lined ponds are required for the pregnant solution and barren solution storage. Sometimes a slime settling pond will precede the pregnant solution pond. Adequate size for these ponds is necessary for efficient and safe operation, and several factors must be considered in sizing them. The pregnant solution pond must generally be designed to receive runoff from a 100 year maximum flood occurring in the total heap area being drained, and it must handle runoff from the annual spring snow melt. The pregnant solution pond will continue to receive heap drainage during an extended shutdown of the metal recovery plant. Alternatively, the plant can be bypassed sending pregnant solution to the barren solution pond and back to the leaching heaps.

The barren solution pond must be able to accumulate solution during a shutdown of heap leaching under an emergency or during planned changes in the heap. Generally, several ore heaps are operated concurrently at each mine with some being loaded or unloaded, while others are being leached in a planned schedule of rotation.

An aerial photograph of a small gold mine is shown in Fig. 3.16. The leaching heaps are in the foreground. The two large solution storage

Figure 3.16. Heap leaching showing leaching area and solution ponds. The open-pit gold mine is in the background.

ponds are readily seen in the middle while the open-pit mine is in the background.

SOLUTION CHEMICAL CONTROL FOR GOLD/SILVER LEACHING

In most closed cycle hydrometallurgical systems, impurities accumulate and must be purged by bleedinxg and treating a small stream prior to discharge. This is usually not necessary in gold (silver) heap leaching practice. However, exceptions may occur if there are large amounts of cyanicide minerals in the ore.

As little as 100 ppm of HCN gas in air is fatal upon breathing. Sufficient protective alkalinity prevents formation of HCN,

$$CN^- + H_2O = HCN(g) + OH^- \qquad (3.1)$$

and usually lime is added to the ore or NaOH is added to the barren solution to maintain a high pH. Gold and silver cyanide leaching is inhibited by a pH above 12 because of competition of OH^- with CN^- for attachment to the leaching sites on the gold particle surface. At pH 10 the equilibrium partition between dissolved HCN and CN^- ion leaves 18.6% of the total free cyanide as HCN. The optimum pH is 10.3 and, consequently, alkaline additions are usually controlled to obtain a pH between 10 and 11.

Dissolved oxygen, in equilibrium with air, has a solution concentration of only 7 ppm, but this usually is not a limiting factor for gold (silver) heap leaching with shallow lifts. Sprinklers saturate the distributed solution with oxygen. Unlike copper dump leaching, the heaps are often shallow and the oxygen demand is low. Oxygen in the air inside shallow heaps is rarely decreased significantly during cyanide heap leaching. But when high heaps or multiple lifts are used, heap drainage oxygen concentrations as low as 1 ppm have been measured. Under these circumstances oxygen starvation can slow the gold (silver) leaching rate.

Sodium cyanide is usually selected because it is the least expensive cyanide, but potassium cyanide is equally effective. A strong master solution is usually made, either by dissolving NaCN briquettes into leach solution or into a preliquor with high pH. In some mining districts, a master cyanide solution is delivered from a vendor by tanker truck. This strong solution is metered into the diluting barren solution, or when appropriate it is added at the ore agglomerator. The large volume of the barren solution pond acts to mix and level any variations in free cyanide concentration before solutions are returned to leaching.

Sodium cyanide briquette dust is extremely toxic and must be controlled to prevent wind losses that cause environmental contamination. Breathing 180 mg of NaCN dust is fatal to adults. Proper respirators must be worn at all times when handling sodium cyanide. Fortunately, cyanide vapor, even in small concentrations, has a strong distinctive odor that is a useful warning.

The cyanide dosage in leaching solutions will depend primarily on the amount of cyanide consuming minerals present in the ore. Among these are sulfides and arsenides of copper, cobalt, mercury, and nickel, which may be present in significant amounts. Sufficient sodium cyanide must be added to maintain a desired level of free cyanide $(CN^- + HCN)$ in the pregnant solution. Ore testing, in conjunction with free cyanide assays, is used to determine the optimum economic dosage, which for well-oxidized ores is typically about 0.3–0.5 kg/tonne of leaching solution.

The use of a pipeline scale inhibitor is mandatory, even when NaOH is the source of protective alkalinity. Scale inhibitors are supplied by oil well service firms and usually consist of polyphosphates, phosphonates, organic phosphates and polyacrylates. They function by preventing adhesion of calcium carbonate and calcium sulfate (gypsum) precipitates to pipe walls, sprinklers, and to some extent granular activated carbon, which is used to capture gold from the pregnant solution.

Controlled metering of a master scale inhibitor solution into the barren solution is used to maintain scale inhibitor concentration at a few ppm, which is usually sufficient. In order to be most effective, the inhibitor master solution is usually added near the barren solution pump returning solution to the leaching heaps.

Calcium bicarbonate saturates in leach solution as it passes through alkaline rocks in the ore heap. This is the source of most of the scaling problem, because soluble calcium bicarbonate eventually reacts with both caustic and lime, added to maintain protective alkalinity, and precipitates calcium carbonate especially in the barren solution return pipelines:

$$2NaOH + Ca(HCO_3)_2 = Na_2CO_3 + CaCO_3(s) + 2H_2O, \quad (3.2)$$

$$Ca(OH)_2 + Ca(HCO_3)_2 = 2CaCO_3(s) + 2H_2O. \quad (3.3)$$

But from this stoichiometry it is readily seen that less calcium carbonate precipitates when NaOH is used than when lime is used to achieve protective alkalinity.

GOLD (SILVER) RECOVERY FROM
DILUTE LEACHING SOLUTIONS

Two methods of removing gold and silver from pregnant solutions are used: (1) zinc precipitation and (2) adsorption on hard-shell granular activated carbon.

The chemistry of zinc precipitation (Merrill–Crowe process) is a simple metathesis, or cementation, reaction:

$$2Au(CN)_2^- + Zn = 2Au° + Zn(CN)_4^{-2}. \tag{3.4}$$

Prior to zinc precipitation, dissolved oxygen must be stripped in a de-aeration tower to prevent reoxidation of the gold in the continued presence of cyanide ion. The gold (silver) precipitate must be completely recovered in a clarifying filter press using a filter aid to prevent any particulate loss. The flowsheet of the Merrill–Crowe process is shown in Fig. 3.17. A disadvantage of this process is the cost of clarifying and de-aerating the solutions. Copper ions, when present, are serious interferences.

Gold and silver cyanide complexes are readily adsorbed from dilute solutions at room temperature onto granular activated carbon leaving a low concentration barren solution, if the activated carbon is not overloaded. Because the residual solution is recycled, ultimate metal losses are minimal. Partial

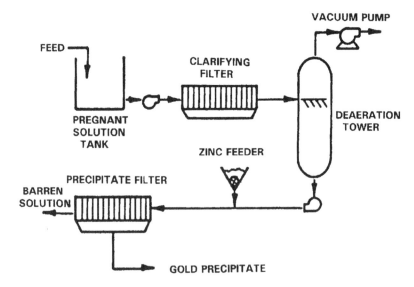

Figure 3.17. Flowsheet of the Merrill–Crowe zinc precipitation process.

Table 3.2 Stripping Solution Compositions.

Strip Solution	Temperature (°C)	Strip Time (hr)
1% NaOH + 0.1% NaCN	88	52
1% NaOH + 0.5% ethanol	120	9
1% NaOH	85	24
0.1% NaCN + 20% ethanol	77	24

stripping is accomplished by removing and heating a batch of loaded carbon to near boiling temperatures in a very strong caustic solution. More complete stripping can occur by heating in caustic-alcohol mixtures or by stripping in pressure vessels at temperatures above the normal boiling point. Some of the strip solution compositions are shown in Table 3.2. However, the alcohol and pressure stripping methods are less commonly practiced.

Granular activated carbon is supplied by several manufacturers. Activation occurs when organic materials are pyrolized, usually with steam in the absence of oxygen. This provides a structure with high microporosity and very high internal surface area. Stripped carbon is usually regenerated in the same manner by heating wet carbon in the absence of air, to between 600°C and 800°C. Accumulated calcium carbonate can be removed after reactivation by rinsing with acid.

Gold and silver are separated from the concentrated strip solution by electrowinning, using stainless steel anodes and steel wool as a cathode, or by zinc precipitation. In either case the resulting solids are smelted to a doré bullion.

With pregnant liquors obtained from heap leaching, carbon loading often occurs in a series of three or more tanks through which the pregnant solution flows. Upward flow in each tank fluidizes a bed of granular activated carbon of carefully controlled particle size, usually about -6 to $+14$ mesh. The bulk density of activated carbon is less than $1 \, g/cm^3$, so it is easily fluidized. Overflow screens prevent carbon loss, provided undersize carbon is not present, but carbon particle attrition does occur and leads to minor losses of carbon containing gold. A conical bottom tank for fluidizing activated carbon is shown in Fig. 3.18.

Periodically, the activated carbon is advanced from tank to tank countercurrent to the solution flow, and carbon from the first tank is sent to stripping. A flowsheet is shown in Fig. 3.19. Granular activated carbon manufactured from coconut shells is usually used because of its good adsorption properties combined with excellent attrition resistance.

Various types of commercial anion exchange resins have been evaluated with respect to their ability to extract gold from cyanide leach solutions

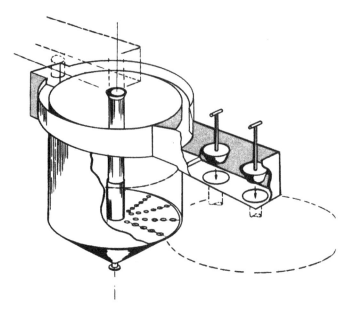

Figure 3.18. Activated carbon adsorption tank with solution bypass arrangement (McQuiston and Shoemaker, 1981).

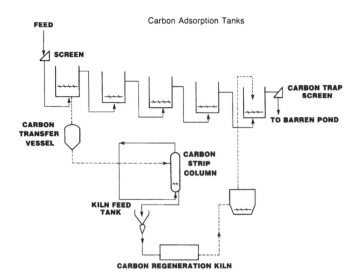

Figure 3.19. Countercurrent activated carbon adsorption circuit and carbon stripping flowsheet.

(Riveros, 1990; Fleming and Cromberge, 1984). Thus far, inadequate chemical properties have prevented substitution of resin beads for granular activated carbon. Weak base resins will not extract the gold cyanide complex at pH much above 9, while strong base resins are difficult to elute. However, development of a resin based on guanidine functionality by Mintek and Henkel (Mackenzie, 1993) shows promise. Elution would be simpler, and the thermal regeneration required with granular carbon would not be needed.

TESTING GOLD (SILVER) ORES FOR HEAP LEACHING

Adequate ore testing is critical to the success of a new heap leaching operation. Several sequential, and often iterative, steps in ore testing are usually linked with the exploration and engineering feasibility activities for the prospective mine. It is necessary to obtain and assay the most representative ore samples available. These may be drill core, drill cuttings (preferably from reverse circulation drilling) or a bulk sample.

Ore amenability to cyanide leaching is determined by agitation leaching, typically a 24 hr leach of -100 mesh ore. Good aeration needs to be promoted either with a bottle roll test or with stirring, using vortexing to inject air. These simple tests are used to determine cyanide and lime (or caustic) requirements as well as the extent of gold and silver extractions (total extractability).

If cyanide extraction is feasible and heap leaching is being considered then column leach tests, typically with about 50 kg of ore, are conducted to simulate on/off heap leaching. Typically, these tests are run in vertical columns approximately 150 mm diameter by 2 or 3 m high, with ore crushed to about -20 mm. Continued success leads to scaled-up column leaching at the planned operating ore size, using columns preferably with heights close to the planned lift height in the ore heap. The column is intended to simulate a vertical section through the eventual ore heap from top to bottom, as shown in Fig. 3.20. These larger column tests may involve up to a 500 kg of ore per column.

To prevent the column wall effects from distorting the results, the column diameter should be at least 10 times the largest rock size. For crushed ore, this often equates to a column of about 250–300 mm in diameter. Large commercial plastic irrigation pipe is often used to construct these larger columns.

A column for ore testing is illustrated in Fig. 3.21. Often several column tests of different ore samples from the same deposit under study are run concurrently. For each column, solution is continuously drained into a

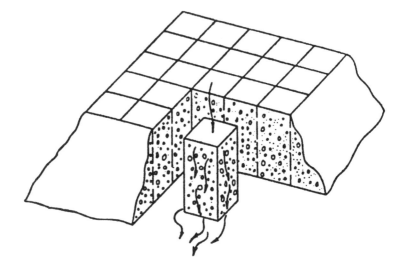

Figure 3.20.   Schematic diagram of a leach heap cut out section.

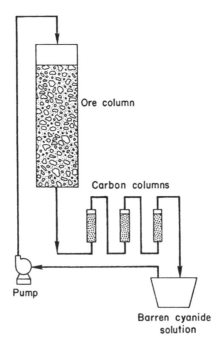

Figure 3.21.   Ore testing in a leaching column to simulate heap leaching extraction.

small sump and recirculated, at typical commercial leaching application rates, with a peristaltic pump. Chemical reagents are periodically dosed and sump samples are collected for chemical analyses. Representative samples of the ore mass are obtained and assayed before and after completing leaching. Materials balances on gold and silver are obtained daily from the solution assays and corroborated, upon completion of the test, with the ore residue fire assay.

Materials balances are used to generate cumulative extraction curves, as shown in Fig. 3.22. These are either plotted against the leaching time or the cumulative leach volumes that are circulated. Both sets of data are made available and used to design the commercial heap leaching operation.

The result is either a prescribed time of leaching before either removing the ore, or stacking a new ore lift, or a prescribed amount of leaching solution to be applied before removing the ore, or beginning a new lift. For a truly continuous operation at a constant solution application rate followed by immediate replacement of fresh ore at the end of a leaching cycle, leaching operations based on either time or the amount of leaching solution applied are equivalent. However, many interruptions usually occur in both ore materials handling and solution application. Therefore, it is

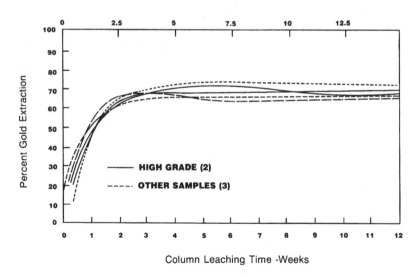

Figure 3.22. Metal extraction curves obtained from ore testing in leach columns.

common for operations to be based on a prescribed amount of solution to be applied before the ore is either removed from the heap or a new ore lift may be installed. For example, at one large Nevada gold mine heap leaching ore in a multilift operation, the operating rule is to apply 1.2 tonnes of leaching solution to each tonne of ore before a new lift may be started. Drip emitters are used predominately at this heap leaching operation. At typical drip emitter application rates and an expected range of internal rock microporosity, this is equivalent to about one week of leaching before a new lift is installed, based on information in the previous chapter.

Rock microporosity and variability of rock microporosity in an ore is generally unknown. Consequently, leaching operations must be designed on the basis of experimental information. Column leaching tests usually include *all* of the factors that may effect leaching extractions for a particular disseminated gold ore. Column leaching tests are a minimum requirement for designing a commercial heap leaching operation. Nevertheless, evaluating the experimental leaching results with methods of analysis provided in the previous chapter is recommended. This is critical when the ore particle size distribution is different than the crushed ore used in the column leaching tests. A theoretical perspective can often illuminate aspects of the leaching of a particular ore that are not obvious from the experimental test results alone, and this can lead to changes or further inquiries that may improve the ultimate heap leaching operation.

Ore testing should include composite ore samples, representing average characteristics. However, samples taken from various parts of the ore deposit or having different characteristics, such as variations in *host rock type*, should also be tested to determine if these differences affect leaching results. To the extent possible, the effect of *rock size* on mineral accessibility should be determined because mineral accessibility often increases as rock size decreases. Consequently, crushing the ore further than otherwise planned may be indicated by ore testing results. Mineral accessibility as a percent of *grade* may decrease with grade. An irreducible minimum amount of gold, or other ore mineral, that cannot be extracted may exist in an ore. This is known as the "constant tailing" effect.

Conversely, very high grade ore samples may result from a nugget effect. Since the nuggets are little dissolved in normal leaching times, the ore will appear to have a low ultimate leach accessibility. One way to get around this is to cap the gold grade of ore samples when making extraction forecasts. Very high grade samples are down graded to the cap.

Considerable **judgment is needed in interpreting the results of column leaching tests** because of these and other ore non-homogeneities, which may cause anomalous extraction behavior. Statistical methods and statistically based procedures using replicate samples have been used to improve the interpretation of leaching test results from non-homogeneous ores (Cavender and Granger, 1995).

Pilot heaps are expensive and usually not necessary, but they are occasionally operated when either the mine project economic feasibility is marginal or uncertainties require them, and especially when planning to use run-of-mine ore or limited crushing. Pilot heaps may be (1) large columns on the order of 2 m diameter, (2) concrete or wooden square cribs holding the ore mass—a box of ore, and (3) small unconfined heaps. Because of the angle of repose of an unconfined heap, some ore cannot be adequately exposed to the leaching solution. This presents metal accounting difficulties unless the unconfined ore heap exceeds several thousand tonnes. More precisely, the dead volume in the side slopes of heaps decreases as the ratio of the heap length to heap height, L/H, increases. Computed results are shown in Fig. 3.23 for a square shaped heap in plan view and a rectangular or longitudinal heap with its length four times its width. These results were computed from the geometry of the heaps and

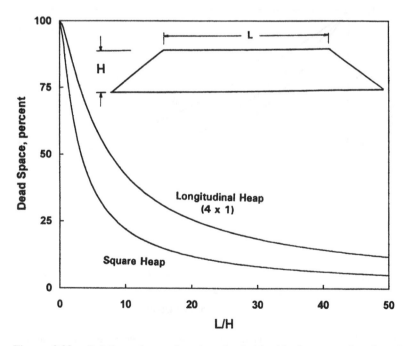

Figure 3.23. Relative volume of ore in unleachable side slopes as a function of heap length, L, and height, H; computed with an angle of repose of 37°.

an angle of repose of 37°. Note that L/H must exceed about 25 before the unleached dead space becomes less than ten percent of the ore. For a single lift heap stacked to 9 m, which is typical when using radial belt stackers, this requires a length in excess of about 228 m (750 ft). For a square shaped heap, it would necessarily amount to nearly one million tonnes. Multilift heaps with setbacks between lifts to ensure heap geotechnical stability have an effective angle of repose less than 37°, and a larger percentage of the ore contained in the unleached dead space.

Another factor that may cause losses or deductions from column results that need to be considered when estimating commercial recoveries are reduced solution/ore contact, for example when caused by truck compaction during ore emplacement. This may amount to five percent of the ore.

A significant postponement in recovering metal from the solution inventory, either in the heap or otherwise located can have a significant negative economic impact. Fig. 3.24 shows how the present value of

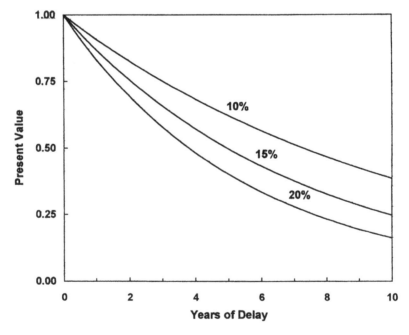

Figure 3.24. The decline in the present value of a deferred revenue, such as metal recovery and sale, as a function of three selected discount rates and the deferred time.

recoverable metal decreases with the years of delay after ore emplacement and leaching begin before the metal is actually recovered. Present value versus delay time results are plotted for three economic discount rates, 10%, 15% and 20%.

ALTERNATE PROCESSES FOR GOLD (SILVER) LEACHING

The most frequently encountered alternative process for oxidized gold (silver) ores amenable to cyanide leaching is grinding (milling) followed by agitation leaching. Cyanide leaching and metal recovery variants of milling include:

1. Leaching followed by solid/liquid separation and metal stripping:
   a. countercurrent decantation and Merrill–Crowe precipitation,
   b. filtration and Merrill–Crowe precipitation,
2. Leaching followed by Carbon-in-Pulp (CIP) metal recovery,
3. Carbon-in-Leach (CIL).

Milling always entails higher capital and operating costs than heap leaching, but it gives more certain and predictable recoveries upon which to base an economic feasibility analysis. Milling often, but not always, yields higher recoveries that may offset its additional cost. Justification of mill construction in lieu of heap leaching normally requires the discounted value of the additional gold and silver recovered by milling to equal or exceed the additional operating cost. This calculation requires a forecast of the future prices of gold and silver during the expected operating life. The additional cost should also include an amortization charge, but only for the mill construction capital required beyond the capital needed for heap leaching.

These factors lead to a choice of heap leaching for lower grade ores and smaller operations (small mines and/or limited reserves and short mine lifetimes). Average ore grade and ore reserves for fourteen western U.S. disseminated gold deposits, that were being mined by open pit mining methods in the 1980s, are shown in Fig. 3.25, Van Zyl et al. (1988). A simple diagonal line divides these mines into an upper group that uses milling and agitation leaching, and a lower group that uses heap leaching. Many in the lower group of mines could not exist without development of heap leaching as an effective solution mining technology.

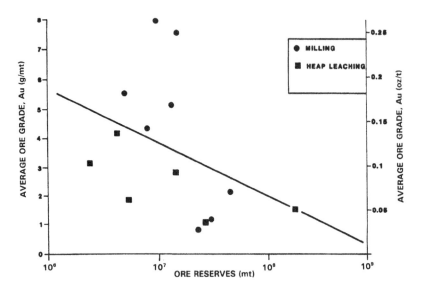

Figure 3.25. Gold ore grade versus ore reserves showing extraction processes chosen for 14 mines (Van Zyl et al., 1988).

REFRACTORY ORE

Most disseminated gold ores are not amenable to flotation, and when these ores have low cyanide extractability, they are considered to be "refractory"— meaning difficult to treat. These ores are usually refractory because they are not thoroughly oxidized. Often they contain small amounts of pyrite and/or arsenopyrite, up to a few percent, that surround the micron sized gold particles and obscure contact with the cyanide leach solution. These ores often contain organic material, exhibiting a black or gray color, and are called "**carbonaceous**". The organic material is frequently activated with respect to the gold cyanide complex, and any soluble gold is quickly adsorbed onto the organic matter within the ore. Mixing oxidized ore with carbonaceous ore during processing will often transfer gold by solution to the refractory carbonaceous ore. Hence, such ores are called "**preg robbing**".

Removing the gold containing sulfide minerals by flotation is effective when possible, but usually the western disseminated refractory ores must be oxidized by either roasting or pressure oxygen hydrometallurgical treatment before cyaniding. Typically, these oxidation pretreatment processes increase both capital and operating costs to twice those required for milling oxidized ore. Consequently, these processes can only be used on high grade ores. A low cost step is needed to oxidize low grade refractory ores prior to heap leaching.

Research is being conducted in laboratories with chemical oxidants. Both laboratory and field tests are being conducted on ores that are inoculated with bacteria and aerated over extended periods of time. This concept (Bartlett, 1990) of heap biooxidation or biocuring the ore is directed toward inexpensively oxidizing the sulfide minerals and possibly also the organic material. In practice, after a suitable biooxidation pretreatment period in a large wetted stockpile or heap, the ore will be reclaimed, neutralized with adequate lime to obtain protective alkalinity, and conventionally heap leached with a cyanide solution.

Laboratory results by Burbank et al. (1991) using this heap biooxidation process on refractory ores show considerable promise. Cyanide leach extractions before biooxidation and after biooxidation, for a period of 3 months and a period of 6 months, are shown in Fig. 3.26. Results from the same ore samples without any biooxidation pretreatment are also shown. These results are from tests conducted by Prisbrey (1991) on refractory gold ores from five widely scattered U.S. gold mines. Heap biooxidation pretreatment of sulfidic refractory gold ore is covered in more detail in Chapter 9.

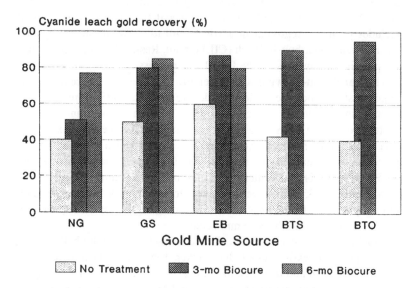

Figure 3.26. Laboratory results from −19 mm ore biooxidation tests in inoculated wet column tests followed by cyanide extraction (Prisbrey, 1991).

EXAMPLES OF GOLD ORE HEAP LEACHING OPERATIONS

The **Round Mountain Mine** in central Nevada is a large gold operation, mining 70,000 tpd of ore and an equal amount of waste rock, or more. Several ore processing methods are used, depending on the ore grade and whether the ores is refractory. Both reusable pad and valley-fill heap leaching methods are used. Approximately 25,000 tpd of oxide (non-refractory) ore, assaying above 0.65 g/t (0.02 opt), is crushed in a tertiary crushing plant to − 19 mm (− 3/4 in) and stacked with a conveyor belt system on reusable leach pads. Because the ore has a low extraction rate (low internal microporosity), the reusable pad leaching cycle is 150 days. During the first half of this cycle, ore receives lean recycled pregnant solution assaying 0.25 ppm gold. During the second half, the ore is leached with barren solution returned from a conventional carbon adsorption plant. Wobblers are used to distribute leaching solution on the ore heap. The reusable leaching pad is nearly a mile long and is lined with asphalt. Lower grade ore down to the cut-off grade of (0.012 opt) is trucked to a dedicated (valley-fill) leaching heap and end-dumped into 15 m (50 ft) high lifts. This heap is legally permitted to grow to a height of 105 m (350 ft).

Refractory ore, above a grade of 1 g/t (0.03 opt), is concentrated in an 8,000 tpd mill that uses gravity methods (spirals and tables) to treat ore that has been ground, with a semi-autogenous mill, and wet screened to pass 20 mesh. Nugget gold is separated and the gravity tailing, which includes pyrite, is reground in a Tower mill to − 400 mesh and cyanide leached using a carbon-in-leach (CIL) circuit. Residual cyanide in the tailing pulp is destroyed using the INCO SO_2 process.

The **Zarafshan–Newmont** joint venture in Uzbekistan processes a very large existing stockpile of low grade ore (Suttill, 1995). Ore is loaded directly into a mobile jaw crusher set at − 200 mm and transported by portable conveyor belts to a coarse ore stockpile at a central, fixed location, where three more crushing stages are located. Each crushing stage is preceded by "banana-deck" screens. The final screen passes ore through 3.35 mm, which is much finer than usual for heap leaching. There are thirteen parallel 3.35 mm screens needed to accomplish this for a daily throughput of 38,000 tonnes of ore. Ore discharged from the quaternary crusher is recycled to the 3.35 mm screens.

Crushed ore is stacked on a dedicated pad in 10 m high lifts using grasshopper conveyors and a radial stacker. The multilift heap will eventually be 80 m high and over one mile long. Leaching occurs in two sequential steps similar to that practiced at Round Mountain. The primary leach, on fresh ore, uses a lean pregnant solution resulting from the secondary leach of old ore. Barren solution from the Merrill–Crowe metal recovery plant is used in the secondary leaching step. The leaching time for each step is 49 days. After the two leaching steps are completed, the ore is over-dumped with a new ore lift.

PROBLEMS

1. Refer to the five column leach tests extraction curves shown in Fig. 3.22. Why do the extractions not exceed about 65%? What leaching duration (weeks) would you recommend for this ore in a commercial reusable pad heap leaching operation? Explain your answer.

2. Calculate and plot on graph paper your estimated fractional extraction curve (as a percent of the accessible gold) versus time in days for the crushed ore with the sieve analysis shown in Fig. 3.27. The measured rock microporosity is 1.0%.

3. A 4.01 bulk sample of crushed ore with specific gravity $2.65\,g/cm^3$ is weighed and has a mass of 8.30 kg. After drying the sample mass is 7.42 kg. Compute the following: (1) weight percent moisture content, (2) wet bulk density, (3) dry bulk density, (4) void space as a ratio of total sample volume and (5) percent water saturation.

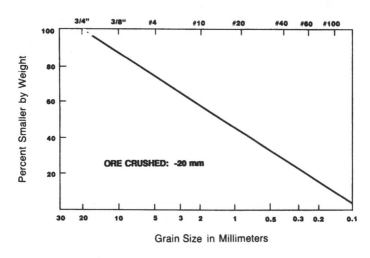

Figure 3.27. Ore rock size analysis for Problem 2.

4. The feasibility of heap leaching a newly discovered, low-grade, disseminated gold ore deposit is being considered. Two-month column tests of
the ore crushed to $-19\,mm$ $(-3/4\,in)$ show excellent ultimate gold
extractions, 90% of the gold contained in the ore. Seven days of column leaching provides 67% extraction, but 14 days of leaching provide
77% extraction.

The ore is fairly soft and the mining plans are to rip it with a bulldozer followed by scrapers for haulage. The ore will pass through a
15 cm (6 in) grizzly on the way to being stacked in the leaching pad.
Oversize, expected to be only about 5% will be broken with an impact
crusher. After gold heap leaching, the ore will be removed for use in
another process. (a) Recommend a heap leaching cycle time and the
expected percent recovery of the gold contained in the ore if on/off
leaching with a reusable pad is used. (b) Recommend, with an explanation, whether this ore should be leached in multiple lifts (valley-fill) or
as on/off ore heaps.

REFERENCES AND SUGGESTED FURTHER READING

Anderson, P.A. (Dec 1989). Heap leach solution management at Coeur-Rochester. *Mining Engineering*, pp. 1186–1188.
Bartlett, R.W. (Feb 1990). Aeration pretreatment of low grade refractory gold ores. *Minerals and Metallurgical Processing*, pp. 22–29.
Bernard, G.M. (1993). Transporting and stacking ore for heap leaching. *Randol Gold Conference 93*, Randol Int'l. Ltd., Golden, CO, pp. 275–277.
Burbank, A., Choi, N. and Prisbrey, K. (1990). Biooxidation of refractory gold ores in heaps. *Advances in Gold and Silver Processing*. Soc. for Mining, Metallurgy and Exploration, Littleton, CO, pp. 151–159.
Cavendar, B.W. and Granger, B.L. (1995). Column leaching procedures for non-homogeneous ores. *XIX Int'l. Mineral Processing Congress, Vol IV.*, Soc. of Mining, Metallurgy, and Exploration, Golden, CO, Chap. 36.
Fleming, C.A. and Cromberge, G. (1984). The extraction of gold from cyanide solutions by strong and weak-base resins. *J.S. Afr. Inst. Min. Metall.* (5) pp. 125–137.
Gross, A.E. and Strominger, M.G. (1990). Development of a polymeric agglomer-ation aid for heap leaching. *Randol Gold Conference, 90*, Randol Int'l. Ltd., Golden, CO, pp. 205–208.
Heinen, H.J., McClelland, G.E. and Lindstrom, R.E. (1979). Enhancing percola-tion rates in heap leaching of gold–silver ores. *U.S. Bureau of Mines Report of Investigation RI 8358*.
Kim, Y.C., Cervantes, J.A. and Farmer, I.W. (1989). Predicting *in situ* rock fracture parameters using soft kriging. Weiss, A. (Ed.), *Proc. 21st APCOM Symp.* Society of Mining Engineers, Littleton, CO, pp. 237–252.

Li, T.M. and Plouf, T.M. (Eds.) (1987). *Small Mines Development in Precious Metals*. Society of Mining Engineers, Littleton, CO, Chap. 24.

Mackenzie, J.M.W. (1993). Henkel IX resins and Henkel LIX79 solvent for gold recovery from alkaline cyanide leach solutions. *Randol Gold Conference 93*, Randol Int'l. Ltd., Golden, CO, pp. 287–291.

McClelland, G.E. and Eisele, J.A. (1982). Improvements in heap leaching to recover silver and gold from low-grade resources. *U.S. Bureau of Mines Report of Investigation RI 8612*.

McQuiston, F.W. and Shoemaker, R.S. (1981). *Gold and Silver Cyanidation Plant Practice*. Society of Mining Engineers, Littleton, CO, Vol. 2.

Milligan, D.A. and Engelhardt, P.A. (1984). Agglomerated heap leaching at Anaconda's Darwin Silver Recovery Project. Hiskey, J.B. (Ed.), *Gold and Silver Heap and Dump Leaching Practice*. Society of Mining Engineers, Littleton, CO, pp. 29–39.

Prisbrey, K. (1991). Unpublished study sponsored by US Bureau of Mines.

Riveros, P.A. (1990). Evaluation of ion exchange resins from the extraction of gold from cyanide solutions. *Randol Gold Conference 90*, Golden, CO, pp. 257–261.

Roman, R.J. and Poruk, J.U. (1996). Engineering the irrigation system for a heap leach operation. Preprint 96–116, Soc. for Mining Metallurgy and Exploration, Littleton, CO.

Schlitt, W.J., Larson, W.C. and Hiskey, J.B. (Eds.) (1981). *Gold and Silver Leaching, Recovery and Economics*. Society of Mining Engineers, Littleton, CO.

Suttill, K.R. (1995). Zarafshan–Newmont joint venture. *Eng. & Min. J.*, Sep., pp. 29–31.

Van Zyl, D.J.A. (1987). *Geotechnical Aspects of Heap Leach Design*. Society of Mining Engineers, Littleton, CO, Chap. 10.

Van Zyl, D.J.A., Hutchison, I.P. and Kiel, J.E. (Eds.) (1988). *Introduction to Evaluation, Design, and Operation of Precious Metal Heap Leaching Projects*. Society of Mining Engineers, Littleton, CO.

FOUR

Environmental Control in Percolation Leaching

ENVIRONMENTAL CONCERNS

This chapter focuses on precious metal heap leaching regulations and practices, which have evolved rapidly because of the recent expansion of the industry coupled with the extreme toxicity of cyanide. However, the developed technology and approach are generally applicable to copper ores and other applications of percolation leaching. Much of it is derived from Van Zyl et al. (1988).

Cyanide is also toxic to fish and wildlife. Consequently, preventing bird poisoning on wet heaps and solution storage ponds is a concern, especially in desert climates where wetlands are rare and the ponds attract migrating waterfowl. Wind blown dust and other air contaminants can be a serious environmental problem. Collapse or other geotechnical failure of improperly designed or constructed heaps is a potential earth movement hazard. Failure is more likely when the heap is wet and particularly so if there is a water saturated zone (perched water table) within the heap, due to a region of low permeability that cannot drain the leach solution fast enough.

However, the major concern with heap leaching, copper mine waste dump leaching, and most other applications of percolation leaching, is contamination of surface and ground waters with chemicals, including cyanide, and heavy metals dissolved in the leaching solutions.

The key to effective environmental control of leaching solutions is nearly always **containment** of the solutions under a worst case scenario of possible emergencies, so that environmental contamination is prevented. Clean-up of solution spills and leaks is usually only partially accomplished, is often infeasible, and nearly always proves to be much more expensive than prevention. The key to containment is proper design and construction of the leaching system, coupled with an adequate **monitoring** system to give early warning of any failures so that little leaks and problems can be corrected before they become big leaks and disasters.

This chapter will focus on containment, prevention of contamination and cyanide detoxification. Remediation of ground and water contaminated with other substances will be covered in Chapter 15.

PERCOLATION LEACH PAD DESIGN

Because flow is downward in percolation leaching, the leach pad is the most important element of the solution containment system. The leach pad supports the ore heaps, collects solution flowing through the heaps, transports the solution laterally to drainage pipes or ditches, and prevents toxic or hazardous chemicals from penetrating into the ground. The leach pad site and its topography should be selected to be free from flooding or other natural hazards. The foundation must be stable to prevent movement or cracking of the pad liner under the weight of the ore heaps, which may eventually reach heights of nearly 60 m at some multilift cyanide heap leaching operations.

The minimum leach pad system consists of: (1) a prepared foundation and bedding layer, (2) an impervious liner, and (3) a covering layer of coarse rock (drain blanket) to facilitate drainage of percolating solutions and prevent rupture or tearing of the liner. Usually the drain blanket will contain a network of plastic pipes, constituting a French drain, to facilitate removal of the leachate. The general features of a heap leach pad are illustrated in Fig. 4.1. The heap leach pad is protected from extraneous water by diversion ditches or berms and on the down slope end it has a solution collection ditch, which also has an impervious liner. The slope and permeability of the drain blanket, which must be adequate to carry away percolating solution, are important factors in determining the actual rate at which leaching solutions penetrate the liner. However, if the slope is too great the heap may be unstable and slump or slide laterally. Proper foundation preparation is necessary to prevent movement and tearing of the liner. The drain blanket will also prevent polymer liner damage by sunlight, prevent evaporation and cracking of clay–soil liners, and protect the liner during heap construction.

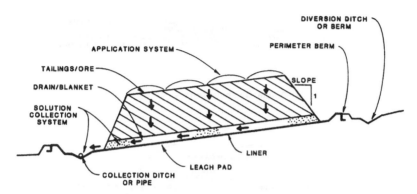

Figure 4.1. Leach pad and heap in cross section, slope = 2–6°.

Factors to be considered in liner selection and in leach pad and settling pond design are: seepage rate, settlement of the foundation, slope stability for heaps on slopes, weathering, solar radiation, and regulatory requirements.

Two types of impervious liners are in use: (1) clay and clay-amended soil liners and (2) polymeric sheet materials, sometimes referred to as **"geomembranes"**. Combinations of both types are in use and often multiple liners are required. Examples of typical geomembrane impervious liners sections are illustrated in Fig. 4.2. Asphalt road paving material has also been used in impervious liner systems, especially with reusable leaching pads.

Clay liners and **soil liners**, with or without clay additives, consist of selected materials placed in thin layers and compacted to prescribed moisture content and density specifications. The soil borrowed to produce the liner must be tested and the optimum additions of water and imported clay need to be determined in advance. The materials must be thoroughly mixed prior to introducing water, followed by further mixing and compaction. Permeabilities of liners are often specified by regulation with the maximum allowable **hydraulic conductivity** in the range of 10^{-5}–10^{-6} mm/s (1–0.1 ft/yr). Generally, the higher the amount of clay and plasticity of the clay–soil mixture, the lower will be its ultimate permeability for a constant liner thickness (see Fig. 4.3). However, highly plastic soils are difficult to mix because they tend to form clumps. Usually 5% bentonite, either naturally present or added to the soil, is sufficient to obtain the required low hydraulic conductivity.

Soils are tested by a standard Proctor Compaction Test. The relationships between optimum water content, dry density, and the resulting hydraulic conductivity are illustrated in Fig. 4.4. After compaction, ideally, all of the clay/soil void space will be filled with water but without any excess water.

a) Single Geomembrane Liner.

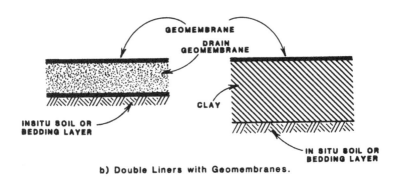

b) Double Liners with Geomembranes.

Figure 4.2. Examples of typical geomembrane liner sections (Van Zyl, 1987, Chapter 6).

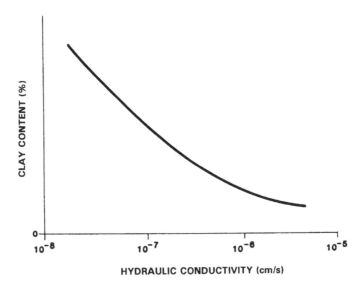

Figure 4.3. Typical relationship between hydraulic conductivity and clay content in leach pad liner.

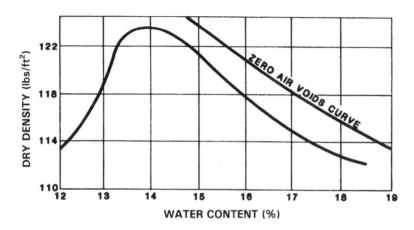

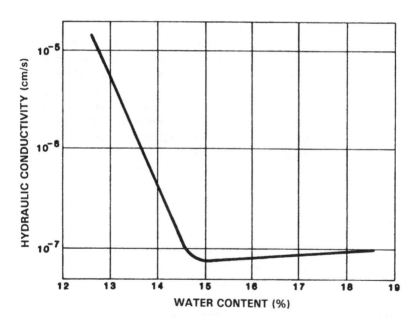

Figure 4.4. Relationships between density, moisture content and hydraulic conductivity for clay based liners.

Clay and clay-amended soils are dry mixed using a road grader or agricultural disc. Incremental amounts of water are then sprayed in with a water truck while mixing continues, using the same equipment until the optimum predetermined water content is achieved. Vibratory rollers or "sheeps foot" rollers are used to compact the liner to its maximum density.

In constructing clay and clay-amended soil liners, a series of compacted layers are sequenced. Each layer thickness is controlled by the effectiveness of the compaction equipment, and usually does not exceed 150 mm. A layer is spread and compacted followed by a second layer, up to the number of layers required to meet the specified hydraulic conductivity. Generally, total thickness after compaction should not be less than 300 mm of clay or clay-amended soil. When clumps are present in the soil, the total liner thickness should be increased to about 12 times the diameter of the largest clumps (sieve size that will pass 95% by weight of the soil). Once properly compacted, a clay or clay-amended soil liner should be kept moist to prevent desiccation and microcracking. After the drain blanket is installed and leaching begins, the liner will retain its moisture.

Organic polymer **geomembranes** come as sheet rolls and are laid out in slightly overlapping strips or panels that are subsequently bonded to each other by a solvent cement, vulcanizing tapes or adhesives. Various thicknesses of the geomembrane sheet are used depending on the application and regulatory requirements, but a 1 mm (40 mils) to 2 mm (80 mils) sheet thickness is common. Theoretically, these sheets are impervious, but tiny pin holes yield effective hydraulic conductivities in the range of 10^{-8}–10^{-9} mm/s.

In recent years high density polyethylene (HDPE) has been favored for its resistance to ultraviolet radiation and because it does not require reinforcement for adequate tear resistance. Polymer geomembranes shrink and expand with temperature changes. This factor and weather conditions must be considered when installing and joining the sheet panels.

Most heap liners are hybrids, involving both clay and a polymeric geomembrane. Typically, it is a compacted clay, or amended clay soil overlain by a **friction layer** of crushed rock about 5–10 cm thick followed by the geomembrane. The friction layer prevents lateral movement of the heap when on a slope and it also acts as a secondary drain that can be used both to detect and remove leakage that may occur through the primary geomembrane liner. Sometimes a **geotextile** sheet is used in place of the friction layer as a secondary drain.

A liner cross section of the leach pad for a cyanide leaching system using multiple lifts on low grade mine waste at Mercur, Utah (Brewer, 1986) is shown in Fig. 4.5. This heap is located in a valley with drainage to a central collection pipe in the bottom of the valley. The liner consists

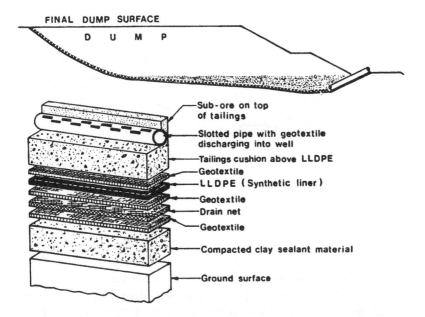

FINAL DUMP SURFACE

D U M P

Sub-ore on top of tailings

Slotted pipe with geotextile discharging into well

Tailings cushion above LLDPE

Geotextile

LLDPE (Synthetic liner)

Geotextile

Drain net

Geotextile

Compacted clay sealant material

Ground surface

Figure 4.5. Liner materials in cross section for Barrick–Mercur dump leach (Brewer, 1986).

of seven layers below the drain pipe. Geotextile layers strengthen the liner and are included to prevent tearing. This complex liner, involving both polymeric materials and compacted clay, was specified because the spent ore will remain permanently stored on the leach pad.

Bedding materials covering the impervious liner need high permeability to serve as an adequate drain blanket for the pregnant liquor. They should be free of fine particles and not contain angular rocks that are likely to tear a geomembrane liner. Washed stream-rounded gravel is ideal. Sometimes the first ore layer itself can be an adequate drain blanket if crushed and free of fine particles. For reusable leach pads, a few feet of ore are left over the liner at all times to protect it.

LEACH PAD SITE SELECTION

Site selection is a compromise between a number of factors, including haulage distance from the mining areas, environmental suitability, and the amount of earth work that must be done to provide grading of the slope for proper drainage of the heap. Figure 4.6 illustrates an equal balance between cut and fill to construct a nearly flat but adequately sloping leach pad.

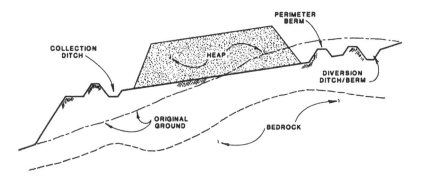

Figure 4.6. Illustration of cut and fill excavation required for leach pad construction.

The pregnant solution pond is invariably located down slope from the heap so that solution will flow to it by gravity. The barren solution pond should either be at the same elevation or slightly below the pregnant solution pond. This allows the metal separation plant to be by-passed during a prolonged shutdown. Preferably, in an emergency, the pregnant solution pond can overflow directly to the barren solution pond, without requiring pumps.

Additional factors to be considered in site selection are: flood potential, unacceptable subsoil conditions, and future prospecting potential for economic mineralization. A sufficient number of tests of the soil or rock under the proposed heap leach site are necessary to ensure that the ground remains stable under the weight of the heaps.

Some exploration drilling may be appropriate to determine whether the selected site contains economic mineralization.

In some situations it is necessary to construct a leach pad over an earth-filled gully. Settlement, especially differential settlement, of this fill can lead to rupture of the impervious base or geomembrane resting on the fill. Settlement can be minimized by the choice of earth-fill material, earth placement and compaction method, but settlement of 1–2% may still occur. HDPE geomembranes are less tolerant of multiaxial strain than the alternative geomembrane materials, very low density polyethylene (VLDPE) and polyvinyl chloride (PVC).

SOLUTION COLLECTION AND POND DESIGN

The solution collection system consists of a series of lined ditches and other components to collect solutions from the heap and convey them by

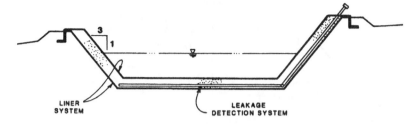

Figure 4.7. General features of a solution pond design.

gravity to the pregnant solution pond. The collection system must be designed to accommodate the rate of flow of solutions from leaching plus the runoff from major storms. The drain blanket is an important component and may contain highly permeable gravels, drain pipes (French drains) or merely crushed ore, if it is sufficiently permeable.

The necessarily large solution storage ponds will also be earthworks with impermeable liners. As shown in Fig. 4.7, solution ponds often have a double liner system for monitoring leaks. A highly permeable material separates the two liners and leaked solution, if detected, can be removed by a pump.

The pond side slopes, typically 3:1, are dictated by the liner materials and the stability of the underlying earth materials. The liner must be resistant to damage from sunlight, temperature variation, wave action, wind pressure and in-flow from ditches and pipes.

SEEPAGE AND FRENCH DRAIN PIPE SPACING

Ore in the leaching heap will usually have a hydraulic conductivity preferably in the range of 10^{-1}–1 mm/s so that the solution infiltration rate is well below the permeability limit and solution saturation will not occur within the heap.

When solution trickles down to the impervious liner, it must flow laterally over relatively long distances. Consequently, free standing saturated solution will exist at the bottom of the heap. Under steady state heap leaching conditions, the lateral drainage rate must be equal to the cumulative infiltration rate, W, through the heap over the area being leached. The drainage rate will also depend on the pad slope, drainage path length, L, hydraulic conductivity of the drain blanket, K_{DB}, and the water head, h, of the free standing solution lying immediately over the liner. This situation is illustrated in Fig. 4.8, where the drainage path length is the distance between equally spaced drain pipes in a French drain field.

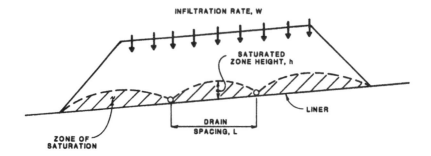

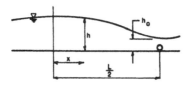

Figure 4.8.   Leach pad and heap with equally spaced drain pipes.

The water head at the drain pipe is h_0, so the lateral driving force for flow is related to $(h - h_0)$. For a given heap construction all the parameters are fixed except the water head, h, in the saturated zone above the liner. Although there are more rigorous expressions based on finite element computations, the "agricultural drain equation" is sufficiently accurate and conservative to select drain pipe spacings and materials for a drain blanket to adequately minimize the height of the saturated zone for a level pad:

$$h^2 = h_0^2 - \frac{WL^2}{4K_{DB}} - \frac{Wx^2}{K_{DB}}. \qquad (4.1)$$

For $h_0 = 0$ and $x = 0$, then:

$$\frac{L}{2}\left(\frac{W}{K_{DB}}\right)^{1/2} \simeq h. \qquad (4.2)$$

For the usual case of a sloping pad, the variation in head with slope distance must be included. With a modest slope and good K_{DB}, drain pipes may not be required, but most commercial leaching ore heaps are so extensive in area, that a network of plastic drain pipes are used within the drain rock.

From the height of the saturated region, h and h_0, and the permeability of the liner it is possible to estimate the seepage rate through the liner. This situation is illustrated in Fig. 4.9, where h_{avg} is the average head of the saturated zone and the liner has thickness, ΔX_L, and hydraulic conductivity, K_L. According to Darcy's equation, the seepage flux, loss through the liner per unit area, Q_L/A, is:

$$\frac{Q_L}{A} = K_L \frac{h_{avg}}{\Delta X_L} \quad [Lt^{-1}], \tag{4.3}$$

where K_L is the hydraulic conductivity of the liner.

Note that the solution flux is a volumetric flow rate per unit area which is identical with a (superficial) solution velocity or **specific discharge**, and it has the same dimensions as the hydraulic conductivity, Lt^{-1}. This relationship will be encountered frequently throughout this text. Various dimensions for the seepage flux are commonly used, e.g., gallons/day per acre, mm/s and feet per year.

Examples of leach pad solution seepage (losses) through a clay-amended soil liner, that is 250 mm thick with a typical value for the average head of the overlaying saturated zone, follow in Table 4.1. There are usually regulations on both the maximum allowable seepage rate and the maximum

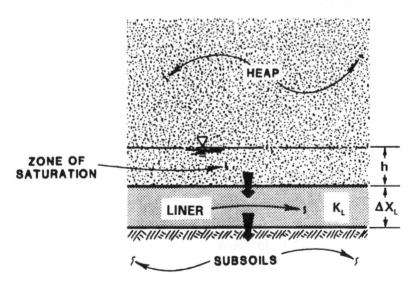

Figure 4.9. Seepage through the leach pad liner from the saturated layer of water immediately over the liner.

Table 4.1   Examples of Leach Pad Solution Seepage.

Avg Head, h_{avg} (mm)	K_L (mm/s)	Seepage Flux, Q_L/A	
		(mm/s)	(gal/day per acre)
50	1×10^{-5}	2×10^{-6}	1.85
50	1×10^{-4}	2×10^{-5}	18.5
50	1×10^{-3}	2×10^{-4}	**185**

allowable head, h_{avg}, on the liner. For example, in Nevada, h_{avg} must not exceed 24 in (0.6 m). This regulation requires precise grading during heap pad construction to prevent low spots that will accumulate solution at depths exceeding the regulation.

Because of the large size of leach pads, which may cover several acres, the seepage flux expressed as gallons per day per acre may more strongly impress the reader. It is clear that clay liner hydraulic conductivities should be 10^{-5} mm/s or lower for both economic and environmental purposes.

EXAMPLE PROBLEM I

For a drain blanket hydraulic conductivity of $K_{DB} = 0.2$ cm/s, a maximum head of stagnant solution on the heap pad liner of $h = 30$ cm, and a solution distribution rate using impact sprinklers of 0.005 gal/(min ft^2) of heap surface area, compute the maximum allowable distance separating drain pipes using eqn 4.2.

ANSWER

(1) The infiltration rate W will be equal to the sprinkler distribution rate as long as flooding on the top or within the heap does not occur. Converting the distribution rate of 0.005 gal/(min ft^2) to cm/s yields:

$$W = \frac{(0.005 \text{ gal/(min ft}^2) \times (3785.4 \text{ cm}^3/\text{gal})}{(60 \text{ s/min}) \times (30.48 \text{ cm/ft})^2}$$

$$= 3.4 \times 10^{-4} \text{ cm/s}$$

(2) Rearranging eqn 4.2 provides:

$$L = 2h(K_{DB}/W)^{1/2}$$

(3) Inserting the values of the parameters into this equation yields the maximum value of the separating distance, L:

$$L = 2(60)(0.2/3.4 \times 10^{-4})^{1/2} = 2910\,\text{cm},$$

$$L = 29\,\text{m}, \qquad L = 95\,\text{ft}$$

EXAMPLE PROBLEM II

At a large gold mine in Nevada, the heap area covers two million square meters and it is underlain with a 2 mm (0.80 in) thick geomembrane with a hydraulic conductivity of 1×10^{-11} cm/s. The average thickness of the saturated layer on the geomembrane is 60 cm. The average concentration of the pregnant solution or leachate is $0.41\,\text{g/cm}^3$. Calculate the annual seepage volume of pregnant solution in cubic meters, and the annual gold loss both in grams and in dollars at a gold price of $400/oz. Note that gold is priced in troy ounces and not in avoirdupois ounces.

ANSWER
Using eqn 4.3:

$$K_L = 1 \times 10^{-13}\,\text{m/s}, \quad h_{avg} = 0.6\,\text{m},$$

$$\Delta x_L = 2 \times 10^{-3}\,\text{m}, \quad A = 2 \times 10^6\,\text{m}^2,$$

$$Q_L = (1 \times 10^{-13}\,\text{m/s})(0.6\,\text{m})(2 \times 10^6\,\text{m}^2)/(2 \times 10^{-3}\,\text{m})$$

$$= 6 \times 10^{-5}\,\text{m}^3/\text{s}$$

$$Q_L = 5.2 \ \textit{cubic meters per day}$$

gold (g/day) = (solution volume/day) × (soluble gold concentration)

$$\text{Au/d} = (5.2\,\text{m}^3/\text{day})(0.41\,\text{g/m}^3) = 2.13\,\text{g/day}$$

$$\text{Au/d} = 2.13\,\text{g/day} \times 0.032151\,\text{oz Tr/g} = 0.0685\,\text{oz Tr/day}$$

dollar loss = gold price × daily gold loss × 365 day/yr

$$\$/\text{yr} = \$400/\text{oz} \times 0.0685\,\text{oz/day} \times 365\,\text{day/yr} = \$10,000/yr$$

EVIDENCE OF SEEPAGE FROM SOLUTION CHEMISTRY

A chemical inventory of the leach solution can sometimes be used to approximate total leakage rates from the circulating solution system. For example, in copper mine waste leaching, aluminum and magnesium are dissolved from mica and possibly other gangue minerals by the acidic leaching solution. There is apparently no precipitation or other chemical sink for magnesium cations, and they accumulate in the leach solution over time. However, the magnesium cations are also being bled away by seepage and their bleed rates are proportional to their increasing concentrations in the circulating leach solution. Eventually, a steady state condition is reached when the magnesium removal rate by seepage balances the magnesium generation rate, and at that time the solution concentration of these cations ceases to increase.

Achieving steady state magnesium (and aluminum) solution concentrations in a copper mine waste leaching system may take several years.

A seepage rate of the order of one percent of the recirculating solution flow rate was estimated with this chemical balance method for a copper dump leaching system at a major copper mine. This estimate seems reasonable, because the mine waste was dumped on adjacent sides of mountains (valley-fill) without using impervious liners or otherwise making preparations to eliminate seepage. Leaching began years after mine waste dumping occurred so that intervention with a liner to prevent seepage was not possible. Subsequently, this copper mine has undertaken a major ground water reclamation program to decontaminate ground penetrated by the dump seepage.

HEAP GEOTECHNICAL FAILURES

Heap movement can occur with high heaps or heaps located on steeply sloping sites. This problem can be particularly acute if perched water tables are located at the edge of the heap because of slow percolation (i.e., infiltration rates that are less than the solution application rate causing local saturation). Furthermore, the normal angle of repose for dumped material at the edge of a heap, 37 degrees, may be too steep to prevent a slumping failure in high heaps. Edge slumping failure surfaces are illustrated in cross section in Fig. 4.10. The probability of edge slumping failures can be minimized by stair-stepping back some of the lifts, as shown in Fig. 4.10, to provide a flatter average angle of repose than the normal dumping angle of repose. However, the amount of ore lost to leaching in lower lifts will increase.

Heaps may also fail by sliding along a high slope leach pad because the saturated solution layer in the drain blanket over the impervious liner lubricates motion. This is also illustrated in Fig. 4.10. Designing with a leach pad slope less than 6 degrees is generally considered adequate to prevent liner slippage. When slippage along the liner contact is a concern, sand or

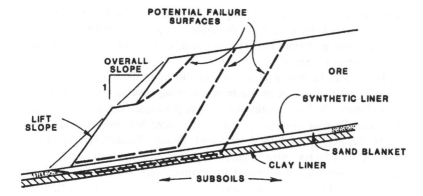

Figure 4.10. Potential ground failures in a wet ore heap.

other materials can be emplaced during construction of the heap to increase the coefficient of friction along this surface. Minimizing the height of the saturation zone on the liner is also important to preventing slippage.

A containment dam of stronger material can also prevent down-flow slippage of a heap in steep terrain (see Fig. 4.11). Concerns about the stability of leaching heaps should be addressed by geotechnical engineering consultants.

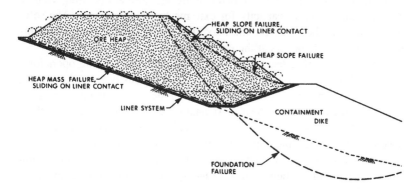

Figure 4.11. Failure modes in a contained heap in steep terrain.

WATER BALANCE

The importance of water balance is related to avoiding spills of leaching solutions or other contaminated water. Through a typical year, the water circulating through a heap leach operation is stored in different facilities. During start-up water is added to the system to partially fill the ponds and start the leaching operations. Once the heap is soaked to its fill capacity, solution starts to drain out.

A key aspect in designing a leach facility is to determine the overall average water balance. A **continuously gaining system**, such as those in wetter climates or those requiring extensive post-leach rinsing will require a solution disposal system to achieve balance. Solution disposal may include chemical or biological treatment. Desert climates, typical of the western United States, often require a large input of make-up water.

Solution ponds must be sized to accommodate large fluctuations in water volume. This involves meteorological water in-flow and out-flow, as well as the process circuit (Fig. 4.12). These fluctuations can involve: (1) short periods, such as intense storms, (2) annual accumulations, and (3) long-term climatic changes, which can be critical if the solution mining location is marginal with respect to either evaporation water loss or net water accumulation. Arid climate locations at high elevations or northern latitudes can encounter a large spring snow melt, which must be taken into account in planning the size of solution ponds.

An illustration of a one-year fluctuation in solution pond volume requirements is given in Fig. 4.13. In this case, the mine is closed during winter and the volume of heap drainage during the first quarter is large. Superimposed on the winter drawdown are the normal operating volume and additional reserve capacity for extreme hydrologic events. The solution pond must be sized for the maximum pond volume, which for this particular example occurs at the end of March.

In the interest of reducing the capital cost of heap leaching operations, there has been a recent trend in the gold mining industry away from using a barren solution pond. Barren liquor leaving the separation plant is immediately returned to the ore heaps. At established leaching operations there are usually several ore heaps or cells and almost always one that can accept returned leaching solution. Very recently, in valley fill heap leaching operations in desert climates, where there is a large net water evaporation, consideration is being given to using one small pond or eliminating the pond altogether and substituting the storage capacity in the ore heaps themselves as an emergency storage reservoir. Large ore heaps and dumps have an extensive volume of void space and well-drained heaps that have been out of service can act as a sponge to accept additional water in an emergency.

However, when this is done, it is better to either divide the pregnant liquor pond into two separate units or construct an overflow pond, so that

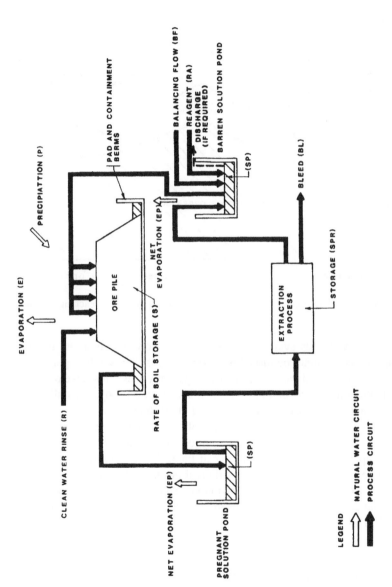

Figure 4.12. Water flows in connection with a heap leaching circuit.

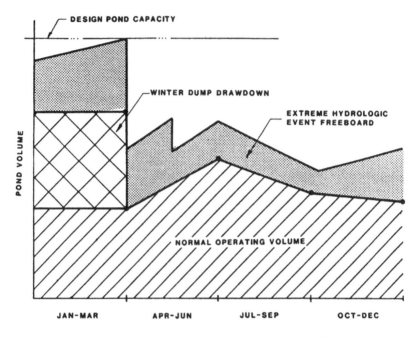

Figure 4.13. Example of annual water fluctuations in the heap leaching system.

there are two ponds available. Leaks may develop in a pond, but the ore heap drainage cannot be turned off. When at least two solution ponds are available, one pond can be drained to repair any leaks that develop in it, while the other pond continues to function and prevent solution spillage.

Another source of leaching solution spills is the occasional blow out of perched water tables within ore heaps, often causing a large ore landslide and a temporary spring of excess solution running down the sides of the ore heap and beyond the leaching pad liner. While rare, these spills can generate environmental citations with stiff penalties. Perched water tables, regions of saturation, result from inhomogeneities in the distribution of fine and coarse ore and local areas of very low permeability underlying the perched water table.

CYANIDE DETOXIFICATION

Some cyanide species are highly toxic, whereas others are relatively inert and harmless. Hydrogen cyanide is the most toxic form of cyanide, either in air or dissolved in solution. Cyanide ion is also extremely toxic. Cyanide in solution is generally classified as: (1) free cyanide, both dissolved HCN and CN^- ion, (2) total cyanide, which includes free cyanide plus various cyanide metal complexes, and (3) weak acid dissociable (WAD) cyanide.

Free cyanide concentrations greater than 0.2 mg/l (0.2 ppm) can kill sensitive species in fresh water or marine environments. Because WAD cyanide will convert to free cyanide as the ore pH decreases, the pore water in spent ore should have a WAD cyanide concentration that does not exceed this amount. The simplest detection method for free cyanide is the specific cyanide electrode technique. WAD cyanide concentrations can be determined by this method after weak acid additions to the sample are made.

Several natural processes occur in heap leach tailings that affect free cyanide concentration over time. Hydrolysis of cyanide ion occurs as carbon dioxide enters the material and eventually lowers the pH. Hydrolysis occurs according to the following reaction:

$$CO_2 + CN^- + H_2O = HCN + HCO_3^-. \tag{4.4}$$

The pH relationship between HCN and cyanide ion is shown in Fig. 4.14. Once HCN is formed, it is quite volatile and will eventually diffuse as a gas out of the spent ore and be air-oxidized to benign products.

Several other natural reactions can occur, including direct oxidation and aerobic biodegradation of HCN. For most host rocks, the minerals will react with the pore water and its pH will drift to values below pH 9.5. So, the free cyanide concentration will generally decline over time by natural processes. Thus, one method for cyanide destruction is **passive abandonment**, coupled with monitoring and/or fencing to prevent contact with wildlife or humans until the spent ore is adequately detoxified.

Another approach is to wash the spent ore with water to flush out cyanide and also lower the pH. However, large volumes of rinse solution that must be detoxified will result.

Research conducted on agglomerated mill tailings that had been heap leached, at a Darwin, California desert site, showed that passive abandonment lowered the cyanide concentration by a pseudo first order rate, i.e., the rate of elimination of free cyanide was proportional to the free cyanide remaining at any time. Cyanide concentrations were reduced from 0.6 g/l to 0.1 g/l in 18 months (see Fig. 4.15 from Van Zyl et al., 1988).

Results from water washing heap leach tailings originally containing 133 ppm of total cyanide are shown in Fig. 4.16. Cyanide is lowered to

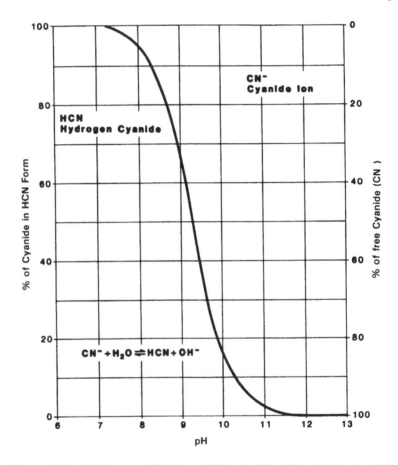

Figure 4.14. Solution equilibria for HCN and CN⁻ (Van Zyl et al., 1988, Chapter 13).

less than 1 ppm in 5 effluent pore volumes and acceptable detoxification was reached in these experiments in 7 pore volumes. These data were presented as a semi-logarithmic plot in Fig. 4.16 because it is widely believed that detoxification is a first order chemical rate process,

$$\text{Rate} = -k[\text{CN}^-] \tag{4.5}$$

and where the integrated form of the rate expression shows that the concentration declines in a straight line when $\log[\text{CN}^-]$ is plotted versus time.

Actually, the data in Fig. 4.16 are a poor fit of a straight line. There is no reasonable explanation of the cyanide destruction rate being dependent on

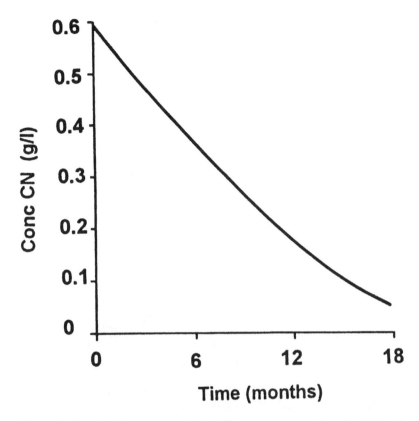

Figure 4.15. Cyanide concentration in tailings pore water at Darwin, California (Van Zyl et al., 1988, Chapter 14).

a molecular chemical reaction, when a chemical degradation reagent is not present, which is implied by the first order rate equation. It may be more reasonable to expect that the rate of cyanide destruction is controlled by its slow diffusion out of the rock micropores—similar to soluble gold extraction occurring in the heap earlier during leaching prior to detoxification. A semi-logarithmic plot of fraction extracted versus dimensionless time for diffusion–extraction from porous spheres is given in Fig. 4.17. This plot is based on eqn 1.2 and essentially equivalent to Fig. 1.1 but carried out further in dimensionless time. Computed plots are shown for a Gates–Gaudin–Schuhmann distribution of ore particles (expected from normal rock breakage) and for monosize spherical ore particles. The data point trend of Fig. 4.16 should be compared with the normal rock breakage

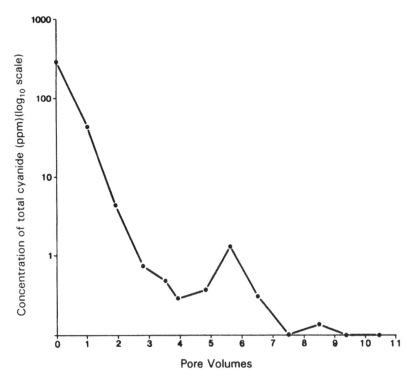

Figure 4.16. Semi-log data plot of cyanide concentration versus water wash pore volumes (Van Zyl et al., 1988, Chapter 14).

computed curve in Fig. 4.17. Over at least three orders of magnitude of cyanide removal, the extraction plots are very similar.

Dixon et al. (1994) have developed a similar diffusion based mathematical model for rinsing spent ore heaps, except that diffusion occurs out of "stagnant zones" to the downward flowing washing liquid, rather than out of individual ore particles. The stagnant zones include both ore particles and a nearby region of interstices that are filled with stagnant liquid because of capillary action. The dispersion pathways vary in length and cross-section and a Gates–Gaudin–Schuhmann distribution of stagnant zone size was assumed. The model verified the time dependent experimental cyanide concentration decrease over several orders using reasonable parameter values, and it fairly accurately predicted the effect of different rinsing solution flow rates.

Chemical treatments are more expensive but allow for quicker detoxification and usually better control of the final result. The following

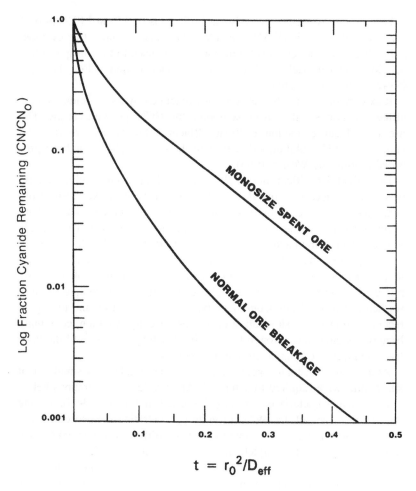

Figure 4.17. Semi-log plot of diffusion–extraction (cyanide) from spherical ore rocks.

processes have been used or proposed in connection with detoxifying cyanide in spent ore: (1) alkaline chlorination, (2) sulfur dioxide/air oxidation, (3) ammonium bisulfite (in place of SO_2), (4) hydrogen peroxide oxidation, (5) Caro's acid, (6) direct acidification, and (7) ferrous sulfate. Caro's acid (H_2SO_5) is produced in a rapid reaction on-site by mixing sulfuric acid and hydrogen peroxide. It has been used primarily to detoxify tailings slurries and is claimed to be faster and less expensive than using hydrogen peroxide alone (Castrantas et al., 1993).

Presently, most states have effluent limitations of 0.2 mg/l (0.2 ppm) of weak acid dissociable (WAD) cyanide, which also includes free cyanide. All of the preceding chemical processes can be used to treat heap effluent solution and can yield a WAD cyanide concentration below 0.2 mg/l. However, they are relatively expensive.

Recovery of cyanide for reuse is also practiced by direct acidulation of effluent solutions followed by scrubbing the HCN gas with caustic. The use of solvent extraction to treat effluent solutions and recycle the enriched cyanide solution stripped from the organic phase has been proposed (Virnig and Weerts, 1993).

Biological detoxification of CIP cyanide tailings using aerobic bacteria was first practiced near the Homestake Mine in South Dakota. Spent ore heaps have also been detoxified using chemolithotrophic bacteria. Many bacteria species are capable of aerobic oxidation of cyanide. The cyanide provides both carbon and nitrogen that are needed for cell metabolism.

The spent ore heap detoxification procedure is to add a bacteria culture to the barren solution pond as soon as additions of cyanide to the spent ore pile cease. The bacteria begin to oxidize the auricyanide complex as it is washed out of the heap. Consequently, both detoxification and final gold recovery from the wash solution occur concurrently. Solution, containing bacteria, is circulated through the existing leaching system until the heap leachate meets the cyanide effluent regulations.

At a 1.2 million tonne spent ore heap, which is relatively small, about six months were required to achieve WAD cyanide concentrations below 0.2 mg/l using bacteria oxidation (Thompson and Jones, 1993). During the treatment period, the WAD cyanide concentration ramped down at a constant rate. However, at a much larger, multilift heap it may take much more than one year to complete detoxification using bacteria.

The potentially long time to detoxify cyanide in large spent ore heaps has stimulated a concept for passive bio-detoxification by Mudder et al. (1995). The approach is similar to passive detoxification of acid mine water using anaerobic bacteria, such as Pseudomonas. Machinery is not used and human attendance is not required. This procedure is as follows. At a closing gold heap leaching operation, one of the former solution storage ponds, located at an elevation below the spent ore heaps, becomes a collector of heap seepage. This storage pond is filled with a permeable bed of inexpensive organic material, such as manure and straw, and covered with vegetation, thereby becoming an anaerobic cell. This cell has a lower and upper drain field. Heap effluent enters the bottom drain field of the cell and passes upward and out through the upper drain field pipe system. Laboratory column tests with cyanide heap effluent solutions have shown that WAD cyanide concentrations below 0.2 mg/l can be obtained with 20–30 days of effluent solution residence time in the cell. A contractor used this anaerobic method to successfully detoxify cyanide at Newmont Gold Company's Rain Mine.

While cyanide is the major problem in gold heap leaching, some leachates contain mercury and selenium in excess of environmental standards. Mercury is removed by activated carbon and should not be a problem with continued circulation of the leachate through a carbon adsorption plant, if available. The mercury follows gold to the gold refinery where mercury is captured. A large variety of methods for capturing selenium have been explored, principally by the Bureau of Mines, and reviewed by Brierley (1992).

PERMITTING

The primary goal of environmental permitting is to avoid contamination of surface and groundwater resources with solutions. This generally means a zero discharge philosophy and requirements for leach pad liners with very low hydraulic conductivity. The emphasis is on an appropriate environmental assessment, coupled with designing the project to prevent environmental emergencies and also manage them, if they occur.

The primary regulatory agencies in the U.S.A. are the Bureau of Land Management, Forest Service, state and county governments and sometimes the Environmental Protection Agency (EPA). Permitting is required under the National Environmental Policy Act (U.S.A.) of 1969 (NEPA) and requires an Environmental Assessment (EA). Heap leaching projects in sensitive areas require a full Environmental Impact Statement (EIS).

Preparing an Environmental Assessment requires a number of base-line environmental studies in the region of the proposed site. These studies generally address geology, hydrology, soils, vegetation, wildlife, aquatic features, history and archeological aspects of the area.

Other concerns such as air quality, noise, visual resources, and socioeconomics may be included. Bird (waterfowl) suppression at cyanide ponds and heaps may be required for precious metal heap leaching operations. Noisemakers have been tried but are rarely successful deterrents. Nets stretched over the top of a cyanide pond can be used if the dimensions of the pond are not so great as to make this impractical.

Hollow plastic spheres, approximately 10 cm in diameter, float on the solution contained in a pond. A huge number of these **birdballs** must be dumped into the pond to provide, as a minimum, full coverage of the solution at is highest level and greatest surface area. At low solution levels, the birdballs override each other. There must also be enough birdballs, with adequate physical properties, to deter wind blown accumulations of bird-balls from occurring at one side of the pond, leaving some of it open to bird landings.

A preferred plan of operation and a discussion of how the plan mitigates any environmental impacts is essential in an environmental assessment. A reclamation plan for mine closure is usually required. Alternative plans should also be presented and evaluated in the EA report. A forum for public notice and comment is included in the overall permitting process. The EA plan must submit information on the geotechnical, hydraulic and engineering design to prevent solution contamination during operations and in the event of an emergency, such as the largest expected flood during a 100 year period. Solution mining air quality permits focus primarily on particulate emissions from crushing, agglomerating and other dust produc-ing steps in the overall operation.

Mining wastes are not currently considered by EPA to be hazardous. But EPA is required to make a final classification later and cyanide leached spent ore may become hazardous, unless detoxified to a specified level of residual free cyanide. Mineral wastes containing: (1) a high acid generation potential, (2) radioactivity, and (3) asbestos type minerals may eventually be declared hazardous by EPA.

Several states have already promulgated cyanide regulations. Effluents are usually limited to a maximum of 0.2 mg/l (0.2 ppm). In desert climates, adequately sized storage ponds, coupled with summer evaporation of excess water, can be used to provide for zero discharge from the solution mining system.

PROBLEMS

1. A **level** reusable heap leach pad is planned to cover 375,000 m² (nearly 100 acres). The drain blanket has a hydraulic conductivity of 5 mm/s and the impact sprinkler solution application rate will be 0.0025 l/s m². Does this heap leach pad need drain pipes if the saturated solution layer thickness, h, is restricted to 50 mm, and if so how far apart should they be spaced? Explain your answer with the necessary calculations.

2. The clay amended soil leach pad liner of problem 1 has a hydraulic conductivity of 1×10^{-5} mm/s and a thickness of 300 mm. Calculate the daily (24 hour) solution seepage from this leaching heap in cubic meters, using your design information from Problem 1, including $h = 50$ mm.

3. The gold content of the pregnant solution in Problems 1 and 2 is 4 ppm (by weight). What is the annual gold seepage financial loss if the gold price is $350 per **troy** ounce. Show calculations.

4. What percentage of the initial soluble free cyanide is present after the residual pH in the heap leach residue (water) is lowered to: (a) pH = 8, (b) pH = 7 and equilibrium has been achieved?

REFERENCES AND SUGGESTED FURTHER READING

Brewer, R.E. (1986). The Barrick–Mercur dump leach project. *Technical Paper A86–36*, The Metallurgical Society, Warrendale, PA.

Brierley, C.L. (1992). Selenium in mine and mill environments. *Randol 92*, Randol Int'l. Ltd, Golden, CO, pp. 175–179.

Castrantas, H.M. et al. (1995). Caro's acid, the low-cost oxidant for CN detoxification attains commercial status. Preprint 95–153, Soc. of Min. Metall. and Exploration, Littleton, CO.

Dixon, D.G., Dix, R.B. and Comba, P.B. (1994). A mathematical model for rinsing of reagents from spent heaps. Hager, J.P. et al. (Eds.) *Extraction and Processing for the Treatment and Minimization of Wastes*. TMS, Warrendale, PA.

Mudder, T. et al. (1995). Lab evaluation of an alternative heap-leach pad closure method. *Mining Engineering, Nov.*, pp. 1007–1014.

Thompson, L.C. and Jones, E. (1993). Bio-detoxification of spent ore and heap leach solutions. *Randol Gold Forum 93*, Randol Int'l. Ltd., Golden, CO, pp. 343–346.

Virnig, M.J. and Weerts, K.E. (1993). Cyanomet™ R-A process for the extraction and concentration of cyanide species from alkaline liquors. *Randol Gold Forum 93*, Randol, Int'l. Ltd., Golden, CO, pp. 333–336.

Van Zyl, D.J.A. (1987). *Geotechnical Aspects of Heap Leach Design*. Society of Mining Engineers, Littleton, CO, Chaps. 1, 3, 6.

Van Zyl, D.J.A., Hutchison, I.P. and Kiel, J.E. (1988). *Introduction to Evaluation, Design and Operation of Precious Metal Heap Leaching Projects*. Society of Mining Engineers, Littleton, CO, Chaps. 3, 10–14.

FIVE

Percolation Leaching Oxidized and Secondary Sulfide Copper Minerals

INTRODUCTION

Nearly 30% of domestic copper mine production is derived by solution mining methods. An important copper contributor is percolation leaching of huge waste rock dumps, resulting from the stripping of overburden at existing open pit copper mines. Most of the copper in mine waste is in the form of the primary copper sulfide mineral, chalcopyrite. Because of its greater complexity, copper mine waste leaching will be discussed in the next chapter. Percolation leaching of oxidized copper ores is also important and will be covered in this chapter. Leaching secondary copper sulfide minerals will also be covered in this chapter.

Natural occurring copper ores usually do not have adequate permeability for percolation leaching without fragmentation (rubblizing). Copper in sandstone is an exception but of minor economic importance. Percolation leaching of copper can occur in heaps, mine waste dumps, and caved mine workings above the water table. Large explosive blasts of ore have been conducted to adequately rubblize copper ore prior to percolation leaching.

An important difference between copper percolation leaching and gold percolation leaching is the much greater concentrations of copper metal in both the ore and the resulting pregnant solution. Typical ore and solution grades are compared in Table 5.1. The copper grades in both cases are

Table 5.1 Typical Ore and Solution Grades.

Percolation Leaching	Solids Grade (ppm)	Solution Grade (ppm)
Gold	1–3	0.2
Silver	20–75	5–40
Copper	2000–4000	1000
	(0.2–0.4%)	(2 lb/ton)

between 1,000 and 10,000 times greater than the gold grades. This increases the time required to complete leaching extractions of copper for similar rock sizes and microporosities, because more material must be transported out of the rock through its solution-filled micropores.

COPPER ORE DEPOSIT GEOLOGY

Copper ore deposits are sedimentary, stratiform or porphyritic. In porphyry deposits, copper is disseminated either in fine grained igneous intrusions or adjacent host rock. This is the dominant type of copper ore deposit in the western hemisphere. Figure 5.1 illustrates an idealized vertical section through a porphyry copper ore deposit, showing primary sulfide ore at

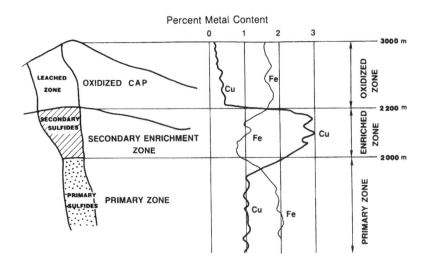

Figure 5.1. Copper porphyry ore deposit types showing primary, secondary enriched and oxide zones (Wadsworth, 1987).

depth, an intermediate secondary enrichment zone and, at the top, a naturally weathered (leached) zone above the water table where all of the copper has been oxidized, given sufficient time. Typical copper concentrations (weight percent) in the ore in each zone are also shown in Fig. 5.1. Because of subsequent erosion, all of these zones may not be present and sometimes there are mixed oxide/sulfide zones in a copper ore deposit.

The primary sulfide mineral is chalcopyrite ($CuFeS_2$) and there may be several secondary sulfide minerals. Pyrite (FeS_2) is always present in primary mineral zones and it is usually present in secondary mineral zones. Oxidized iron minerals are usually present in the oxidized zone (see Fig. 5.2). Water flooding excludes air (oxygen) and, consequently, the oxidized ore zone cannot extend below the water table, at least as it existed historically over the life of the mineral deposit.

Oxidized mineral zones are often depleted by the leaching action of percolating supergene water. If pyrite was previously abundant, acidic solutions were generated that solubilized copper and transported it downward. Precipitation of the soluble copper below the water table causes

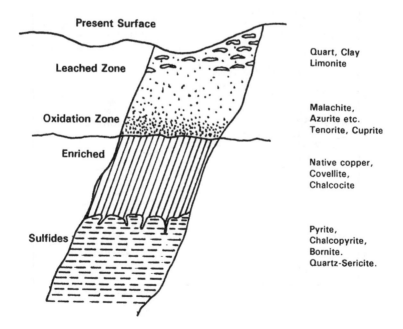

Figure 5.2. Upper portion of copper ore deposit showing zonation, water table, and typical copper minerals (Dudas, 1974).

secondary copper enrichment, as the oxidation potential drops and chalco-pyrite reacts with the infiltrating soluble copper to form secondary miner-als, e.g., chalcocite (Cu_2S), bornite (Cu_5FeS_4), and covellite (CuS).

If acidic conditions cannot be maintained in the oxidized zone, copper carbonate minerals may precipitate. Hence, there are several possible oxi-dized copper minerals, many of which are shown in Fig. 5.2. Ores contain-ing these minerals as the predominant copper source are referred to as "oxide copper ores."

Figure 5.3 illustrates the E_h–pH relations for the copper–iron–sulfur–water system (Wadsworth, 1987). The secondary enriched zone corresponds to the E_h region of approximately 0.2–0.6 V, at low pH, with the oxidized zone above this E_h range and the primary mineral zone below it. Autotrophic bacteria, such as *Thiobacillus ferrooxidans*, catalyze both sulfide mineral oxidation and oxidation of dissolved ferrous ions to ferric ions. This is very important to the percolation leaching of copper sulfide minerals, and it will be discussed in the next chapter. From Fig. 5.3, note that the bacteria are

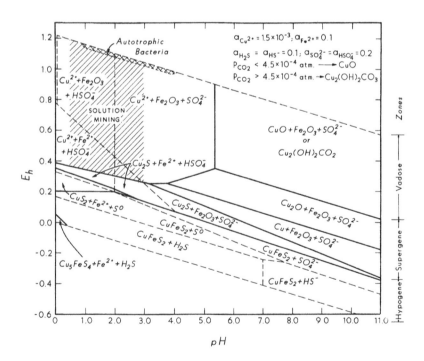

Figure 5.3. E_h–pH diagram for the copper–iron–sulfur–water system showing zoning, solution mining region and region of bacteria action (Wadsworth, 1987).

only active at high oxidation potentials, corresponding to a minimum of 2–3% oxygen in air, and about 1 ppm of dissolved oxygen in the leachate.

CONSEQUENCES OF DIFFERENT COPPER ORE TYPES

As a practical matter, it is useful to distinguish oxide copper ores, both with and without secondary sulfides discussed in this chapter, from primary ores and primary mineralization mine waste, which are considered in the next chapter. Oxide copper minerals leach rapidly in acid, when compared with the other copper minerals, while secondary sulfide minerals require oxidation and leach at a slower rate. Nevertheless, the secondary copper sulfide minerals oxidize and dissolve much more rapidly than either chalcopyrite or pyrite, and they are often called "leachable sulfides." Also, much less oxidizing reagent (dissolved oxygen or ferric ions) is required to leach them than to leach primary copper ore. Chalcopyrite in primary ore (or mine waste) leaches very slowly, and at about the same rate as pyrite, which is usually the most prevalent sulfide mineral in primary ore. Consequently, when leaching these primary ores, most of the oxidizing reagent is consumed to oxidize pyrite rather than chalcopyrite and very large amounts of oxygen are required relative to the amount of copper extracted.

The different ore types cause significant differences between the heap leaching practice for oxide copper minerals and for primary copper ore (chalcopyrite). Some of these differences in the criteria for percolation leaching practice, which are discussed in greater detail in this and the subsequent chapter, are listed in Table 5.2.

Table 5.2   Effect of Selected Criteria on Copper Ore Heap Leaching.

Criteria	Primary Ore	Leachable Sulfides	Oxide Cu
Reaction Kinetics	very slow	fast	very fast
Acid Pretreatment	harmful[1]		important
Soln Irrigation Rate	low	high	high
Ferric Concentration	limited by pH	limited by pH	limited by pH
Bacterial Activity	yes	yes	no
Heap Depth	uncontrolled	important	critical (low)
Aeration Required	high	moderate[2]	none

[1]Concentrated acid introduced to the top of a heap reacts rapidly with the gangue minerals causing excessive decrepitation of the rocks and lowered hydraulic conductivity; see Chapter 6.

[2]See a later section on the ferric cure leaching process.

Ore from secondary enriched zones and oxide copper ore are often present in the same mine and mixed during mining production. With appropriate planning, considerable extraction of copper from secondary sulfide minerals can usually be obtained.

MINERALOGY AND LEACHING CHEMISTRY OF OXIDE COPPER ORE

The more important oxidized copper minerals are malachite ($CuCO_3 \cdot Cu(OH)_2$), chrysocolla ($CuSiO_3 \cdot 2H_2O$), and azurite ($2CuCO_3 \cdot Cu(OH)_2$). A typical leaching dissolution chemical reaction is:

$$CuSiO_3 \cdot 2H_2O + H_2SO_4 = CuSO_4 + 3H_2O + SiO_2. \qquad (5.1)$$

Note that sulfuric acid is a necessary purchased reagent. Generally, the individual oxide minerals are not determined in an ore; rather, the copper content of the ore that is "acid soluble" is determined by a standard sulfuric acid laboratory extraction test. Acid consumption by the ore is also determined by laboratory testing.

Carbonate rocks, e.g., limestone and dolomite, are often significant in oxidized copper ores and will consume additional acid:

$$CaCO_3 + H_2SO_4 + H_2O = CaSO_4 \cdot 2H_2O + CO_2(g). \qquad (5.2)$$

Consequently the cost of purchased acid can be a major economic factor in leaching oxide copper ores. Carbonate rocks also cause gypsum to precipitate in the ore heap during leaching, which can very adversely affect heap permeability and solution percolation. The poor percolation caused by gypsum precipitation in ore heap leaching tests at the Twin Buttes Mine in Arizona is an example. It forced the adoption of agitation leaching and solids/liquid separation by countercurrent decantation, using the world's four largest thickeners. This was a much more costly processing option than heap leaching would have been if it had been feasible.

Unlike the leaching of copper sulfide ores, oxygen and bacteria are not required to leach oxide copper ore. The copper carbonate minerals dissolve rapidly if sufficient acid is present. Copper silicate minerals dissolve more slowly, but this is usually not a kinetically constraining feature of oxide copper ore heap leaching.

BLUEBIRD MINE OXIDE COPPER ORE
HEAP LEACHING PRACTICE

The Bluebird Mine near Miami, Arizona is an excellent example of oxide ore heap leaching practice (Power, 1970). Although the mine has been depleted and closed, this operation is historically significant as the location of the world's first commercial copper solvent extraction and electrowinning plant (SX/EW). Construction of the project was initiated in 1968.

The Bluebird ore body consisted of fractured granite and schist into which copper solutions had been geochemically transported and precipitated. The major mineral was chrysocolla and the average ore grade was 0.5% copper. A small amount of the copper was not acid soluble, but since only a trace of sulfide mineral was present most of the nonsoluble copper was probably occluded by silica or clay present in the host rock. Mining was carried out at the rate of 5,400 tonne/day and since the ore was friable, bulldozer ripping followed by scraper loading and haulage was used, rather than drilling, blasting and shovel loading of ore haulage trucks. This mining method fragmented the ore to an effective size of $-150\,mm$ ($-6\,in$).

Ore deposited by the scrapers was windrowed sideways to the crest of an advancing lift with a road grader. This procedure avoided compaction of the heap by the scraper wheel pressure. After each ore lift was completed, it was furrowed with a bulldozer ripper shank and the top 300 mm (1 ft) was removed and redeposited at the edge of the heap. This practice, common when leaching mine waste dumps in the copper industry, helped to provide uniform distribution of the leach solution and more uniform infiltration of the solution through the heap.

Each ore lift was 5–6 m high and leached for approximately 180 days (half a year) before being overdumped with a new lift. Spent ore was not removed, so leaching of the larger rocks at depth continued for years. Because the rates of copper extraction and acid consumption were greatest during the initial leaching period and declined with time during the 180 day primary leaching period, acid strength was reduced incrementally during this period following the schedule in Table 5.3.

Sufficient heaps were available and staggered in sequence so that a new ore lift could begin leaching every 20 days. This procedure provided a fairly uniform generation of total leachate volume and a fairly uniform leachate copper grade for feeding the SX/EW recovery plant.

Sprinklers could not be used to distribute the acidic barren solution because of its high salt content. Evaporation at the sprinkler head caused salt precipitation and plugging. The major dissolved salts were ferrous sulfate, ferric sulfate, aluminum sulfate, magnesium sulfate, copper sulfate

Table 5.3　Solution Acid Strength Leaching Schedule.

Period (days)	Barren Solution Strength
0–10	50 g/l H_2SO_4
10–30	30 g/l H_2SO_4
30–60	20 g/l H_2SO_4
60–180	7 g/l H_2SO_4 (SX raffinate)

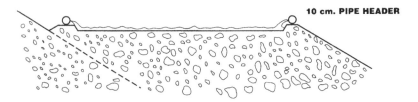

Figure 5.4. Sketch of solution application by discharge from valves in a header pipe at the edge of the heap—method used at Bluebird and many copper mines.

and sulfuric acid. Freezing of the leaching solution in winter did not occur because of the high concentration of its dissolved salts. This is generally true of copper heap leaching in severe winter climates.

At the Bluebird Mine, solutions were trickled from a row of valves onto the leveled heap from 100 mm diameter header pipes that were located on berms surrounding the heap (illustrated in Fig. 5.4). This ponding method is a common practice in copper dump leaching. It is similar to flooded basin irrigation of furrowed panels in farming. However, the rate of solution infiltration is controlled by heap permeability, which is variable, rather than being controlled by a water sprinkler or other metering device. Consequently, infiltration rates are also variable and usually too high. The leaching solution travels through the heap rapidly, yielding a low grade pregnant liquor, and large solution flows must be circulating through the ore heap and processed through the metal recovery plant. These large solution volumes increase metal recovery costs.

COPPER SOLVENT EXTRACTION/ELECTROWINNING

Solvent extraction involves the transfer of cupric ion from clear, dilute pregnant liquor to an organic molecule dissolved in a highly purified kerosene solvent. Actually, cupric ion is exchanged for hydrogen ion during extraction. The loaded organic solvent is stripped into a highly concentrated sulfuric acid solution, with the reverse exchange of cations. The strip solution is a much more concentrated copper solution than the pregnant liquor, and it is substantially free of impurities. Copper is electrowon from this solution using conventional technology. Electrowinning generates additional acid in the electrolyte (from water electrolysis at the anode), which is recycled as the solvent extraction stripping solution and provides the needed two H^+ ions for solvent extraction exchange with each Cu^{2+} ion.

A solvent extraction process flowsheet is illustrated in Fig. 5.5. Copper is transferred through the system from the leach solution to electrowon copper, while hydrogen ions are transferred countercurrently to the leach solution. Consequently some, but not all, of the acid needed for ore leaching is generated by the SX/EW process itself. Simplified reactions are:

Extraction

$$CuSO_4 + 2HR = CuR_2 + H_2SO_4, \qquad (5.3)$$

Stripping

$$CuR_2 + H_2SO_4 = CuSO_4 + 2HR, \qquad (5.4)$$

Electrowinning

$$CuSO_4 + H_2O = Cu^0 + H_2SO_4 + 0.5O_2(g). \qquad (5.5)$$

Both the extraction and stripping processes are conducted continuously in mixer-settlers (Glassner et al., 1976). The organic and aqueous phases are brought together and thoroughly mixed for a sufficient time, typically only a few minutes, to complete the ion exchange transfer. Then, the mixture enters the settler where the two immiscible phases disengage under the influence of gravity. It is important to keep the leachate free of particulate matter because particles tend to stabilize an emulsion in the mixer, commonly referred to as "crud." Copper concentrations in the aqueous and organic phases nearly attain those permitted by equilibrium. Multiple mixer-settler stages for both extraction and stripping may be required. At the Bluebird Mine three extraction stages were used, with countercurrent flow of the organic and aqueous phases between stages. Two stripping stages were used.

The barren solution from extraction, termed the raffinate, is returned to leaching, as is done in other closed-cycle heap leaching systems coupled with a metal separation plant. At the Bluebird Mine this raffinate contained about 7 g/l free sulfuric acid. Some of the raffinate was enriched

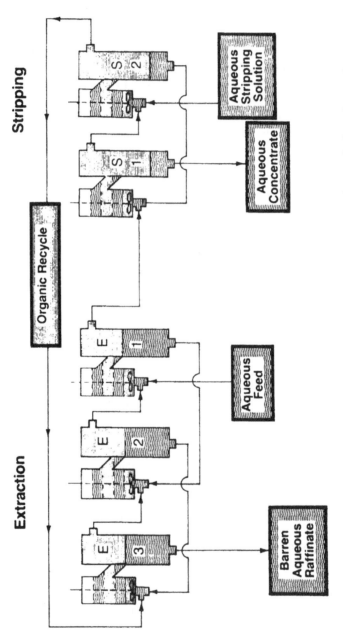

Figure 5.5. Solvent extraction flowsheet with three extraction mixer-settler stages and two stripping mixer-settler stages.

with purchased sulfuric acid for use with new ore lifts as previously described in Table 5.3.

The development of organic extractants selective for copper, beginning with the alkyl–aryl hydroxy oxime reagents that chelate cupric ions (developed by General Mills, now Henkel Chemicals), has been one of the most important twentieth century developments in copper extractive metallurgy (Agers and De Ment, 1972; Flett, 1974).

There are now more than thirty five copper SX/EW plants and several organic extractants tailored to different applications. In 1995 world-wide copper production from SX/EW was about 1,200,000 tonnes and may reach 2,500,000 tonnes by the end of the century. Most leaching pregnant liquors have a pH from 1.5 to 2.5. Strip liquors for electrowinning are typically from 30 to 36 g/l of copper and 130 to 190 g/l of sulfuric acid. Kordosky (1992) recently reviewed the status of copper solvent extraction.

UNDERGROUND COPPER LEACHING PRACTICE

Underground leaching of copper ores has been practiced on both oxide and sulfide ores that have been rubblized. Block caving is a classical method of underground mining of large disseminated ore bodies such as porphyry copper deposits. Blocks of ore are undercut with "finger" raises above a haulage drift. Each raise is opened sufficiently to stimulate caving of the ore above the raise. Caved ore is removed and more caving occurs to fill the open void space of the raise. Eventually a large vertical column of ore is rubblized and available for either further mining or leaching in place, as shown in Fig. 5.6. Leach solutions enter, via upper drifts in the ore or by injection wells located above the ore, percolate through the fragmented ore and are collected on haulage drifts below the ore block. Rubblization requires removal of about 15–25% of the ore with simultaneous creation of an equivalent void volume in the ore rubble.

Leaching of block-caved remnants began in 1941 at the Miami Mine in Arizona (Fletcher, 1971). Commercial production of percolation leached copper from rubblized ore in underground workings is occurring at Magma's San Manuel Mine (Arizona), at Cyprus Minerals' Casa Grande Mine (Arizona) and at the El Teniente Mine (Chile).

Examples of underground percolation leaching of remnant copper ore rubblized by large scale explosive blasting include the Old Reliable Mine in Arizona (Longwell, 1974) and the Big Mike Mine in Nevada (Ward, 1974).

The remaining deposit at the Old Reliable Mine was a breccia pipe exposed along the side of a steep ridge that contained about four million

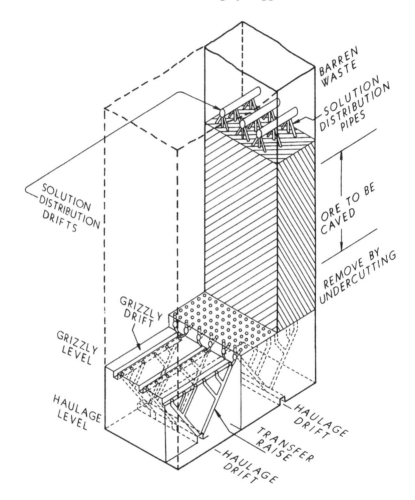

Figure 5.6. Underground percolation leaching of block-caved ore (Wadsworth, 1987).

tonnes of low grade copper ore, mostly as oxide copper minerals. Existing and new underground tunnels were used to emplace 2,000 tonnes of ammonium nitrate (ANFO) explosive that were detonated in one of the largest commercial blasts every conducted. The average ore fragment size was 270–300 mm. The rubblized hillside was terraced into narrow 6 m high benches and leached for a few years with dilute sulfuric acid using impact sprinklers. Copper was recovered from the pregnant liquor at a location below the ore deposit by scrap iron cementation in launder cells.

The small, high-grade Big Mike deposit had been previously open-pit mined down to an economic pit limit. Then, approximately 500,000 tonnes of pit wall and pit bottom mixed oxide and secondary sulfide ore were blasted with explosives. The bottom was fragmented from numerous blast holes, drilled from 12 to 60 m deep on a 3 m × 3 m grid. After leveling and benching, copper was leached with dilute sulfuric acid using impact sprinklers for uniform distribution of the downward percolating leach solution. Pregnant liquor was collected at the bottom of the fragmented ore and was removed by a single well, using a downhole turbine pump. Copper was recovered by iron cementation.

LEACH EXTRACTION FROM OXIDE COPPER ORE ROCK FRAGMENTS

Some theoretical aspects of leaching oxide copper ores will now be considered. The accepted mathematical model for leaching individual ore fragments will be considered first for monosize ore and then extended to a distribution of ore particle sizes. This will be followed by a discussion and model for the entire ore heap along its vertical axis, which describes changes to the solution that occur as it percolates down through the heap while being depleted in acid.

Consider a typical ore rock containing oxidized copper minerals, a copper concentration of 0.5% by weight and an internal micropore volume of 5%. With a rock specific gravity of 2.8 g/cm³, the internal micropore volume of 1 kg of rock is 18 cm³ (0.018 l). At 0.5 wt pct, the copper in the micropore solution, if fully dissolved, is 5 g and the concentration would be 278 g/l. Hence, with the Bluebird Mine's average pregnant liquor copper concentration of 2.9 g/l, nearly 100 pore volumes would be required to dissolve all of the copper in the rock. Conversely, only about 1% of the copper will be dissolved by the initial solution injection into the rock micropores. Recall that one pore volume is capable of dissolving all of the gold in typical disseminated gold ores as discussed in the previous chapters.

No pumping mechanism will remove solution from the micropores once, let alone 100 times. The leachate enters once by capillary forces and drainage of the ore will not remove the micropore water, unless the ore mass is dried. Consider this matter from another perspective with the following problem.

EXAMPLE PROBLEM

Granted that the copper solution concentration in the rock micropores may be greater than 2.9 g/l average at the Bluebird Mine, what amount and percent of the copper contained in the ore as chrysocolla will be dissolved by the initial penetration into the rock micropores of the strongest solution strength used at Bluebird, 50 g/l H_2SO_4?

If this acid reacts quickly with the available chrysocolla as it enters the rock micropores, as expected, how far will it penetrate into the ore rock before all of the acid is consumed?

ANSWERS

(1) From eqn 5.1, 1 mol of H_2SO_4 will dissolve 1 mol of copper from chrysocolla. Therefore, the copper dissolved per liter of acid solution at a concentration of 50 g/l H_2SO_4 is:

$$Cu_{diss} = 50 \text{ g/l} \times (64.54 \text{ g/mol})/(98 \text{ g/mol})$$

$$= 32.42 \text{ g-Cu/l-acid soln.}$$

(2) With a pore volume of only 0.018 l/kg-ore, from the preceding discussion, the copper in ore dissolved from the initial acid penetration is:

$$Cu_{diss} = 0.018 \text{ l/kg-ore} \times 32.42 \text{ g-Cu/l-acid soln.}$$

$$= 0.58 \text{ g-Cu}_{diss}/\text{kg-ore.}$$

(3) At 0.5 wt pct, the total copper in the ore is 5 g-Cu/kg-ore

$$\%Cu_{diss} = 100(0.58 \text{ g-Cu}_{diss}/\text{kg-ore})/(5 \text{ g-Cu/kg-ore}) = 11.6\%$$

$$11.6 \text{ wt pct.}$$

(4) The unreacted rock core is $100\% - 11.6\% = 88.4\%$ of the rock. Therefore the rock radius, r_0, and the core radius, r_c, are:

$$r_c^3 = 0.884 r_0^3; \text{ so } r_c = (0.884)^{1/3} r_0 = 0.96 r_0.$$

(5) Penetration and rim thickness $= r_0 - r_c = (1 - 0.96) r_0$;
so *Penetration* $= 0.04 r_0$.

The preceding problem is instructive. It shows that the initial penetration of even a strong acid solution will produce a **leached rim or shell** surrounding an unaffected core of the rock. And, in the example problem, using the typical conditions of heap leaching at the Bluebird Mine, **the initial leached rim thickness is only about 4% of the radius of the rock**. Contrast this situation with disseminated gold ore where **all** of the submicron gold accessible to the micropores is quickly dissolved with the injected leachate without consuming all of the cyanide reactant.

So, what is the mechanism for extracting copper from a simple ore fragment during leaching? As it dissolves, copper leaves the rock by diffusing through the stagnant pore liquor. Although additional acid **cannot flow** into the rock micropores, acid (hydrogen ions) **can diffuse** into the rock micropores under an acid **concentration (or chemical activity) gradient**, dA/dt. Hence, countercurrent chemical diffusion of hydrogen and cupric cations occurs through the stagnant liquid solution in the rock micropores. This mechanism is illustrated in Fig. 5.7. As more hydrogen ions diffuse

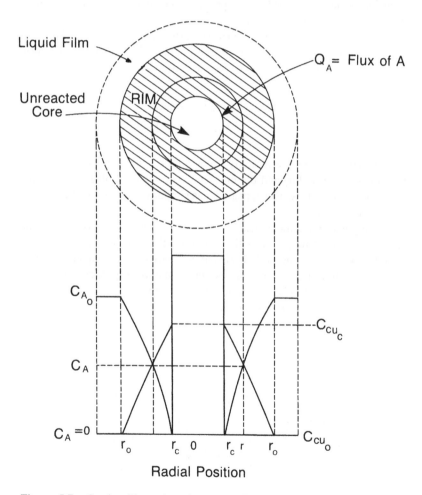

Figure 5.7. Section illustration of concentration gradients for a single leaching oxide copper ore rock, showing a leached rim surrounding an unleached core.

into the rock, the leached rim widens and the unleached core shrinks until eventually the entire rock is leached.

Observe from Fig. 5.7 that the geometry of the diffusion process differs markedly from the non-steady state diffusion occurring in the extraction of soluble gold, as described in Chapter 2 using equations from Chapter 1. In the present copper case, transport of hydrogen ions and cupric ions occurs between two boundaries: (1) the rock surface at r_0 and (2) the leached rim/core interface at r_c. Because these interfaces separate very slowly, diffusion is nearly a steady state process; and, it is called "quasi-steady state", countercurrent diffusion.

As the leached rim widens, the **rate** of extraction will decrease because, (for constant external solution concentrations of hydrogen and cupric ions), the concentration gradients decrease; for example:

$$\frac{dc_{Cu}}{dr} \simeq \frac{(\Delta c_{Cu} \text{ [constant])}}{(\Delta r \text{ [increasing])}}.$$

Fracturing or sawing a partially leached rock readily shows the leached rim (bleached of color) and the unleached core colored blue-green by the copper oxide minerals. The appearance is similar to the rock section model illustrated in Fig. 5.7.

In computer modeling the leaching process, it is necessary to consider the various rock sizes weighted by their mass fraction. The appearance of sectioned rocks of differing particle size after a period of incomplete leaching is illustrated in Fig. 5.8. The smallest rock is completely leached while the larger rocks show leached rims and progressively smaller fractions of the copper extracted as rock size increases.

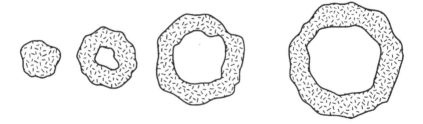

Figure 5.8.   Section illustration of partial leaching of different sized rocks.

THE SHRINKING CORE MODEL OF
MONOSIZE OXIDE COPPER ORE ROCK LEACHING

The shrinking core leaching model, sketched in Fig. 5.7, is valid for oxide copper ores because the dissolution reactions are rapid and can be ignored in the rate determining sequence of reaction and transport steps. Hence, all of the copper is dissolved within the leached rim, and hydrogen ion must diffuse all the way through the rim to the rim/core interface at r_c before it can react with copper oxide mineral. Because the dissolution reaction at this interface is rapid, diffusion of hydrogen ion beyond the interface into the unleached core does not occur. All of the diffusing acid is consumed at the internal interface, r_c. The shrinking core model is not valid for leaching rocks containing dispersed sulfide copper minerals because the heterogeneous dissolution reactions of the sulfide grains are not sufficiently rapid to be ignored in determining the rock leaching rate. Two models for this more complex leaching behavior will be described in Chapter 12.

In developing equations to express the rate of leaching for the shrinking core model, following Wadsworth and Miller (1979), a spherical geometry is assumed. The total number of copper moles, n_{Cu}, in the unreacted sphere is

$$n_{Cu} = \frac{4}{3}\pi r_0^3/V_{Cu}. \tag{5.6}$$

Differentiating yields

$$\frac{dn_{Cu}}{dt} = \frac{(4\pi r_c^2)}{V_{Cu}}\frac{dr}{dt}, \tag{5.7}$$

where V_{Cu} is the molar volume of copper in the rock, defined as, $V_{Cu} = MW_{Cu}/(\rho_{ore}G)$, where MW_{Cu} is the copper molecular weight, ρ_{ore} is the ore rock density and G is the fractional copper grade in the ore. The unreacted core radius is r_c.

If r_0 is the initial radius of the spherical rock and F is the fraction reacted (or the fraction of copper extracted), from geometry it may be shown that

$$F = 1 - \left(\frac{r_c^3}{r_0^3}\right) \tag{5.8}$$

and

$$\frac{r_c}{r_0} = (1-F)^{1/3}. \tag{5.9}$$

Upon differentiation of eqn 5.8 with respect to time,

$$\frac{dF}{dt} = -\left(\frac{3r_c^2}{r_0^3}\right)\frac{dr}{dt}. \tag{5.10}$$

The rate of reaction may be represented by the equation:

$$\frac{dn_{Cu}}{dt} = -(4\pi r_c^2)D_{eff}\frac{dc}{dr}, \tag{5.11}$$

where c in this example is the copper concentration in the micropore solution (moles/cc). This equation can be integrated, assuming steady state conditions for all values of r between the shrinking core interface radius, r_c, and r_0, resulting in the expression

$$\frac{dn_{Cu}}{dt} = -(4\pi D_{eff})(C_c - C_0)\frac{r_c r_0}{(r_0 - r_c)}. \tag{5.12}$$

The units of time are those expressed in the effective diffusion coefficient, D_{eff}.

A similar expression to eqn 5.12 can be written for the consumption of acid except that these two expressions are coupled by the leaching chemical reaction stoichiometry. Hence, the rates of accumulation in the solids are related:

$$B\frac{dn_{Cu}}{dt} = \frac{dn_A}{dt}, \tag{5.13}$$

$$\frac{dn_{Cu}}{dt} = \left(\frac{1}{B}\right)(4\pi D_{eff})(A_c - A_0)\frac{r_c r_0}{(r_0 - r_c)}. \tag{5.14}$$

In principle, the chemical leaching reaction stoichiometry, such as from eqn 5.1 for chrysocolla, can be used to obtain B; but, because of several reactions and possibly competing reactions from minerals other than copper minerals, it is best to determine this relation and the value of B experimentally for each ore.

For the limiting condition, $(A_c - A_0) = $ constant, then substitution and integration can yield explicit expressions for F_{t,r_0} and $r_{c,t}$. Typically, $A_c = 0$ and A_0 is the acid concentration entering the top of the leaching heap. However, this condition can only be met for shallow or "thin" heaps where the acid concentration in the solution percolating through the heap remains constant (i.e., there is little consumption of the entering acid as it passes through the heap). Under this thin layer leaching condition, including $A_c = 0$, substituting eqn 5.7 into eqn 5.14 yields:

$$\frac{dr}{dt} = -\left(\frac{V_{Cu}D_{eff}A_0}{B}\right)\left(\frac{r_0}{r_c(r_0 - r_c)}\right). \tag{5.15}$$

Substitution of eqn 5.10 yields

$$\frac{dF}{dt} = \left(\frac{V_{Cu}D_{eff}A_0}{B}\right)\left(\frac{3r_c^2}{r_0^3}\right)\left(\frac{r_0}{r_c(r_0 - r_c)}\right). \tag{5.16}$$

Upon substitution of eqn 5.9 this integrates to an explicit expression of the fraction reacted (copper leached) as a function of time:

$$1 - \frac{2}{3}F_{t,r_0} - (1 - F_{t,r_0})^{2/3} = \left(\frac{2V_{Cu}D_{eff}A_0}{Br_0^2}\right)t \tag{5.17}$$

and from eqn 5.9, upon substitution

$$r_{c,t} = (1 - F_{t,r_0})^{1/3}r_0. \tag{5.18}$$

For ease in solving problems involving eqns 5.17 and 5.18, a plot of F versus time, t, based on eqn 5.17 is shown in Fig. 5.9. Note that when the abscissa quantity (right side of eqn 5.17) reaches 0.33, $F_t = 1$ and the reaction is completed.

It should be emphasized that these results are only valid for thin layered heaps where A_0 is approximately constant from the top to the bottom. In copper oxide ore leaching of *high* heaps, it is not possible to substitute and solve directly for the fraction extracted over time, F_t, because A_0 and c_0 vary with time and depth in the heap, as will be shown in a later section of this chapter. Hence, a finite difference computer solution must be used with the following expression for the incremental change in the amount (decreasing) of unreacted copper in the ore particle:

$$\Delta n_{Cu} = -(4\pi D_{eff})\left[(C_c - C_0)\left(\frac{r_c r_0}{r_0 - r_c}\right)\right]_t (\Delta t). \tag{5.19}$$

A similar expression to eqn 5.19 can be written for the consumption of acid:

$$\Delta n_A = (4\pi D_{eff})\left[(A_c - A_0)\left(\frac{r_c r_0}{r_0 - r_c}\right)\right]_t (\Delta t), \tag{5.20}$$

so that,

$$\Delta n_{Cu} = \left(\frac{4\pi D_{eff}}{B}\right)\left[(A_c - A_0)\frac{r_c r_0}{r_0 - r_c}\right]_t (\Delta t). \tag{5.21}$$

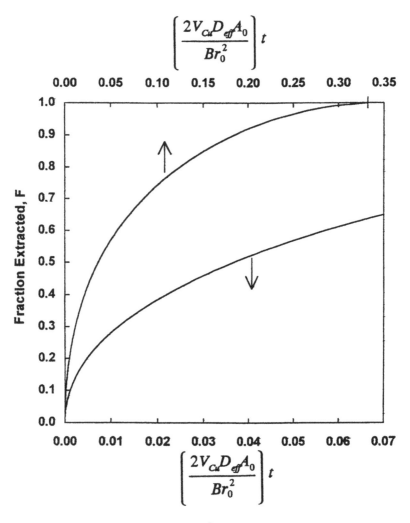

Figure 5.9. Plot of F vs $(2V_{Cu}D_{eff}A_0/Br_0^2)t$, computed for shrinking core extraction of monosize ore particles with diffusion through the rim controlling the rate, eqn 5.17.

EXAMPLE OF THE SHRINKING CORE MODEL
FOR MONOSIZE OXIDE COPPER ORE LEACHING

A comparison was made of the experimental copper extraction curves from oxide ore leached with dilute sulfuric acid and the simulated copper extraction curves generated using the shrinking core model by Roman et al.

(1974). Ore samples from the same source were leached in two different columns under different conditions. The ore was a well-cemented sandstone containing malachite evenly distributed in the rock matrix. Table 5.4 shows the different rock size distributions of the samples in the two columns. Table 5.5 lists the leaching conditions and physical parameters of the two columns. For each column, the volume of effluent solution was measured daily and chemically analyzed to obtain materials balances, which were used to provide the experimental copper extraction data shown as circles in Fig. 5.10. Computed results from the shrinking core simulation model are also shown, as the curves in Fig. 5.10, and there is very good agreement between the simulated and experimental results. F_t was computed by summing F_{t,r_0} for the various particle sizes according to eqn 2.3.

Table 5.4   Rock Size Distribution of Sandstone Ore with Malachite (Roman et al., 1974).

Size (mm)	Column 1 (wt pct)	Column 2 (wt pct)
−76+38	58.1	—
−38+25	15.8	—
−25+19	10.3	—
−19+13	4.1	29.8
−13+9.5	3.2	22.9
−9.5+6.7	1.8	15.0
−6.7+2.4	2.9	15.1
−2.4	3.8	17.2

Table 5.5   Physical Properties of Sandstone Ore and Operating Conditions (Roman et al., 1974).

	Column 1	Column 2
Weight of ore, kg	121.1	74.4
Column Height, mm	1765	3048
Column Diameter, mm	254	143
Solution flow rate, mm/s	0.08	0.034
Acid concentration, g/l	48.8	69.7
% voids	49.8	42.3
Acid consumption, g/g-Cu	3.6	3.6
Copper grade, %	1.9	1.9
Rock density, g/cm^3	2.7	2.7

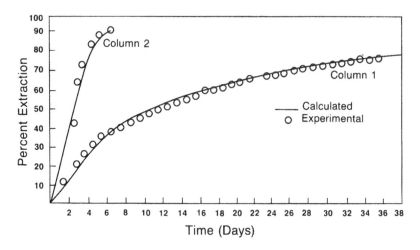

Figure 5.10. Copper extraction curves for a sandstone ore containing malachite (oxide copper) (Roman et al., 1974).

The shrinking core model can be useful for ore leaching systems other than oxide copper. It is applicable whenever the limiting (slowest) step in metal extraction is diffusion of reactant or product dissolved species through micropores in an expanding layer, or shell, surrounding a shrinking core of rock where the mineral is unreacted. It usually pertains to most base metals where the mineral is in a readily soluble form in the lixiviant, but the amount of metal exceeds the solubility limit of the leaching solution contained in the internal micropore space. Gold and silver are exceptions only because of the stoichiometrically very small amount of metal in the ore, which allows essentially complete dissolution with the contained micropore water, as was discussed in Chapter 2.

Readily leachable minerals include many sulfosalts as well as most oxides and carbonates, but the more refractory sulfide minerals, including pyrite and chalcopyrite, are not readily soluble in acid. These ores require a mixed kinetics model wherein both diffusion and the intrinsic heterogeneous mineral chemical oxidation reaction kinetics are important and must be considered simultaneously. This type of leaching is covered in the next chapter of this book, with mixed kinetics models for it described in Chapter 12.

EXAMPLE PROBLEM

After crushing and screening a copper ore containing 0.8 wt pct copper as a 50/50 mixture of malachite and chrysocolla, the $-1+3/4$ inch sieve fraction (nominal size) was leached in a laboratory column with a percolating solution containing 9.8 g/l of H_2SO_4. The ore is devoid of other acid consuming minerals and the leach extractions obtained were 40% in 80 hr and 83% in six weeks. (a) estimate the copper extraction at two weeks and at four weeks. (b) estimate the copper extraction at two weeks resulting from doubling the acid concentration to 19.6 g/l.

ANSWER (a)
(1) From the longest (six weeks) experiment, it is deduced that the accessible copper is only 83%. Therefore, the percent extraction of accessible copper in days is: 40%/0.83 = 48%.
[Note: rounding to accuracy of any of these column leach experimental data.]
(2) With $F_t = 0.48$, Fig. 5.9 gives an abscissa value for this monosize ore, and from eqn 5.17 the following working equation:

$$1 - \frac{2}{3}F_t - (1-F_t)^{2/3} = 0.034 = (2V_{Cu}D_{eff}A_0/Br_0^2)t.$$

(3) The following parameters can be computed from the given data, including the stoichiometry of leaching both malachite and chrysocolla:

$$B_{Chrysocolla} = 1, \quad B_{Malachite} = 1, \quad so \ B_{ore} = 1,$$

$$A_0 = 9.8 \ (g/l)/98 \ (g/mol) = 0.1 \ M = 1 \times 10^{-4} \ mol/cm^3,$$

$$G = 0.008,$$

$$\rho_{ore} = 2.6 \ g/cm^3, \ \text{which is an assumption,}$$

$$V_{Cu} = MW_{Cu}/(\rho_{ore}G),$$

$$V_{Cu} = 63.54/(2.6 \times 0.008) = 3054.8 \ cm^3/mol,$$

$$r_0 = \frac{1}{2}(2.31 \ cm) = 1.155 \ cm \ \text{[from Table A-2 and sieves]},$$

$$t = 80 \ hr = 80/(24 \ h/d) = 3.33 \ days.$$

(4) Substituting these parameters in the working equation in Step 2 above yields a value for the only unknown parameter, D_{eff} in units of cm²/day:

$$0.034 = [2(3054.8 \ cm^3/mol)(1 \times 10^{-4} \ mol/cm^3)D_{eff}(3.3 \ days)]/[1.155 \ cm^2],$$

$$D_{eff} = 0.0157 \ cm^2/day.$$

(5) The value of F_t after leaching four weeks is provided by using this value of D_{eff}, inserted back in the working equation with a leaching time of 14 days (two weeks) to obtain the abscissa value of Fig. 5.9, but since the only variable that is changed is time:

$$1 - \frac{2}{3}F_t - (1 - F)^{2/3} = 0.0333(14/3.33) = 0.140$$

and the value of F_t is read from the graph of Fig. 5.9:

$$F_t = 0.81.$$

However, only 83% of the copper is accessible, so the predicted copper extraction *from the ore* is:

Two week extraction $= 0.81 \times 83 = 67.2\%$ *at two weeks.*

(6) Similarly for copper extraction from the ore at four weeks:

$$F = 0.985$$

Four week extraction $= 0.985 \times 83 = 81.7\%$ *at four weeks.*

ANSWER (b)

(1) Doubling the acid concentration will double the abscissa from a value of $0.144 - 0.288$ with a corresponding F of 0.98 and an expected extraction of *81.7% of the copper in the ore.*

ESTIMATING METAL EXTRACTION VERSUS LEACHING TIME FOR MULTISIZE ORE USING DIMENSIONLESS DESIGN CURVES BASED ON THE SHRINKING CORE MODEL

The heap leaching process is treated as a batch process—all ore rock particles experience the same residence time and lixiviant composition. As will be discussed in the next section, constant lixiviant concentration is not strictly true for a high heap, but the approximation is useful for estimating necessary leaching times to obtain desired metal extractions.

The time for 100% extraction ($F = 1$) of the accessible mineral in a single (monosize) rock particle, t_{100}, can be obtained from eqn 5.17; hence,

$$t_{100} = \left[\frac{B\rho_{\text{ore}}G}{6(\text{MW}_{\text{Cu}})D_{\text{eff}}A_0} \right] r_0^2 \tag{5.22}$$

and all of these parameters must be in consistent dimensions.

With the computed value of t_{100} it is easy to compute a value of t_F necessary to obtain a given extraction percent or conversely estimate the percent extraction of metal that will result from a specified leaching time, t_F, using Fig. 5.9. But, this is only valid for monosize ore.

The shrinking core model can be extended to a distribution of rock particle sizes resulting from crushing or other breaking of the ore with the largest rock having radius r_*:

$$t_{100} = \left[\frac{B\rho_{ore}G}{6(MW_{Cu})D_{eff}A_0} \right](r_*)^2. \tag{5.23}$$

The Gates–Gaudin–Schuhmann multisize rock particle distribution has been combined with the shrinking core model to provide a design curve for multisize broken ore that is pertinent to oxide copper ore leaching (Bartlett, 1972). The computed results are indexed to the largest ore particle radius in the distribution, r_*. However, as a useful approximation, $d_{80}/1.6$ has been substituted for r_*, according to eqn 2.6. The resulting design curve for multisize ore, plotted in Fig. 5.11, expresses F_t versus the dimensionless time variable, t/t_{100}, where t_{100} is the time required for complete conversion (leach extraction) of *all* of the ore particles by the shrinking core model, which is defined by the following modified version of eqn 5.23:

$$t_{100} = \left[\frac{B\rho_{ore}G}{6(MW_{Cu})D_{eff}A_0} \right](d_{80}/1.6)^2. \tag{5.24}$$

The use of d_{80}, determined from a sieve analysis on a representative ore sample, is preferred over using the largest ore particle size because it compensates automatically with sufficient precision for variations in the rock breakage function, m, in the Gates–Gaudin–Schuhmann relationship. However, if d_{80} is unknown and cannot be obtained, the best estimate of r_* should be used, e.g., from eqn 2.12. Refer to the discussion of this topic in Chapter 2.

Figure 5.11 and Fig. 2.6 (gold leaching) are superficially similar. Both processes depend on diffusion in the micropore solution of the rocks and an effective diffusivity, D_{eff}, that includes the rock microporosity. The time required for a specified fractional leaching extraction increases with the rock radius squared for both processes. However, when leaching acid soluble copper minerals, the leaching time to achieve a given value of fraction extracted, F_t, also depends on the acid concentration, A_0, and is inversely proportional to the copper grade in the ore, G. For gold/silver leaching, the amount of metal is so small in relation to the amount of lixiviant in the rock micropores that these are not leaching rate parameters.

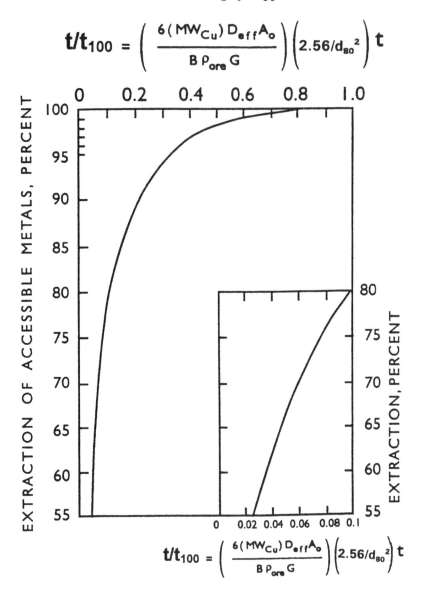

Figure 5.11. Mineral extraction from broken ore versus dimensionless time for shrinking core model and a multisize GGS rock size distribution indexed to d_{80}.

Another important use of Fig. 5.11 is to empirically determine a value of t_{100} from experimental data on fraction extracted versus leaching times. This can be done using crushed ore particles and the F_t versus t_F/t_{100} curve in Fig. 5.11. This is best accomplished with several experimental measurements, for example by using column leaching tests. For each measured value of F_t, the corresponding value of t_F/t_{100} is obtained from Fig. 5.11, and compared with the actual (measured) leaching time, t_F, to obtain an estimate of t_{100}. Several such derived estimates of t_{100} are averaged to obtain a "best" estimate of t_{100}. This experimentally determined t_{100} can then be used with the design curve of Fig. 5.11 to estimate necessary leaching times to obtain a desired extraction percentage. Later, if ore is crushed to a different top size, a new value of t_{100} can be determined from the old value of t_{100} from the radius squared relationship of eqn 5.23,

$$t_{100(b)} = t_{100(a)}[r^2_{*(b)}/r^2_{*(a)}] \tag{5.25}$$

or the d_{80} squared relationship of eqn 5.24,

$$t_{100(b)} = t_{100(a)}[d^2_{80(b)}/d^2_{80(a)}]. \tag{5.26}$$

If some of the mineral contained in the ore is unaccessible and cannot be extracted, then the fraction extracted scale must be adjusted to coincide with $F = 1$ at completion of extraction of the accessible mineral, as was done when using Fig. 2.7 for gold and silver extraction estimates, as described in Chapter 2.

EXAMPLE PROBLEM

Continuing with the copper ore from the previous example problem, estimate the leaching times, using Fig. 5.11, to obtain 60%, 70% and 80% extraction of the copper in the ore if it is crushed with a primary crusher without screening. The estimated maximum ore particle size (diameter) is 175 mm, and the acid concentration is maintained constant at 4.9 g/l during the leaching cycle.

ANSWER

(1) With a maximum diameter of 175 mm, the maximum radius of the ore particles is:

$$r_* = 175/2 = 87.5 \text{ mm} = 8.75 \text{ cm}.$$

(2) Using the data generated in the previous problem, including the value of the effective diffusivity of $D_{eff} = 0.0157 \text{ cm}^2/\text{day}$, and the changed

value of acid concentration, A_0, the time for complete extraction of the accessible copper in the largest ore particle, t_{100}, is calculated using eqn 5.23:

$$t_{100} = \left[\frac{1(2.6\,g/cm^3)}{6(63.54\,g/mol)(0.0157\,cm^2/d)} \right] \left[\frac{0.008(8.75\,cm)^2}{5 \times 10^{-5}\,mol/cm^3} \right]$$

$$= 5321 \text{ days.}$$

(3) Beginning with the 80% extraction goal, the desired value of F to obtain 80% extraction of copper *in the ore* is found as follows, using the value of the accessible copper in the ore at 83%:

$$F_{80\%} = 0.80/0.83 = 0.964\ (96.4\%).$$

(4) From reading the graphed band in Fig. 5.11, the value of t/t_{100} corresponding to $F = 96.4$ is determined to be:

$$t/t_{100} = 0.34 - 0.40.$$

A conservative design is to assume the worst rock breakage function which corresponds to the largest value of t/t_{100}, namely:

$$t/t_{100} = 0.40.$$

(5) Hence, the recommended leaching time is:

$$t_{80\%} = 0.40\ (t_{100}),$$

$$t_{80\%} = 0.40\ (5321),$$

$$t_{80\%} = 2128 \text{ days } (\textbf{5.8 Years!!!!!}).$$

(6) For the other values:

Ore extraction (%)	F	t/t_{100}	Leaching time, t
60	0.723	0.06	319 days, 10.5 months
70	0.843	0.13	692 days, 1.9 years
80	0.964	0.40	2128 days, 5.8 years

Note: This ore would require either multilift heap leaching or further crushing if on/off leaching with a reusable pad is preferred.

VERTICAL SOLUTION VARIATIONS IN HEAP LEACHING OXIDE COPPER ORES

In the previously discussed chapters on gold, the percolating solution concentrations of gold, cyanide and dissolved oxygen were assumed to be invariant from top to bottom of the heap. This approximation is not valid for percolation leaching of oxide copper ores when the heap height is appreciable. As the leaching solution percolates down through the heap it becomes depleted in acid while its cupric ion concentration strengthens. Hence, computer modeling of the process must consider these changes, as well as the chemistry and transport within ore fragments themselves. The heap solution chemistry variation with depth from top to bottom through an ore heap can be determined experimentally using a well instrumented ore column leach test.

Using an analogous approach, computer simulation can occur by taking a vertical block of ore with a unit cross sectional area through the heap from top to bottom. As illustrated in Fig. 5.12, this column can be sectioned into a vertical stack of cubic ore boxes, each of equal unit volume. In computer modeling, a materials balance must be determined for: (1) each rock size, in (2) each unit volume at (3) each increment of time using a suitable finite difference approximation to the governing transport equations. Furthermore, the boundary conditions for the downward flowing leach solutions

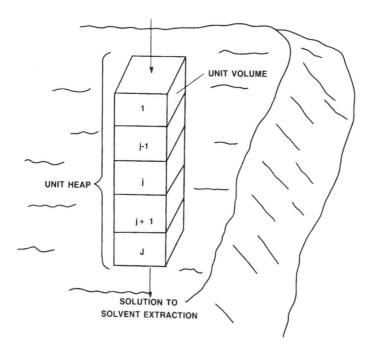

Figure 5.12. Schematic diagram of a heap leach vertical section: stacked cubic boxes of equal unit volume (Roman and Olsen, 1974).

must be consistent for adjacent boxes at each increment of time. Input and output of hydrogen ions and cupric ions in the flowing solution in each unit volume must be balanced for each time increment.

The variations in percolating solution acid concentration and copper concentration with heap depth, at four progressively increased leaching times, are shown schematically in Fig. 5.13. Note that at t_1 all the acid is consumed by the bottom of the first unit volume, but at t_4 considerable acid concentration remains as the solution exits the heap. Average copper recovery from the ore in each unit volume is also shown in Fig. 5.13. Copper recovery is nearly complete in the first unit volume at t_4 but only about half complete in the bottom unit volume.

In this copper oxide heap leaching model the concentrations of copper and acid in the percolating solution entering unit volume j are c_{j-1} and A_{j-1}, respectively. The concentrations leaving unit volume j are c_j and A_j, respectively. The material balance equations for copper and acid include the flow terms, based on the superficial liquid velocity, u_l, of the solution percolating downward in and out of the unit volume and the contributions from the ore contained within the unit volume: copper extraction and acid

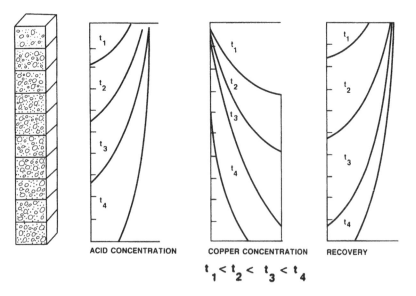

Figure 5.13. Schematic diagram of an oxide copper leaching heap section showing percolating solution concentrations and average recovery of copper with time and heap height (Roman and Olsen, 1974).

consumption. The material balance is illustrated in Fig. 5.14 and the copper balance equations for the jth cubic unit volume of length, L_h, are

$$c_{j-1}(u_1 L_h^2 \Delta t) - c_j(u_1 L_h^2 \Delta t) + \sum_{y=1}^{Y} (\Delta n_{Cu})_{y,j,t} N_y$$

$$= (c_j - c_{j-1}) L_h^3 (v_1), \tag{5.27}$$

where $\sum (\Delta n_{Cu})_{y,j,t} N_y$ is the sum of the extraction of copper into the percolating solution from all of the y rock particles in the jth volume during the time increment Δt, based on eqn 5.19 for a single rock particle. N_y is the mass fraction of y rock particles in the j unit volume, but it is assumed that the ore is well blended so that N_y is invariant with the j unit volumes in the heap.

The left-hand side of eqn 5.27 are the sources and sinks while the right-hand side is the accumulation within the percolating solution in unit volume j. The term "v_1" is the volume fraction occupied by percolating solution (outside the rocks), which is typically about 20% of the total volume in a percolating heap system.

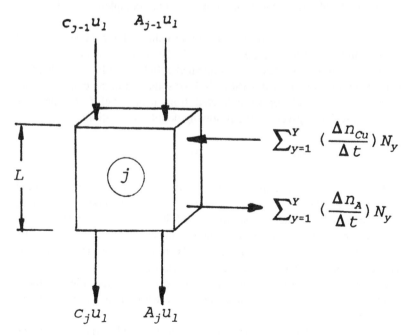

Figure 5.14. Percolating solution copper and acid balances around unit volume j.

Similarly for acid,

$$A_{j-1}(u_1 L_h^2 \Delta t) - A_j(u_1 L_h^2 \Delta t) + \sum_{y=1}^{Y} (\Delta n_A)_{y,j,t} N_y$$
$$= (A_j - A_{j-1}) L_h^3 (v_1), \tag{5.28}$$

where $-\sum(\Delta n_A)_{y,j,t} N_y$ is the sum of the acid absorbed from the percolating solution into all of the y rock particles in the jth volume during the time increment Δt.

A computer algorithm for copper oxide percolation heap leaching proceeds as follows. Beginning with the initial acid concentration, A_0, entering the top of the heap and the first unit volume, $j=1$, Δn_A and Δn_{Cu}, and the products $\Delta n_A N_y$ and $\Delta n_{Cu} N_y$, are computed for each of the y rock particle sizes (using eqns 5.20 and 5.21) and summed. Then new values of $A_{j=1}$ and $c_{j=1}$ are computed and stored, using their initial (or previous) values (zero for the "normal" initial situation at $t=0$), and eqns 5.27 and 5.28, respectively.

The new $A_{j=1}$ is used to repeat the process for the next unit volume, $j=2$. This computation process continues, iteratively, through $j=J$, the bottom unit volume of the heap column, which completes the first time period, Δt.

The entire process is repeated beginning with $j=1$ for the second time period, and so forth for I iterations until the desired leaching duration, $t_I = I \Delta t$, is completed. This is accomplished with "nested loops". It is often convenient to take $\Delta t = 1$ day ($86,400$ s) for solution mining computational programs. The total copper extracted from heap level j at time t_I is $(\Delta n_{Cu})_{j,t_I}$, which is given by the following double summation:

$$(\Delta n_{Cu})_{j,t_I} = \sum_{i=1}^{I} \sum_{y=1}^{Y} (\Delta n_{Cu})_{y,j,t} N_y. \tag{5.29}$$

This example of modeling leaching reactions through a vertical section of the heap was given, in part, because it is a common method of approaching the numerical simulation of heap and mine waste dump leaching in the vertical direction. The simulation results can be tested by comparing them with experimental results from leaching ore in columns. However, as a practical matter columns are limited in height; whereas, computer simulation can easily be extended to full scale mine systems. Some copper and gold mine heaps and mine waste dumps range from 30 to 150 m in height—equivalent to office buildings from 10 to 50 stories.

Using this method of approach, Roman and Olsen (1974) developed one of the earliest simulation models for a vertical heap section. They used it to estimate copper extraction and heap effluent (pregnant liquor) concentrations over 600 days, for three heap heights: 15, 22.5 and 30 m (see Fig. 5.15). The specific conditions used as input data are given in Table 5.6. The percolation flow rate and initial acid concentration were identical in all cases.

The preceding discussion and figures explain why the Bluebird Mine used a more concentrated acid solution (50 g/l) at the beginning of a heap leaching campaign and gradually reduced the acid strength over the 180 day campaign before overdumping the heap with a new lift. Acid consumption is very high at the beginning when the rock's leached rims are thin but slows down as the leached rims thicken and the smaller rocks become completely leached. Stronger acid concentrations toward the end of a campaign would result in acid passing through the lift. Acid would be wasted and the acid concentration entering the solvent extraction plant would be too high, causing problems with extraction. The SX process relies on the thermochemical drive of extraction from a dilute acid solution followed by stripping into a concentrated acid solution.

More recently, Dixon and Hendrix (1993) developed a mathematical model for heap leaching monosize porous ore particles when the mineral is readily soluble in the lixiviant without requiring aeration for mineral

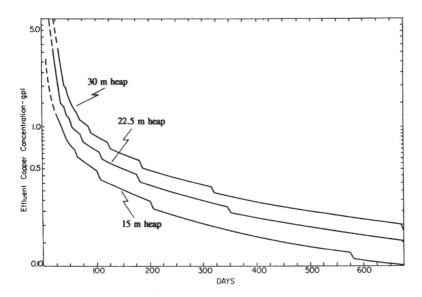

Figure 5.15. Computed copper effluent concentrations for conditions of Table 5.6 and the heap heights shown (Roman and Olsen, 1974).

Table 5.6   Input Data for Computing Results of Fig. 5.15
(Roman and Olsen, 1974).

Copper content: 0.6% weight
Voids in heap: 45%
Maximum acid concentration fed onto heap: 50 g/l
Solution flow rate: 0.03 mm/s
Acid consumption: 3.6 kg acid per kg Cu
Effective diffusivity of ore: 3.2×10^{-5} mm^2/s
Ratio of surface area factor to volume factor: 2.65
Saturation in heap: 35% (percentage of pore space filled with solution)
Maximum concentration of acid in SX Feed: 5.0 g/l
Size Analysis:

Diameter (cm)	*% weight*
51	23.7
36	18.1
26	13.8
18	10.5
13	8.0
9	6.1
6.5	4.7
4.5	3.6
3.2	2.7
2.3	2.1
1.6	1.6
1.1	1.2
0.8	0.9
0.6	0.7
0.4	2.3

oxidation. Model results were successfully validated by column leaching experiments using artificial ore particles made by agglomerating sand, portland cement and very fine silver powder. A dilute cyanide solution was used as the lixiviant. The following dimensionless parameter was used to characterize heap leaching behavior, and it has practical implications for optimizing heap height:

$$\omega = \frac{3v_s D_{eff} H}{u_1 r_0^2}, \tag{5.30}$$

where v_s is the volume fraction of the ore heap occupied by the solids (rocks), u_1 is the downward percolating lixiviant superficial velocity and H is the height of the ore heap. It was shown that when $\omega \ll 1$, the heap is being leached uniformly from top to bottom ("homogeneously"), with the ore particle leaching kinetics all that needs to be considered. However, when $\omega \geqslant 1$, the heap is leaching in a zone wise manner beginning at the top, and the global heap behavior and heap height must be considered. Computing the value of ω from its parameters will indicate whether a planned heap height is too great ($\omega \geqslant 1$) and should be shortened to decrease the overall leaching time needed to obtain the desired level of metal extraction throughout the ore heap. This leads to the next topic, thin layer or shallow lift ore heap leaching practice.

SHALLOW LIFT (THIN LAYER) LEACHING AND ACID CURE LEACHING

Note that the preceding modeling results and supporting column leaching tests indicate that increasing heap height in oxide copper ore percolation leaching will not increase the extraction **rate** proportionately. Doubling the heap height will not double the copper concentration in the effluent solution, even if the acid strength entering the heap is doubled. Also, very strong acid solutions will react with gangue minerals after the copper oxide minerals are dissolved in the upper part of the heap, which wastes acid and causes decrepitation of the ore mass with attendant loss of permeability to the percolating leach solution.

The use of shallow heap lifts avoids the problem of acid depletion before the bottom of the lift is reached by the downward percolating acidic leaching solution. Also, size segregation, with boulders at the bottom and fines at the top of the lift, is avoided if the heap lift is kept shallow when dumping end over from haulage trucks.

Thin layer leaching was originally developed by Holmes & Narver and Soc. Minera Pudahuel Ltda. (Domic and Brim, 1980) and first applied at Lo Aguirre in Chile. Ore was mixed with concentrated sulfuric acid, cured for 24 hr and stacked in a "thin layer" heap 2–3 m high. This layer was leached with wash solution from the previous leach yielding a 6 g/l pregnant liquor for an SX-EW plant. After this primary leach, the heap was washed. Extraction of oxide copper was about 95% while extraction of (secondary) sulfide copper was about 55%.

At the Inspiration Mine in Arizona, as-mined oxide copper ore is heap leached using 4.5 m (15 ft) lifts. The mine management would prefer to use shallower lifts if the local topography and ore haulage distances from the mine would economically permit shallower lifts spread over a larger area. Leaching at the Inspiration Mine utilizes an "acid cure" process that involves adding or mixing into the ore the required amount of acid needed to leach the oxide copper before leaching begins. Acid in the form of a fairly concentrated solution, up to 200 g/l sulfuric acid, is used.

Two approaches to acid cure leaching have been used at different locations in the United States and Chile: (1) premixing acid and ore before
the lift is constructed and (2) soaking the ore heap with acid distributed
onto the heap surface immediately after constructing a very shallow lift.
Premixing practice is similar to agglomerating gold ores in a mixing appa-
ratus as described in Chapter 3. Acid premixed ore is either directly placed
in the heap or first cured for a period to complete acid penetration into the
rock micropores and begin dissolution of the soluble copper. When the
soaking method of acid curing is used, the lifts must be of very limited
height and the correct volume of acid solution, sufficient to thoroughly
soak all of the ore without an excess, must be distributed onto the constructed lift. The volume of acid needed and applied is usually greater than
the internal rock microporosity, but the additional acid will fill fine capillaries between rocks. As will be discussed further in Chapter 7, the total
heap volume occupied by capillary held liquid can be about 20% and
sometimes higher. Eventually the capillary held acid (H^+ ions) will diffuse
into the micropores during the cure period and will complete the dissolution of the acid soluble copper, or convert the oxide copper minerals
to cupric sulfate salt. The H_2SO_4 concentration in the required soaking
volume of acid solution will depend on the amount of H_2SO_4 required by
the ore to complete dissolution of all of the acid soluble copper. Ore in
each lot or stockpile may need to be sampled and tested to evaluate the
amount of acid it requires during the acid cure.

Following an acid cure period, the solubilized copper is leached
(washed out), using spent raffinate returning from the SX/EW copper
recovery plant as already discussed. However, the time of cure before
washing must be sufficient to complete the copper dissolution reactions,
and the required time may vary between different ores within the same
mine. It may be several days to weeks after adding acid to the ore before
leaching begins.

ACID CURE LEACHING PRACTICE AT THE
EL ABRA COPPER MINE IN CHILE

The El Abra Mine, located at a very high altitude in northern Chile, began
operating in 1996. It is an open pit mine that includes a very large oxide
copper heap leaching operation (Hickson, 1996). Chrysocolla and pseudo-
malachite are the predominant minerals. The operation was planned with
a constant annual production of 225 kt of copper from a fixed electrowinning plant, which represents a major part of the capital investment.

While the average grade of the ore deposit is 0.52% copper, grade will decrease from about 1.0% to 0.4% during the mine's life. Consequently, the ore mining rate will gradually increase from nearly 100,000 t/d to about 170,000 t/d. El Abra's initial daily tonnage is nearly twenty times greater than the mining rate at the Bluebird Mine, the first oxide copper SX-EW operation, which began operations twenty eight years earlier.

Ore is crushed at the open pit mine to $-200\,mm$ and sent 14 km to the leaching plant by an overland conveyor belt. Two further stages of crushing, with the last crusher set at $-11\,mm$, yield an ore feed that is 100% $-18\,mm$. Crushed ore is blended with water and sulfuric acid. The acid-cured ore is stacked to a height of 8 m (26 ft) on a reusable 60 mil HDPE leach pad, covering an area of $1.3\,km^2$. The leach pad slopes at a 5% gradient (about five degrees) and contains drain pipes in the drain blanket to collect leachate.

Ore is leached with raffinate returned from the solvent extraction plant for a period of at least 45 days. Cycle time may vary depending on the ore's response. The operating goal is an average copper recovery of 78%. Total acid consumption is projected at 19 kg/t and most of this will be added before curing.

Pregnant liquor drains to a pregnant liquor storage pond and flows by gravity to the SX-EW plant. As the ore grade decreases and the ore tonnage increases more pregnant liquor will be produced. Consequently, the SX plant began with four parallel trains of mixer-settlers and will eventually expand to seven trains. Each solvent extraction train includes two extraction stages, a water wash/scrub stage and two organic stripping stages.

ADVANCES IN MECHANICAL ORE STACKING SYSTEMS AT LARGE COPPER MINES

While soluble metal recovery is quicker and solution control is usually better with reusable pads, the cost of double handling of ore in a reusable (on–off) pad heap leaching system has been considered a major drawback when compared with multilift heap leaching of either copper or gold ores. However, new large-tonnage copper mines have been installing mechanical stacking systems with reusable pads for heap leaching. While initial capital costs are high, operating costs are very low, typically below $0.10 per tonne of ore, and life cycle costs are beginning to favor mechanical stacking systems. These systems usually consist of crawler-mounted, self-propelled mobile stacking and reclaiming bridge conveyor belts. The crawler tracks create low ground pressures that allow them to walk over rough surfaces on either the drain blanket layer or the leach pile without compacting it. A moving tripper discharges ore anywhere along the conveyor. A mobile bucket wheel excavator, also on crawlers, removes spent ore to a reclaim conveyor and on to the spent ore disposal area. One person can operate the ore loading and unloading system from a central console, which saves the considerable cost of labor for truck drivers and dozer operators that would otherwise be required.

The ore handling system at the El Abra Mine is a good example of the use of mechanical ore systems for heap leaching (Kahrger, 1996). Two parallel leach pads, 400 m wide by 1,600 m long, are separated by an ore delivery corridor. Ore enters and leaves the pad area over two parallel conveyors extending down this corridor. Incoming ore is discharged through a tripper car moving on rails to a mobile stacking bridge conveyor which walks around both sides of the central conveyor on crawlers, building an 8–10 m high ore pile. This stacker works in a high cast mode retreating ahead of the heap.

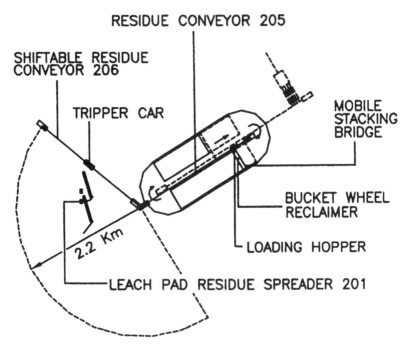

Figure 5.16. Plan sketch of ore heap leaching and residue disposal at El Abra (Kahrger, 1996).

Spent ore is reclaimed with a bucket wheel excavator attached to the end of a mobile conveyor, which is mounted on a crawler. The reclaim conveyor transfers spent ore to the second belt conveyor located in the corridor between the two leach pads. Spent ore is disposed in a large semi-circular area with a 2.2 km long (infrequently moved) shiftable conveyor containing a tripper car that discharges onto mobile residue spreading conveyors. A plan view sketch of the El Abra ore handling system with its spent ore reclaim belts labeled is shown in Fig. 5.16.

LEACHING SECONDARY SULFIDE COPPER MINERALS

Chalcocite (Cu_2S) and covellite (CuS) are prevalent secondary sulfide minerals that are easily leached with sulfuric acid if an oxidant is present. Commonly, the immediate oxidant is ferric sulfate present in the leach solution. Dissolution of these secondary sulfide minerals occurs by reactions typified by the following sequence:

$$Cu_2S + Fe_2(SO_4)_3 = CuSO_4 + CuS + 2FeSO_4, \tag{5.31}$$

$$CuS + Fe_2(SO_4)_3 = CuSO_4 + S^0 + 2FeSO_4, \tag{5.32}$$

$$3Fe_2(SO_4)_3 + 4H_2O + S^0 = 4H_2SO_4 + 6FeSO_4. \tag{5.33}$$

Note that ferric sulfate is consumed, and sulfuric acid is produced by oxidation of these, and most, secondary copper sulfides. Thus, the presence of secondary copper sulfide minerals can generate some of the acid needed to leach oxide copper minerals when a mixed oxide/sulfide ore is being leached, which often occurs. However, acid must be present in sufficient concentration to maintain an adequately low pH to prevent hydrolysis of ferric ions to jarosite, goethite or other insoluble iron hydrolysis products. An "adequately low" pH is one in the range of 2.0–2.8 depending on a variety of factors including temperature, sulfate complexing and the presence of alkali ions needed for jarosite precipitation. When higher ferric sulfate concentrations are needed to chemically oxidize all of the available secondary sulfide minerals, the pH must be suitably lowered by using more sulfuric acid. However, much of the extra acid will be consumed by reactions with gangue minerals if the pH is too low. Hence, as a practical matter there is a limit on the ferric sulfate concentration that can be used, and generally this is less than 10 g/l of dissolved iron.

Two non-sulfide minerals resulting from secondary enrichment, and also requiring oxidation to solubilize the copper, are native copper and cuprite (Cu_2O):

$$Cu^0 + Fe_2(SO_4)_3 = CuSO_4 + 2FeSO_4 \qquad (5.34)$$

and

$$Cu_2O + H_2SO_4 + Fe_2(SO_4)_3 = 2CuSO_4 + H_2O + 2FeSO_4. \qquad (5.35)$$

FERRIC SULFATE GENERATION

Ferric sulfate, needed to oxidize both primary and secondary sulfide minerals, can be generated by the oxidation of ferrous sulfate with dissolved oxygen. This process is kinetically very slow at heap leaching temperatures without the presence of bacteria to catalyze the reaction, which will be discussed in greater detail in the next chapter. Ferric ions, ferrous ions and their complexes will be present in an adequately aerated copper leaching heap. They are ultimately derived from the oxidation of iron bearing sulfide minerals and from the acid leaching of iron oxide minerals such as hematite (Fe_2O_3), magnetite (Fe_3O_4), and goethite ($FeOOH$). Acid chemical weathering of other gangue minerals, e.g., biotite, may provide minor amounts of dissolved iron.

The dissolved iron accumulates in the leach solution until it is removed by ferric ion precipitation reactions. The steady state amount of dissolved iron depends on pH, E_h and other ions, particularly alkali cations.

Equilibrium relations for these iron precipitation reactions are shown in Fig. 5.17 for pH conditions normally encountered in copper heap leaching. The total solubility of ferric ion, Fe^{3+}, plus the principal ferric ion complex, $Fe_2(OH)_2^{4+}$, in equilibrium with goethite is plotted versus pH. As sulfate ion is added to the system, $FeSO_4^+$ or $Fe(SO_4)_2^-$ become important ferric complex ions, as shown in the ferric-sulfate–water system, Fig. 5.18, by McAndrew (1974). With sulfate present, alkali cations, derived from the leaching of silicate minerals, lead to the precipitation of jarosites, $MFe_3(SO_4)_2(OH)_6$, where M is a monovalent cation, rather than precipitation of goethite, $FeOOH * H_2O$.

Potassium jarosite, $KFe_3(SO_4)_2(OH)_6$, is usually precipitated during copper heap leaching because potassium is released by the acid attack of potassium silicate minerals present in the ore matrix. The ferric ion concentration in equilibrium with potassium jarosite depends on the pH, but also on the concentration of sulfate ions and potassium ions in the leaching solution. The Kjarosite line shown in Fig. 5.17 is for a sulfate concentration of

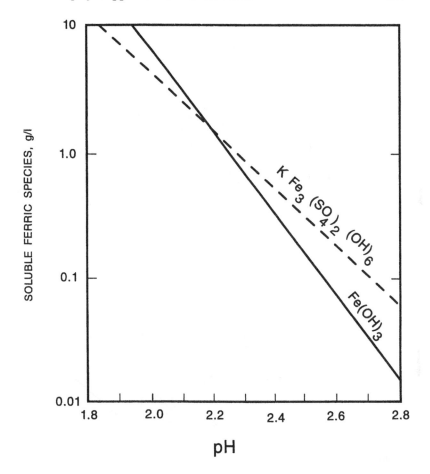

Figure 5.17. pH dependence of ferric ions and ferric ion complexes with hydrolysis precipitation of $Fe(OH)_3$ and jarosite $[KFe_3(SO_4)_2(OH)_6]$, at $[SO_4^{2-}] = 48\,g/l$ and for jarosite, $[K^+] = 0.1\,g/l$.

0.5 M (48 g/l) and a potassium ion concentration of 0.1 g/l (100 ppm). For additional information on iron complexing and precipitation, see the review by Dutrizac (1980).

Regardless of the precipitated iron product, decreasing the pH will tend to increase the amount of ferric ion and its complexes that are available to oxidize copper sulfide minerals. Hence, a reasonably strong sulfuric acid solution is needed for both leaching oxide copper minerals, which are usually also present with secondary sulfides and for maintaining a higher

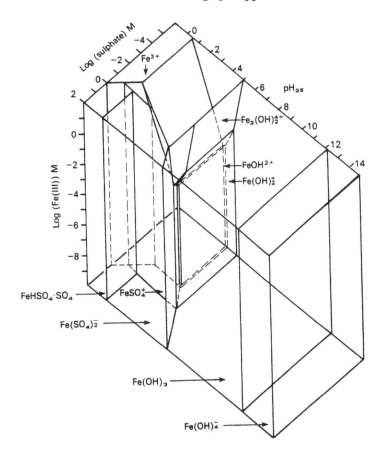

Figure 5.18. Ferric-sulfate–water system at 25°C (McAndrew, 1974).

concentration of ferric ions in solution. However, it can't be overdone. Too
much acid will be consumed by attacking gangue minerals, decrepitating
the host rocks, and ruining heap permeability.

MEASURING THE FERRIC ION CONCENTRATION IN LEACHING SOLUTIONS

Determining the concentration of ferric ions in the leaching solution is often
needed to see if secondary sulfide copper mineral oxidation and leaching
can proceed at an acceptable rate. While the total dissolved iron concentration, $[Fe_{Tot}]$, is easily measured, a direct measurement of the ferrous and

ferric ions concentrations is difficult. The ferrous/ferric ion oxidation/reduction electrochemical half cell reaction, involving one electron, is:

$$Fe^{2+} = Fe^{3+} + e. \tag{5.36}$$

The standard half cell voltage for this reaction (compared with the zero voltage based hydrogen reference half cell reaction) is $E^0 = 0.771$ V. The leaching solution E_h of this reaction couple can be measured with a suitable reference electrode, such as the calomel electrode, and used in conjunction with a measurement of the total dissolved iron concentration to obtain the ferric ion concentration. Doing so requires use of the Nernst equation:

$$E_h = E^0 + \frac{RT}{n\mathscr{F}} \ln \frac{[Fe^{2+}]}{[Fe^{3+}]}, \tag{5.37}$$

where n is the number of electrons involved in the half cell reaction. Consequently,

$$\frac{[Fe^{3+}]}{[Fe^{2+}]} = \exp\left(\frac{-n\mathscr{F}}{RT}(E_h - E^0)\right) \tag{5.38}$$

and because $[Fe_{Tot}] = [Fe^{3+}] + [Fe^{2+}]$, the ferric ion concentration is given by:

$$[Fe^{3+}] = \frac{[Fe_{Tot}]}{\exp(-n\mathscr{F}/RT)(E_h - E^0) + 1}. \tag{5.39}$$

At room temperature, 25°C, $RT/\mathscr{F} = 0.02568$ V.

Leaching solutions are often **NOT in equilibrium** with the adjacent air (oxygen) because of the very slow kinetics of oxidation of ferrous ion to ferric ion at heap leaching temperatures. The low solubility of oxygen in aqueous solutions also impedes reaching redox equilibrium, but this is of lesser importance than the slow reaction kinetics. Consequently, it is much more reliable to measure the solution oxidation potential directly with a suitable electrode, convert this to the E_h and use it, with eqn 5.39, to determine the concentration of ferric ions in the leaching solution.

EXAMPLE PROBLEM

Calculate the Ferric and ferrous ion concentrations for a leach solution with the following measured parameters:

$$E_h = 0.69 \text{ V}, \qquad [Fe_{Tot}] = 5 \text{ g/l}.$$

ANSWER

(1) The half cell reaction is given by eqn 5.36 with a standard potential, referenced to the hydrogen electrode, of 0.77 V and $n = 1$.

(2) Substituting these values into eqn 5.39 yields:

$$[Fe^{3+}] = \frac{[Fe_{Tot}]}{\exp[-(0.69 - 0.77)/0.02568] + 1}$$

$$= [Fe_{Tot}]/23.53 = (5\,g/l)/23.53$$

$$= 0.212\,g/l.$$

(3) $[Fe^{2+}] = [Fe_{Tot}] - [Fe^{3+}]$.

(4) $[Fe^{2+}] = 5\,g/l - 0.212\,g/l = 4.79\,g/l$.

REQUIRED SECONDARY COPPER SULFIDE LEACHING TIME WHEN GOVERNED BY FERRIC ION AVAILABILITY

The ability of air to flow into and through an ore heap to provide oxygen is limited and will be discussed in greater detail in Chapter 8. Large ore heaps tend to have much larger lateral dimensions than the height of a single ore lift. Consequently, natural aeration is limited to the heap edges and, therefore, a relatively small volume of the entire heap. Hence, it is reasonable to expect that most of the re-oxidation of ferrous ions by air to ferric ions in the leachate will occur outside the ore heap as it circulates through storage ponds and a recovery plant. If necessary, re-oxidation can be enhanced by sparging air through the leaching solution, for example, in a solution storage pond.

Calculating the leaching time required to oxidize all of the secondary copper sulfide minerals by the solution's depletion in Ferric ion concentration as it passes through the heap is a conservative estimate of the required leaching time when delivery of ferric ion is rate limiting. Of course, this assumption requires that the ore has been sufficiently crushed so that pore diffusion in the size-limited rock particles is not rate limiting. The leaching time for complete sulfide mineral oxidation will be proportional to the ore grade and the heap height. It will be inversely proportional to the solution application rate (flow rate through the heap) and inversely proportional to the change in Ferric ion concentration, $\Delta[Fe^{3+}]$, where,

$$\Delta[Fe^{3+}] = [Fe^{3+}]_{in} - [Fe^{3+}]_{out}. \tag{5.40}$$

Chalcocite, the secondary copper sulfide mineral, requiring the least amount of ferric ion for oxidation still requires two mols of Ferric ions for each mol of copper dissolved. For this mineral a calculation of leaching time was made as a function of the ferric ion depletion, measured in g/l, for a solution percolating at an application rate of $2.5\,cm^3\,s^{-1}\,m^{-2}$ (0.216 m/day), which is typical of the percolation flow obtained when using an impact sprinkler to distribute the leaching solution onto the heap. The ore heap bulk density was assumed to be $1.825\,tonnes/m^3$.

Results of these calculations are shown in Fig. 5.19. The leaching time required for complete oxidation of the mineral, including sulfur to sulfate ion, is plotted against the product of copper grade (wt pct) and heap height (meters). At a grade of 1.0% copper, all as chalcocite, and a heap height of 15 m oxidation can be completed in about 75 days with a 2.0 g/l depletion in the ferric ion concentration in the solution each time that it passes through the heap. For a covellite ore, the calculated time to complete

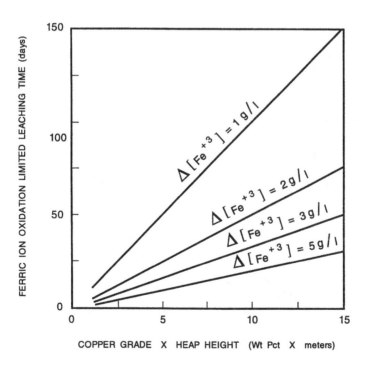

Figure 5.19. Required leaching time for a chalcocite copper ore when ferric ion consumption in the leaching solution applied to the ore heap is controlling the leaching rate.

leaching increases to 120 days for the same depletion in the ferric ion concentration. These calculations assume that there are no other ferric ion consuming reactions, including either hydrolysis reactions or oxidation of other sulfide minerals, such as pyrite.

FERRIC CURE LEACHING PROCESS

The preceding considerations led to the development of the "ferric cure" leaching process for mixed oxide and secondary sulfide copper ores at the Inspiration Mine (Fountain et al., 1983). This process is an extension of the acid cure process in which ferric sulfate is present in the curing liquor soaked into the ore when a new heap lift is constructed. The concentrations of H_2SO_4 and ferric ions depends on the particular ore characteristics but a solution with 200 g/l sulfuric acid and 8 g/l ferric ion would be typical.

The stoichiometric requirement for ferric ion, expressed as concentration in the lixiviant solution soaked into the heaped ore mass, is plotted in Fig. 5.20 for various copper concentrations in the ore present as chalcocite.

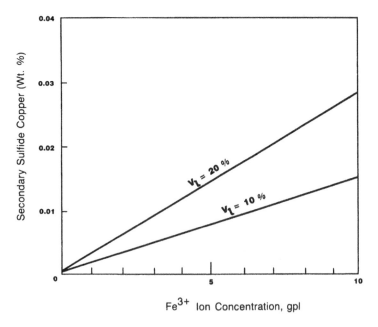

Figure 5.20. Ferric ion concentrations needed in the solution by the Ferric Cure process to dissolve copper in the form of chalcocite, as a function of the solution fractional volume, v_l.

The plotted curves are based on an ore heap with an apparent bulk density of $2.0\,g/cm^3$. Curves are shown for different values of the percentage of total heap volume occupied by the ferric cure solution.

Note that a solution with $8\,g/l$ dissolved ferric iron is inadequate to dissolve more than a small amount of copper in the ore as leachable sulfides, about $0.015\,wt\,pct$ copper. In a typical leaching grade ore this is only about 10% of the secondary sulfides that are present. However, during the cure period the ferrous ions, generated by sulfide mineral oxidation, are reoxidized to ferric ions by dissolved oxygen supplied from air in the heap through the bacteria catalyzed reaction. Air, either blown through pipes under the heap or diffusing into it, is indirectly responsible for most of the oxidation of the secondary sulfide copper minerals in the ferric cure process.

The high concentration of acid that is present with the ferric cure deters precipitation of iron hydrolysis products and allows a relatively high concentration of dissolved iron to be present. This accelerates the copper sulfide mineral leaching reaction process.

ORE TESTING

Ore testing involves: (1) determining the copper and major gangue mineralogy, (2) sulfuric acid leaching of ground ore pulps, and (3) column leaching tests to replicate heap leaching insofar as possible. Ground pulp leaching tests must determine the amount of acid soluble copper and the amount of acid consumed. The discussion in Chapter 3 on testing ores in columns is generally applicable to copper ore column testing. Ore testing should be coupled with an understanding of oxide copper ore leaching theory and the use of Fig. 5.11 in scaling laboratory results to different leaching times and ore particle size distributions.

Both composite samples and individual samples from different parts of the ore deposit or representing different ore characteristics are important. It is very important to determine the amount of acid soluble copper in the ore, which may be substantially less than the total copper concentration. The type and amount of secondary copper minerals must be determined, as well as whether pyrite or other forms of sulfide iron and chalcopyrite are present. The amount and variation with mine location of other acid consuming minerals, such as calcite and dolomite, must be determined. Column tests should be used to determine both copper extraction versus time and adequacy of solution percolation. Precipitation of gypsum resulting from acid reaction with calcite and dolomite can cause percolation problems. Working out the iron balance from analysis of the leaching solution is very important in the leaching of oxide copper ores and in particular ores with secondary copper sulfides.

PROBLEMS

1. An ore with a rock microporosity of 5% contains 0.7% copper, all as chrysocolla. Estimate the percentage of contained copper that will be dissolved by the initial penetration of a 20 g/l sulfuric acid leaching solution. There are no other acid soluble minerals. Ore specific gravity is 2.6 g/cm^3.

2. The ore of Problem 1 is uniformly sized to a radius of 2.3 cm and a very shallow heap of this ore is continuously exposed to 7 g/l percolating sulfuric acid. Estimate the fraction extracted and the leached rim thickness after leaching continuously for a period of: (1) one day, (2) one week, (3) one month, and (4) three months.

3. With the ore of Problem 1, and assuming that no other acid consuming minerals are present, how many tonnes of 100% sulfuric acid will be required for each tonne of ore leached? If sulfuric acid sells for $35/tonne, what is the acid cost per tonne of ore and per kg of copper in the ore?

4. Compare the relative leaching times to obtain 95% extraction of the accessible mineral in a gold ore and a copper ore crushed to the same size, both with a d_{80} of one inch and with the same internal microporosity, two volume percent. The copper ore grade is 1.0% copper, as chrysocolla, and it is leached in a solution with a constant sulfuric acid concentration of 7 g/l, and the ore specific gravity is 2.60 g/cc. There are no other acid consumers than the copper mineral.

5. Several replicate column leaching tests of an oxide copper ore crushed and passed through a 3/4 in sieve (nominal size) averaged 80% extraction after leaching 60 days and 83% extraction after leaching 120 days, both under conditions identical with the planned heap leaching operations, including constant leaching solution acidity, except that the final crusher setting (maximum opening) will be 25.4 cm (1 in). Not all of the copper is accessible, and you will need to determine the amount of accessible copper using the given data by visually interpolating along the curve of Fig. 5.11. (A) What should the on/off leaching time be in the actual heap leaching operation to average 90% extraction of the

accessible copper in the ore? (B) What percent extraction of the *total* copper in the ore do you expect with a leaching time of 45 days? (C) What percent extraction of the *total* copper in the ore do you expect by leaching 45 days if fewer crushers are used and the crusher opening is changed to -50.8 cm (-2 in)?

6. Representative samples taken from different parts of the ore deposit were crushed and passed through a 3/4 in screen, acid cured and column leached (with triplicate columns each sample) under identical conditions for 30 days, with the following results:

Sample #	Volume of the Ore Deposit (%)	Average Copper Extraction (%)
1	40	97
2	20	80
3	25	75
4	15	93

All of the copper in this ore is accessible to leaching.

At the mine to be developed, management plans to crush only to -3 in. What will the *average extraction over the life of the mine* be, if on/off leaching is conducted for 180 days? HINT: First compute t_{100} using Fig. 5.11.

7. Compute the necessary on/off leaching time to extract no less than 65% of the accessible copper for the worst case conditions, including an unfavorable rock breakage size distribution, of the following oxide copper ore undergoing heap leaching with acid enriched SX raffinate, following tertiary crushing to minus 19 cm ($-3/4$ in). Acid curing is not presently planned. The maximum grade will not exceed 2.0% copper; acid consumption may rise to $B = 3$; the rock microporosity may drop to one percent; the leachate acidity may be as low as 1.0 g/l; and the setting of the final (last) crusher may expand by wear to as much as 25 mm (1 in). The ore density is 2.60 g/cc.

8. Re-do Problem 7 with the following amended conditions. Acid curing will be used with the amount of acid added at 110% of the ore's acid demand. Blast hole drill cuttings will be sampled and assayed for both copper content and acid consumption to determine the amount of acid to be added at the cure step. On average very little acid in the raffinate will be consumed and the minimum acidity encountered in the leachate is not likely to be less than 5 g/l.

9. A very representative oxide copper ore sample crushed and screened to $-1/2 + 3/8$ inch provided column extractions of 80% of the accessible

copper in several 10 day long replicate leaching tests. At the mine the ore will be crushed to -1.5 in. Recommend a heap leaching time for this ore and the expected extraction of accessible copper.

10. An oxide copper mine is being planned with in-pit primary crushing to -200 mm (-8 in) followed by conveyor belt haulage to valley-fill leaching heaps (dumps). The average ore grade is 0.8% copper with 85% of this copper accessible to leaching. The mining rate will be 2,000,000 tonnes per year. Leaching will begin one year after mining begins, when the first 2,000,000 tonnes of ore have been deposited. Ore crushed and screened to -1.5 in averages 68% extraction of the total copper in 45 days under simulated replicate column leaching. Compute and plot on graph paper the expected annual copper extraction (tonnes of copper produced per year) beginning with the time leaching begins out to ten years for the following two cases: (A) from only the first year of mine production (first 2,000,000 tonnes of ore), and (B) from the cumulative mine production.

REFERENCES AND SUGGESTED FURTHER READING

Agers, D.W. and De Ment, E.R. (1972). The evaluation of new LIX reagents for the extraction of copper and suggestions for the design of commercial mixer-settler plants. *AIME TMS Paper Selection, Paper No. A72–87*.

Bartlett, R.W. (1972). Conversion and extraction efficiencies for ground particles in heterogeneous process reactors. *Met. Trans. 3.*, pp. 913–917.

Dixon, D.G. and Hendrix, J.L. (1993). A mathematical model for heap leaching of one or more solid reactants from porous ore pellets, *Met. Trans. 24B*, pp. 1087–1101.

Domic, E. and Brim, E.O. (1980). TL leaching: experimental studies for the Lo Aguirre Leach Project. W.J. Schlitt (Ed.), *Leaching and Recovering Copper from As-Mined Materials*. AIME, p. 71.

Dudas, L., Maass, H. and Bhappu, R. (1974). Role of mineralogy in heap and *in situ* leaching of copper ores. Aplan, F.F. et al. (Eds.), *Solution Mining Symposium AIME*, AIME, pp. 193–210.

Dutrizoc, J. (1980). The physical chemistry of Iran precipitation in the zinc industry. Cigan, J.S. et al. (Eds.), *Lead, Zinc, Tin, 80*, The Metallurgical Society, Warrendale, PA, pp. 532–564.

Flett, D.S. (1974). Solvent extraction in copper hydrometallurgy: a review. *Trans. Inst. Min. Metall. (Sect. C: Mineral Process. Extr. Metall.) 83*, p. 30.

Fletcher, J.B. (1971). In place leaching at Miami Mine, Arizona. Aplan, F.F. et al. (Eds.), *Trans. A.I.M.E. 250*, p. 310.

Fountain, J.F., Bilson, E.A. and Timmers, J. (Oct 1983). Ferric cure dump leaching process. *SME-AIME Preprint No 83–104*, Society of Mining, Metallurgy and Exploration, Littleton, CO.

Hanson, R.W. (July 1996). RAHCO analyzes heap leaching costs. *E&MJ.*, p. 75.

Hickson, R.M. (1996). El Abra: World's largest SX-EW mine on track to join copper mining elite. *Mining Engineering*, Feb, pp. 34–40.

Glassner, D., Arnold, D.R., Bryson, A.W. and Vieler, A.M.S. (1976). Aspects of mixer-settler design. *Miner Sci. Eng. 8*, No. 1, 23.

Kahrger, R. (1996). Bulk material handling by conveyor belt at El Abra. Alspaugh, M.A. and Bailey, R.O. (Eds.) *Bulk Material Handling by Conveyor Belt*, SME, Littleton, CO.

Kordosky, G.A. (1992). Copper solvent extraction: the state of the art. *JOM*, May 1992, pp. 40–45.

Longwell, R.L. (1974). In place leaching of a mixed copper ore body. Aplan. F.F. et al., *Solution Mining Symposium*, AIME, New York, Chap. 16.

McAndrew, R.T. et al. (1974). Precipitation of iron compounds from sulphuric acid leach solutions. *Metallurgical Society, CIM. Hydrometallurgy Section*, 4th Annual Meeting, Toronto, pp. 44–53.

McAndrew, R.T., Wang, S.S. and Brown, W.R. (Jan 1975). Precipitation of iron compounds from sulfuric acid leach solutions. *CIM Bulletin*, pp. 101–110.

Power, K.L. (1970). Operation of the first commercial copper liquid ion exchange and electrowinning plant. Ehrlich, R.P. (Ed.), *Copper Metallurgy*, The Metallurgical Society, Warrendale, PA, pp. 1–26.

Roman, R.J. and Olsen, C. (1974). Theoretical scale-up of heap leaching. Aplan, F.F. et al. (Ed.), *Solution Mining Symposium*, AIME, pp. 211–232.

Roman, R.J., Benner, B.R. and Becker, G.W. (1974). Diffusion model for heap leaching and its application to scale-up. *Trans. Soc. Min. Engrs.*, pp. 247–256.

Rossi, G. (1990). *Biohydrometallurgy*, McGraw-Hill, New York, Chap. 4.

Wadsworth, M.E. (1987). Solution mining systems. Szekely, J. et al. (Ed.), *Mathematical Modeling of Materials Processing Operations*, The Metallurgical Society, Warrendale, PA, pp. 311–328.

Wadsworth, M.E. and Miller, J.D. (1979). Hydrometallurgical processes. Sohn, H.Y. and Wadsworth, M.E. (Eds.), *Rate Processes of Extractive Metallurgy*, Plenum Press, New York, pp. 133–244.

Ward, M.H. (1974). Surface blasting followed by *in-situ* leaching: the Big Mike Mine. Aplan, F.F. et al. (Eds.), *Solution Mining Symposium*, AIME, New York, Chap. 17.

SIX

Percolation Leaching Copper Mine Waste (Primary Sulfides)

INTRODUCTION

The vertical zoning of minerals in porphyry copper deposits was discussed in the preceding chapter. Horizontal zoning of the hypogene primary deposit is also typical of these ore bodies, with a protore core surrounded by a mineralogically different halo. The protore core contains a low **pyrite to chalcopyrite ratio** (Py/Cp = 1/1), and most of the molybdenite. The halo contains higher pyrite to chalcopyrite ratios, typically Py/Cp = 10/1, and lower copper concentrations. Mining of ore for milling is usually confined to the protore region where copper concentrations exceed an economic cut-off grade.

With most of the world's copper being produced from open pit mines, large volumes of subgrade overburden rock must be stripped to gain access to flotation mill grade ore (illustrated in Fig. 6.1). Often at mature mines, overburden stripping ratios (mass of overburden/mass of ore) are 3/1 or higher. And often the total copper in the mined waste is equal to the total copper in the ore mined and sent to mills. It is common practice to discard this mine waste in huge dumps surrounding the pit and filling adjacent canyons and slopes, using the shortest available haulage distances to minimize haulage costs.

The volume of the accumulated copper mine waste in the western USA exceeds the volume of all other mining wastes combined, and is measured in several billion tonnes of fragmented rock containing millions of tonnes

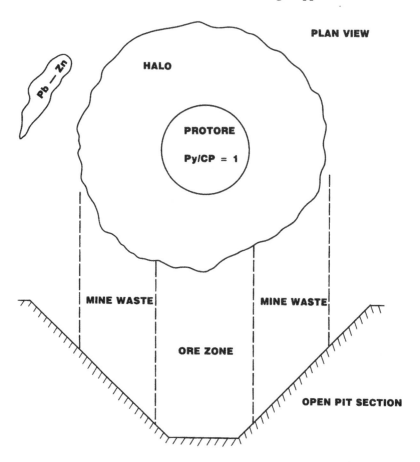

Figure 6.1. Idealized porphyry copper open pit mine showing plan and section views and mine waste (overburden) stripping in the halo around the protore.

of copper. When these dumps overlay rock or other impervious strata they can be percolation leached in place. However, several copper mine dumps in southern Arizona overlay poorly consolidated conglomerates and cannot be leached because of continuing percolation of solutions into the ground with eventual groundwater contamination.

Open pit mining must end when an excessively high and costly stripping ratio is required to continue. The slopes of pit walls cannot be overly steepened during mining or pit walls will collapse. Often conventional mining ceases when less than 50% of the copper has been removed from

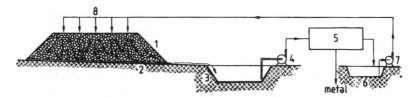

Figure 6.2. Copper mine waste dump leaching process flowsheet: 1 = dump; 2 = draining subsurface; 3 = pregnant solution pond; 4 = pump; 5 = metal recovery; 6 = barren solution pond (aeration optional); 7 = pump; 8 = solution distribution system.

the ore deposit. Pit walls, and other rubblized rock remaining after completion of conventional mining, were the first sources of copper solution mining practice which began in Spain centuries ago. The amount of accumulated leachable mineralized rock rubble is expected to be very great in the future, and solution mining of pit walls and rubble is expected to continue to be important even at closed mines, as it has been at Bisbee, Arizona. Typical copper mine waste averages 0.2% copper and the amount of copper extracted by leaching mine waste is often less than 1 kg/Mg of waste (0.1%). Consequently, a simple, inexpensive process is required. The mine waste dump leaching circuit, shown in Fig. 6.2 includes a pregnant liquor pond, copper recovery plant, and barren liquor pond. Mechanical aerators are infrequently installed in the barren pond to oxidize ferrous ions so that precipitation of ferric salts occurs preferentially in the pond rather than on the dump surfaces where it may otherwise impede percolation and air flow.

Solution mining techniques at surface mines are reviewed by Schlitt (1992). Rossi (1990) provides a review of many aspects of copper leaching, covered in this chapter and in Chapter 5.

OVERVIEW OF COPPER MINE DUMP LEACHING PRACTICE

Dumps, mine waste or excavated overburden, are usually formed by end dumping haul trucks near the edge followed by pushing the muck over the edge with a dozer. These dumps are usually much larger than the ore heaps typically encountered in gold leaching, both in height and lateral extent, some being 150 m high. As dumping is continued, leaching must be confined to inner surface areas of the dump away from its crest in order to not interfere with truck dumping operations. Before leaching, the dump surface is deep ripped and leveled with a dozer. In some cases the top few feet are removed by pushing muck into inner berms or over the dump side. The dumped material remains permanently in place.

Leaching solutions circulate between the mine dumps and a central copper separation plant. Both solvent extraction with electrowinning (SX-EW) and cementation with scrap iron are practiced, but the latter is being phased out. Solutions are introduced onto the horizontal surface of the dumps by one of several methods, most commonly ponding (flooding leveled basins) or using sprinklers. Ponding was the original method, but channels in the dump can permit rapid uneven percolation and short circuiting of the solution. So, the trend in solution application is to use sprinklers and other devices, described in Chapter 3, especially when SX-EW is used for copper recovery. With SX-EW, acidity is increased slightly without increasing soluble iron in the barren liquor returning to the dumps. When scrap iron cementation is used for copper separation, the barren liquor contains additional dissolved iron and a heightened proportion of ferrous iron caused by the chemically reducing conditions of cementation.

Most dumps are located on mountain slopes as shown in Fig. 6.3. Solutions percolate down to bedrock and then flow laterally to solution collection ponds from which they are drained or pumped to the copper recovery plant. Figure 6.3 also illustrates a channel that may form because of cracks developing or because of an unusually low amount of fine particles. Low permeability regions are also shown and occur because of a higher than normal amount of fines. These can lead to locally perched water tables, which occasionally blow out of the side of a dump as a large spring of "rocky mountain water."

Air flows slowly through the dump because of modest buoyant effects in the dump associated with heating from the exothermic oxidation of the sulfide minerals and from an increase in moisture and depletion of oxygen within the air as it passes through the dump. Air convection cells can form inside the dump but are usually confined to regions near a dump surface (illustrated in Fig. 6.3).

Intermittent leaching is practiced, with short leaching periods followed by much longer rest periods. This allows the slow sulfide mineral oxidation process to continue without an excessive circulation of leaching solution. With adequate rest periods, pumping costs are reduced, the pregnant liquor is not overly diluted and the metal separation plant operates more efficiently. Also, excessive circulation of solution transfers energy from the dumps, thereby lowering internal temperature and decreasing air convection.

Even with long rest periods, several dumps are often being leached simultaneously at a large copper mine under a rotation plan. This provides, in aggregate, a fairly uniform supply of pregnant liquor to the recovery

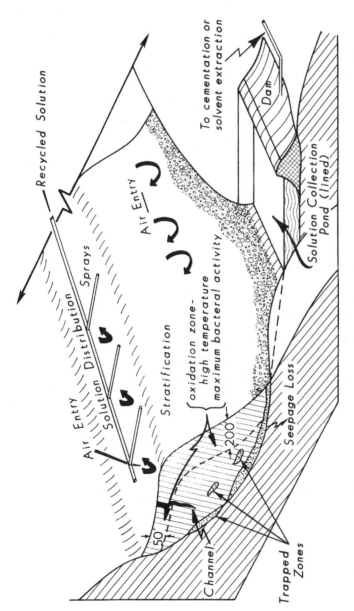

Figure 6.3. Cross-section of copper mine waste dump illustrating solution management during leaching (Wadsworth, 1987).

plant. As with gold ore heap leaching, adequately sized pregnant liquor and barren liquor storage ponds must be provided to handle excess solution from unusual weather and operating emergencies.

Copper dump leaching is carried on for many months and years. Some dumps have been periodically leached for over twenty years, usually with progressively longer rest periods between leaching cycles.

Solution mining of sulfide copper minerals by percolation leaching is also practiced in abandoned underground workings that are located above the water table. Most frequently, these are the subsidence areas located above former block caving mine operations.

Effective leaching depends on a sequence of processes and requires (1) effective air circulation, (2) good bacterial activity and (3) uniform solution contact (good permeability and uniform percolation). Solution flow and air flow phenomena will be considered in greater detail in Chapters 7 and 8.

COPPER SULFIDE MINERALS AND FERRIC SULFATE LEACHING KINETICS

Copper in mine waste is usually present as sulfide minerals that must be oxidized by air to be solubilized by mine water during leaching. The primary mineral chalcopyrite, $CuFeS_2$, is usually predominant and this mineral exhibits oxidation reaction kinetics that are much slower than those of the secondary copper sulfide minerals.

The intrinsic leaching kinetics of copper sulfide minerals have been extensively studied and reviewed (Wadsworth, 1972; Dutrizac and MacDonald, 1974). Mineral oxidation during heap and dump leaching occurs by dilute ferric sulfate leaching at temperatures generally below 45°C. While the leaching kinetics of the various sulfide minerals are quite complex, it is convenient to divide them into three groups according to their oxidative leaching rates in ferric sulfate; these are shown in Table 6.1. Reaction rates are indicated by the leaching times required to complete oxidation (99%) of sulfide mineral grains, all assumed to be spherical and of the same size. Computations were made for a 0.1 molar ferric sulfate lixiviant at two temperatures: 25°C and 45°C, which are typical copper heap/dump leaching conditions. The reaction times were computed with rate equations, from numerous sources, that have been summarized by Paul et al. (1992). To better illustrate the wide differences in oxidation rates under heap and dump leaching conditions, the computed times in Table 6.1 were normalized to the fastest reaction (shortest time), which was that of chalcocite (phase I) at 45°C.

Table 6.1 Copper Sulfide Mineral Grain Oxidative Leaching Kinetics with Ferric Sulfate ($C_{Fe^{3+}} = 0.1M$).

Mineral	Reaction	Relative Time to Complete Reaction	
		$T = 25°C$	$T = 45°C$
Chalcocite (phase I)	$5Cu_2S + 8Fe^{3+} \rightarrow 5Cu_{1.2}S + 4Cu^{2+} + 8Fe^{2+}$	1.35	**1.0**
Chalcocite (phase II)	$5Cu_{1.2}S + 12Fe^{3+} \rightarrow 5S^0 + 6Cu^{2+} + 12Fe^{2+}$	5,000	730
Enargite	$Cu_3AsS_4 + 11Fe^{3+} + 4H_2O \rightarrow 3Cu^{2+} + AsO_4^{3-} + 4S^0 + 8H^+ + 11Fe^{2+}$	9×10^6	2×10^6
Covellite	$CuS + 2Fe^{3+} \rightarrow Cu^{2+} + S^0 + 2Fe^{2+}$	1×10^8	2.5×10^7
Bornite (phase I)	$Cu_5FeS_4 + 4Fe^{3+} \rightarrow Cu_3FeS_4 + 2Cu^{2+} + 4Fe^{2+}$	1.4×10^8	3.1×10^7
Bornite (phase II)	$Cu_3FeS_4 + 8Fe^{3+} \rightarrow 3Cu^{2+} + 9Fe^{2+} + 4S^0$	9×10^{12}	2×10^{12}
Chalcopyrite	$CuFeS^{2+} + 4Fe^{3+} \rightarrow Cu^{2+} + 5Fe^{2+} + 2S^0$	$\mathbf{4 \times 10^{12}}$	$\mathbf{9 \times 10^{11}}$

As indicated in Table 6.1, oxidation of chalcocite and bornite proceed in two steps at different rates, each through an intermediate compound. Elemental sulfur is typically produced by oxidation of these minerals, but sulfur is thermodynamically unstable and slowly oxidizes to sulfate ion during heap/dump leaching. While enargite, covallite and bornite (phase I) oxidize more slowly than chalcocite, they are not nearly as slow as chalcopyrite. In fact, all of these secondary copper sulfide minerals oxidize at rates that are at least four orders of magnitude faster than the rate of oxidation of chalcopyrite. At heap/dump leaching temperatures, both chalcopyrite and pyrite oxidize at similar rates. The practical effect of this is that primary copper sulfide ores and mine waste leach very much slower than ores where secondary copper sulfide minerals and copper oxide minerals predominate.

Primary copper mine waste and some primary copper ores have a further complication caused by the large amounts of pyrite, which consumes oxygen during leaching, concurrently with chalcopyrite oxidation. The amount of oxygen used by pyrite can dwarf the oxygen actually used to leach the copper mineral. With secondary copper sulfide ores, most if not all, of the pyrite has already been oxidized by the natural geochemical weathering (gossan formation) process. And, if pyrite is present, the secondary copper sulfides tend to leach first.

The relative oxygen requirements, or ferric ion requirements, per unit of copper leached are shown in Table 6.2 for the more common copper sulfide minerals and for primary mine waste (chalcopyrite) with different molar ratios of pyrite to chalcopyrite in the mine waste, Py/Cp ratio. The higher Py/Cp ratio in the halo zone of a typical porphyry copper ore deposit, compared with that in the protore, requires very high oxygen consumption per unit of copper leached. At a Py/Cp ratio of 10, which is not uncommon, **thirty one times more oxygen (air) is required than when leaching chalcocite (Cu_2S).**

The combination of slower leaching kinetics and higher oxygen (air) requirements has a profoundly slowing effect on the extraction of copper from primary ore, and especially primary mine waste with a high Py/Cp ratio. Compare oxygen requirements in Table 6.2 with Fig. 5.19 showing the time required for leaching solution passing through an ore heap or primary waste dump to provide sufficient ferric ion to oxidize the copper as chalcocite. With a ferric ion depletion of 2.0 g/l for each solution pass, the required time for a 1.0 wt pct copper ore as chalcopyrite is 240 days (2/3 year). For a mine waste with Py/Cp = 10, the required leaching time jumps to 6.4 yrs. Injecting sufficient ferric ion or oxygen (air) into a copper mine waste dump in a reasonable amount of time is a serious technical and economic problem.

When either oxidized copper or secondary copper sulfide minerals are present in a primary ore or in primary mine waste, copper from these sources will make a significant early contribution to the leaching production.

Table 6.2 Relative Oxygen or Ferric Ion Consumption Requirements per unit of Copper Leached, including Sulfur Oxidation to Sulfate.

Mineral Source	Relative Oxidant Consumption
Cu_2S (chalcocite)	1
Cu_5FeS_2 (bornite)	1.44
CuS (covellite)	1.6
$CuFeS_2$ (chalcopyrite)	3.2
Primary ore or mine waste	
Py/Cp = 1	6
Py/Cp = 2	8.8
Py/Cp = 5	17.2
Py/Cp = 10	31.2

For some mines and mine waste dumps, it can be argued, from comparing production statistics with mineral assay data from mine waste rock samples, that oxide and secondary sulfide minerals are the major source of copper derived from leaching these primary materials. In other words, very little copper is actually derived from leaching chalcopyrite. And, after several months of leaching, the smaller amounts of oxide copper minerals and the more readily leached secondary sulfide copper minerals are usually no longer present in significantly accessible amounts.

During leaching, secondary copper sulfide minerals are often formed from soluble copper reacting with chalcopyrite located deep inside partially leached rocks, where the oxygen chemical potential is low. Refer to the oxygen chemical potential (E_h) versus pH diagram in Fig. 5.3. Usually the amounts are small because the secondary sulfides are themselves being further oxidized and solubilized. In effect, a steady state is reached within the rock for generation and conversion of secondary sulfides in a localized micro-region of each of the large ore rocks. Consequently, the amount of secondary copper sulfides in a mature leaching dump, originally composed of primary mineralization, is usually very small.

BIOLOGICALLY ENHANCED OXIDATIVE LEACHING OF SULFIDE MINERALS

Bacteria are by far the most prevalent and diverse microorganisms on the earth and in the earth. There are over 200 genera and many species in each genera. These single cell organisms range in size from 0.5 µm to 5 µm and take on various shapes. Reproduction is by cell fission. In soils and ores, bacteria populations are typically 0.1–1 billion/g. They tend to adhere to particle surfaces, referred to as substrates by biologists, and migrate slowly, but they are found in smaller numbers at remarkable depths in the earth's lithosphere.

The composition of bacteria is mostly water, but includes carbon, oxygen, nitrogen, hydrogen and phosphorus and minor amounts of trace elements. Nutrients, often naturally available in soils and ores, are required to supply these elements for bacteria cells to reproduce. Environmental factors, such as temperature, humidity and pH, are also important factors in cell repro-duction rates. Various bacteria species generally fall into low, medium and high temperature ranges: psychrophiles, mesophiles and thermophiles, respectively.

There are many bacteria strains within a single species and they readily mutate. If conditions are changed slowly the bacteria will change to fit the new conditions. In solution mining, the gradual adaptation of a bacteria to large concentrations of what had previously been toxic metals is a very important characteristic. While this adaptation is relatively easy if done slowly, adaptation to higher temperatures to accelerate leaching has not been very successful, even though thermophillic sulfide oxidizing bacteria species are found in natural hot springs. Because bacteria occur in communities of different species, successful insertion of a single thermophillic sulfalobus species into an ore heap may not be possible.

Bacteria derive their energy from oxidation–reduction chemical reactions. The energy source, an electron donor, is oxidized transferring electrons to an acceptor and releasing energy. Generally this involves an enzyme reaction. There are three oxidation–reduction mechanisms used by bacteria, but only one is operative for a given species. **Aerobic** respiration processes require oxygen. **Anaerobic** processes rely on reduction of nitrate, sulfate or carbonate anions in the absence of oxygen. Because oxygen, processed to water, has the highest standard reduction potential, $+0.92\,V$ at 25°C and pH 7, the anaerobic reduction of these other inorganic species cannot occur except in the absence of oxygen. **Fermentation**, a third redox process, is not important in mineral processes, but both aerobic and anaerobic processes are important in mine leaching and mine remediation. Aerobic biooxidation is used to oxidize sulfide ore minerals, which can be very desirable, e.g., to provide a source of copper (the topic of this chapter), or it can be very undesirable as a source of toxic acid mine water drainage originating from mine openings, ore tailings piles and mine waste dumps. Anaerobic reduction of sulfate ion can be used as a biochemical remediation treatment for water. The resulting dissolved H_2S or HS^- ions will react with dissolved heavy metals to precipitate metal sulfides, which generally have very low solubilities. Note that this is completion of the mineral sulfur cycle. Anaerobic reduction of sulfate ions in acid mine water will be considered in Chapter 15.

Leaching heaps and mine waste dumps containing copper sulfide minerals depends on natural convection of internal air to provide oxygen for conversion of the sulfide minerals to soluble copper and iron sulfates. Dissolved oxygen in the leaching solution reacts with pyrite and chalcopyrite but the reaction rate is much too slow at room temperature. Similarly, the reaction rate to chemically oxidize ferrous ions to ferric ions is too slow. However, once formed, ferric ions can also oxidize the sulfide minerals.

Bacteria accelerate both the sulfide mineral oxidation and ferrous ion oxidation reactions (Malouf and Prater, 1961). Although these bacteria, shown in Fig. 6.4, require oxygen, carbon (from CO_2 in the air) and nitrogen for cell development, they obtain energy from the oxidation of sulfides and from the oxidation of ferrous ions to ferric ions. Bacteria

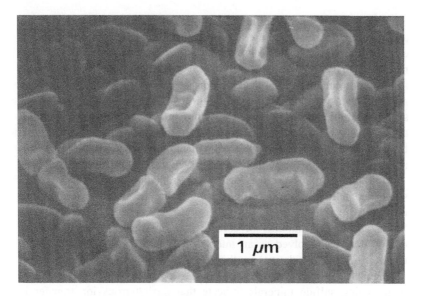

Figure 6.4. *Thiobacillus ferrooxidans* bacteria.

assist the oxidation process through an indirect biochemical mechanism. In effect, the bacteria are catalysts for both the mineral and ferrous ion oxidation processes. In the absence of oxygen from air in heaps and dumps, leaching will not occur whether bacteria are present in large numbers or not. Because of the low solubility of oxygen in leaching solutions near room temperatures, the bacteria must be close to the air source for oxidation of sulfides to occur at an economically significant rate. Oxygen transfer in solution over more than a very few centimeters is too slow, by either flow or diffusion, to have much impact.

Thiobacillus ferrooxidans bacteria are rod shaped, about 0.5 μm in diameter by up to 2 μm long, and multiply by cell division. They are the predominant bacteria catalyzing the leaching of copper sulfide minerals in solution mining and of metal sulfide minerals in nature, including the unwanted production of acid mine water. These bacteria are found naturally when sulfide minerals are exposed to air and water.

T. ferrooxidans is a member of the genus *Thiobacillus*. This aerobic bacterium is acidophilic, operating optimally at a pH near 2.5, but it will metabolize in solutions up to a pH of at least 6 (Rossi, 1990). It is a chemoautotrophic bacterium, deriving its energy from the aforementioned chemical oxidation reactions and its cell carbon only from CO_2.

T. ferrooxidans is a gram-negative procaryotic cell that is divided into a non-compartmentalized interior containing the genetic information, an envelope and a flagellum or tail that is resolvable only with electron microscopy (Rossi, 1990). The cell interior has a neutral pH even in highly acidic environments. The envelope consists of a cytoplasmic membrane and an outer wall that provides rigidity, determines shape and resists swelling from osmotic pressure. Enzymes in the cytoplasmic membrane are responsible for respiration of dissolved oxygen. Ferrous ions diffuse through the cell wall to the cytoplasmic membrane where oxidation occurs and where ferric ions form and then diffuse out of the membrane.

Although morphological evidence suggests direct oxidation of solid sulfur and sulfide minerals by attached bacteria, the mechanism involving transport through the bacterium cell wall is undoubtedly more complex. When substantial amounts of soluble iron are present, the Fe^{2+}/Fe^{3+} couple may be involved as an intermediate accounting for most of the electron transfer (oxidation) by this so-called direct mineral biooxidation mechanism. Recent electrochemical studies of the role of *T. ferrooxidans* in oxidizing sulfide minerals indicate that ferric/ferrous ion transfer is operative, even when the bacterium is **attached** to a sulfide mineral surface. Pesic and Kim (1990) have shown that when a thin reaction product layer forms between the bacteria and the mineral, oxidation continues, apparently by diffusion of soluble inorganic chemical species through the porous layer separating the bacteria and the sulfide mineral surface.

Nyavor et al. (1996) in a detailed study showed that pyrite biooxidation cannot occur by the direct mechanism. Viable cells could not oxidize pyrite when the chemical oxidation is eliminated at pH above 4.0. Ferric ion must always be present as an intermediate. The pitting of pyrite in the presence of *T. ferrooxidans*, which had been used to support the direct contact mechanism is attributed to the higher concentration of ferric ions in the vicinity of bacteria attachment. Furthermore, the rate of pyrite oxidation was independent of an increasing bacteria population above 10^6/ml.

Batch growth of *T. ferrooxidans* populations in the laboratory shows a sequence of phases that are illustrated in Fig. 6.5: (1) lag, which can last for days and weeks with no visible activity, (2) exponential growth, with cells dividing in a fairly uniform mean time interval, (3) a stationary, or steady state phase where the division period is long and new cells are balanced by dying cells, and (4) a death phase when the live cell population is decreasing (Rossi, 1990). Under fairly ideal conditions, *T. ferrooxidans* will divide about once every day. Adverse changes in the environment, including depletion of nutrients, substrates, ferrous ions or oxygen, cause a transition to phases 3 and 4.

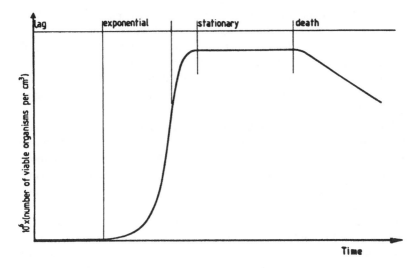

Figure 6.5. Typical bacteria population growth curve for batch cultivation (after Rossi, 1990).

The initial leaching of recently emplaced copper mine dumps may experience lag and exponential growth, but dumps quickly proceed to phase 3, steady state, where oxygen depletion because of inadequate air flow is often the limiting factor. Metabolic activity, and the catalyzing of ferrous ion oxidation and mineral oxidation, does not occur below an oxygen concentration in solution of at least 1 ppm.

The leaching characteristics of *T. ferrooxidans* strains, obtained from different sources, vary considerably. The strain diversity is unusually high for this bacteria species. Also, strains mutate quickly to acquire optimum characteristics for each individual mine site or microenvironment. Although not normally tolerant of toxic metals, these bacteria can be adapted through division and mutation to operate effectively in many chemically adverse environments. This adaptation process may take weeks and months after bacteria are introduced to a new environment.

The bacteria preferably attach to surfaces rather than being suspended in solution. Consequently heaps, dumps and other rubblized rock are ideal hosts for the bacteria.

Most *Thiobacillus* species are mesophiles, with their activity strongly inhibited at temperatures above 45°C. This leaves a sulfide mineral oxidation rate gap between the relatively fast biochemical rate at lower temperatures and the fast chemical (nonbiological) rate at higher temperatures (Malouf and Prater, 1961). This is illustrated schematically in Fig. 6.6.

Sulfalobus acidocaldarius bacteria, found naturally in hot mineral springs, and other similar sulfalobus species also biochemically oxidize sulfur and ferrous ions. These bacteria grow between 55°C and 85°C. However, attempts to adapt these bacteria for accelerated copper dump leaching have not yet been successful.

Ultimately, the rate of copper dump leaching, after removal of oxidized copper minerals, is controlled by the slow rate of air ingress more than any other factor. A high rate of air injection would be required to sustain more rapid oxidation and higher dump temperatures. Dump temperatures have rarely been measured above 45°C, but when thermal excursions above this temperature are found, they occur in new dumps and are usually short lived. And on these occasions the temperature usually is well above 40°C, often as high as 60–80°C, and with a diminished bacteria population (Beck, 1967). Thermal excursions may be associated with either sulfalobus bacteria, or more likely with a transition to non-biological chemical oxidation. With

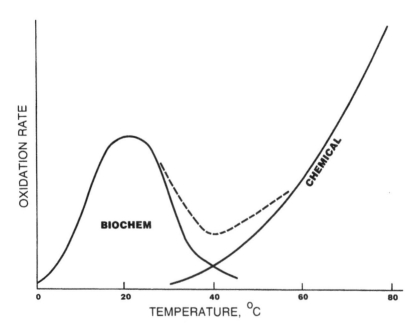

Figure 6.6. Illustration of temperature regimes and comparative rates of biochemical and chemical oxidation of sulfide minerals.

ideal air flow conditions, moderately elevated temperatures will accelerate the air draft through the dump, leading to faster chemical oxidation, still higher dump temperatures, and a thermal excursion with positive feedback.

LIMITING COPPER OXIDATION RATES IN COPPER DUMP LEACHING

Two-phase fluid flow occurs in mine dumps and heaps with the flow of both fluids caused by gravity. Air generally flows up but it also may circulate internally in convection cells. Because downward flowing water is the rock wetting fluid, open micropores within rocks are quickly filled with stagnant solution by capillary action while air is being excluded. Hence, air and oxygen do not enter the micropores of leaching rocks and bacteria catalyzed direct oxidation of sulfide minerals well inside rocks does not occur, even if the micropores are large enough to admit bacteria. Therefore, the rate controlling sulfide mineral oxidation step continues to be transport by diffusion of dissolved oxygen and, primarily, ferric ions from the rock surface to the internal sulfide mineral grains through solution-filled micropores.

Under dump leaching pH conditions, the ferric ion concentration in the percolating leach liquor is usually about 0.25–1 g/l, which is much higher than the oxygen solubility of 7 ppm that exists in equilibrium with air. Oxygen depletion in the air will further lower the oxygen solubility. Since the charge transfer of ferric ion as an oxidant is only 1/4 that of O_2, a greater mass of ferric ion must be transferred, four moles of iron (at. wt. 56) versus one mole of oxygen at (mol. wt. 32). Nevertheless, at 1 g/l, the ferric ion concentration is about 1000 ppm and its diffusion transport advantage over dissolved oxygen is about 20 to 1, as shown by the following proportionality calculation:

$$\left(\frac{1000\,\text{ppm}}{7\,\text{ppm}} \right) \times \left(\frac{32\,\text{g-O}_2}{56\,\text{g-Fe}} \right) \times \left(\frac{1\,\text{e}}{4\,\text{e}} \right) \approx 20,$$

while at 0.25 g/l Fe^{3+} the advantage drops to about 5.

Note that **the significant mechanism for the internal leaching of rocks containing copper sulfide minerals is the countercurrent diffusion of dissolved ferric ions, going in, and product ferrous ions, coming out, through the open rock micropores.** And for dump leaching conditions, **the most important role for bacteria is to catalyze the oxidation of ferrous ions to ferric ions in the capillary held solution outside of the**

rocks, in the near presence of air. Direct oxidation of sulfide minerals by bacteria will only occur for those minerals exposed at the surface of the rocks, which surely is a minor amount compared with the internal sulfide mineral grains.

It is worth mentioning as an aside that "apparently direct" bacteria oxidation of sulfide mineral grains can be important for the bioleaching of flotation concentrates because in that process the individual mineral grains are completely **liberated** and fully exposed to sparged air, leach solution, and bacteria.

The high ratio of pyrite to chalcopyrite present in the halo surrounding the protore of a porphyry copper deposit usually causes the mine waste to contain much more pyrite than chalcopyrite. Consequently, **oxidation of pyrite is usually the dominant oxidant consuming process**, with chalcopyrite oxidation somewhat of a geochemical "sideshow". Oxidation of pyrite generates ferrous sulfate and sulfuric acid by the following net reaction:

$$FeS_2 + \frac{7}{2}O_2 + H_2O = FeSO_4 + H_2SO_4. \tag{6.1}$$

But, the actual route is more indirect as follows:
outside rock, aerobic:

$$14FeSO_4 + 7H_2SO_4 + \frac{7}{2}O_2 = 7Fe_2(SO_4)_3 + 7H_2O, \tag{6.2}$$

inside rock, anaerobic:

$$7Fe_2(SO_4)_3 + FeS_2 + 8H_2O = 15FeSO_4 + 8H_2SO_4. \tag{6.3}$$

Especially, note the production of sulfuric acid by the oxidation of pyrite, eqn 6.1. Acid generation is relatively uncommon among the oxidation of sulfide minerals, but acid generation from pyrite oxidation is very important to most applications of sulfide mineral heap leaching including mine waste dump leaching. Injection of purchased acid is not required, as is the case with oxide copper ore leaching. Most important, the acid is generated *in situ* so that it is **uniformly** generated and consumed throughout the heap. Consequently, the acid concentration (pH) is fairly constant in the percolating solutions from top to bottom of copper mine dumps, even those dumps up to 150 m in height. Contrast this with the depletion of acid as it percolates through the heap in copper oxide heap leaching described in Chapter 5, where thin layer leaching lifts are limited to about 5 m.

IRON REMOVAL FROM COPPER LEACH SOLUTION

The large amount of ferrous sulfate generated by oxidation of pyrite is oxidized to ferric sulfate and hydrolyzed to jarosite. This precipitate forms within the dump, but generally outside the rock micropores, under aerobic conditions:

oxidation

$$2FeSO_4 + H_2SO_4 + \frac{1}{2}O_2 = Fe_2(SO_4)_3 + H_2O, \tag{6.4}$$

precipitation of jarosite

$$3Fe_2(SO_4)_3 + 0.8K_2SO_4 + 12H_2O$$
$$= 2K_{0.8}H_{0.2}Fe_3(SO_4)_2(OH)_6 + 5.8H_2SO_4. \tag{6.5}$$

Hydrolysis and precipitation prevent iron, dissolved from both pyrite and chalcopyrite, from accumulating in the solution beyond the equilibrium value for ferric ion and other ferric complexes. This helps to govern a steady state value for both the ferric ion and ferrous ion concentrations as they are involved in sulfide mineral oxidation. Total soluble ferric iron is increased by sulfate complexing, and mature dump leach solutions have a high concentration of sulfate ions due to the presence of large amounts of magnesium sulfate and aluminum sulfate salts, which have slowly accumulated from acid weathering of gangue minerals.

Precipitation of jarosite outside the rock prevents it from clogging the rock micropores. In fact, the micropores become more open as internal sulfide minerals are dissolved and removed, which is advantageous for leaching.

The potassium consumed in precipitating jarosite results from the acid leaching of various gangue minerals, predominantly biotite mica. If potassium ions are not available, jarosite will not form, but another iron hydrolysis product will be substituted for jarosite. Note from eqn 6.5 that jarosite precipitation releases some sulfuric acid.

Oxidation of the prevalent pyrite and hydrolysis of dissolved iron to jarosite are the major chemical reactions involving ore minerals in leaching copper mine waste. The oxidation of copper sulfide minerals, for example chalcopyrite

$$CuFeS_2 + 4O_2 = CuSO_4 + FeSO_4, \tag{6.6}$$

is relatively minor and does not control overall dump geochemistry. Rather, the copper oxidation reaction and extraction is strongly affected by the pyrite/jarosite chemistry, coupled with the consumption of generated acid by gangue mineral neutralization.

COPPER SEPARATION

The recovery of copper from the leach liquor by solvent extraction and electrowinning, previously described in Chapter 5, also generates some acid:

$$CuSO_4 + H_2O = Cu^0 + H_2SO_4 + \tfrac{1}{2}O_2, \tag{6.7}$$

while extraction of copper from the leach liquor by iron cementation generates more ferrous ion in solution,

$$CuSO_4 + Fe^0 = Cu^0 + FeSO_4. \tag{6.8}$$

Iron cementation also reduces the ferric ions in solution to ferrous ions while dissolving more scrap iron,

$$Fe_2(SO_4)_3 + Fe^0 = 3FeSO_4. \tag{6.9}$$

Because the amount of copper produced is typically small compared with the amount of pyrite oxidized, these processes do not have a major impact on internal dump/heap acidity. However, the incremental dissolved iron, obtained from cementation with scrap, often exceeds the solubility limit in the balanced solution once reoxidation of ferrous ion to ferric ion by air occurs. Consequently, when iron cementation of copper is used, some $Fe(OH)_3$ or other iron salts often precipitate and settle as a red slime deposit on the surface of the dumps, especially when flooded basin leaching is used. This gelatinous precipitate significantly lowers dump permeability near its top where, because of a large accumulation of fines, dump permeability is usually poor to begin with. Jarosite cannot be produced as the iron precipitate in these surface ponds because the pond solution is not in intimate contact with the potassium releasing gangue minerals. The presence of red ponds on flooded basin leached dump surfaces is a very good indication that iron cementation is being used to recover the copper from the leachate. Because this problem does not occur when solvent extraction is used, it is another advantage of SX-EW as a process for the recovery of copper from leach solutions.

Most sulfide ores, other than copper, also contain large amounts of pyrite, which is usually the dominant sulfide present in base metal and precious metal sulfide ores. Therefore, the conclusions about pyrite leaching and iron hydrolysis are likely to be valid for the oxidative leaching of other sulfide ore heaps, although depending on particular conditions, the iron hydrolysis product may be other than jarosite.

ACID WEATHERING OF GANGUE MINERALS

When iron hydrolysis and precipitation are included, nearly two equivalents of H_2SO_4 are generated for each equivalent of pyrite oxidized. This acid in the leach liquor attacks gangue minerals in the region of the dump or heap where it is generated. The rates of attack depend on the local pH and vary among the numerous gangue minerals present. Generally, the rate of acid attack and neutralization increases with increasing acid concentration (lowered pH). A steady state condition is reached in each dump at a pH wherein the net rate of acid generation by the ore minerals balances the rate of acid consumption by gangue mineral neutralization. Since the mineral assemblages, their mineral concentrations, and other leaching conditions differ somewhat among copper mine dumps, even at the same mine, the steady state pH of the leaching solution may also vary somewhat among individual mine waste dumps. Nevertheless, there is enough similarity between mines that this dump leach liquor pH typically ranges between pH 2.6 and 2.9.

The majority of acid consumption in primary porphyry deposit mine waste is usually associated with extraction of magnesium, aluminum, and alkali metal oxides from biotite $[H_2K(Mg, Fe)_3 Al(SiO_4)_3]$ and similar basic silicate minerals. This has been verified chemically, but it can be readily seen in partially leached rocks that have been sectioned. The originally dark (basic) silicate minerals in the leached rim of the rock are bleached of their color because of the reactions with acid. Of course, when carbonate rocks are present in the waste, they will be rapidly attacked by the acid leachant.

Acid attack of gangue minerals also causes rock decrepitation, meaning loss of rock physical integrity. Consequently, the average rock particle size and permeability to both percolating leach solutions and air flow tends to decrease with extended leaching time. This is a major factor preventing adequate aeration and continued economic leaching as the mine dumps age, and it often occurs long before most of the copper has been extracted. Rock disintegration also makes realistic mathematical modeling of the leaching process difficult.

New mine waste dumps are most often gray in color and have coarse, rocky surfaces. Very old dumps, measured in leaching years, are stained yellow from jarosite and often weathered to a smooth, near soil, surface texture.

The acid attack of silicate minerals to release magnesium and aluminum ions is also evidenced by the high concentrations of these cations in dump leach liquors at copper mines where leaching has been established for many years. See Table 6.3 for a pregnant liquor composition that is typical

Table 6.3 Typical Copper
Mine Waste Leaching
Pregnant Liquor.

pH	1.9–3.5
Cu^{2+}	0.2–2.0 g/l
Fe^{3+}	0.2–3.0 g/l
Fe^{2+}	0.01–3.6 g/l
Al^{3+}	up to 12 g/l
Mg^{2+}	up to 7 g/l

of a mature copper mine dump leaching operation. There is apparently no chemical outlet for dissolved magnesium in a closed copper leaching system, without a bleed. A small leakage of solution, with replacement by make up water, will hold concentrations of this cation in check at a steady state concentration dependent on the generation rate from acid weathering and the leakage rate. Most copper mine dumps are not resting on an impervious liner, and some leakage to the underlying ground may occur.

As the major acid consuming gangue mineral, biotite's alteration chemistry is important to an overall dump chemistry balance. Based on structural considerations of the biotite lattice and experimental work on the dissolution of biotite under stronger acid concentrations than are typical of dump leaching, it is expected that biotite alters in the following manner. Potassium is readily exchanged by hydrogen ions or other small cations and is quickly leached from biotite. Ultimately all of the cations from biotite except silicon can be removed, and the final residue would be amorphous silica. Iron, magnesium, and aluminum in biotite are removed at a rate that preserves their original stoichiometric relationship to each other in the biotite lattice. These cations are not leached nearly as rapidly as potassium.

Copper dump leaching solutions are calcium saturated. Carbonates and soluble calcium bearing silicate minerals cause precipitation of gypsum, which is very detrimental to dump permeability because the molecular volume of gypsum is much greater than the minerals it replaces. The mining of relatively small amounts of carbonate rock from dikes and skarns, cutting the ore body and placed in leach dumps, has led to localized high gypsum concentrations and major perched water tables. These perched water tables were caused by gypsum precipitation and sealing that prevented solution flow. Spectacular saturated solution pressure blow outs have occurred at locations where carbonates were previously dumped.

CHARACTERISTICS OF AS-MINED ROCK FRAGMENTS

Copper host rocks vary considerably in size distribution, internal micro-porosity and mineralogy. This makes it difficult to generalize and predict leaching behavior, even at the same mine.

The waste rock size distributions, as mined from three petrologic types of host rocks obtained from two copper mines, are shown in Fig. 6.7. For all three of these samples, approximately half of the rock mass exists at sizes above a diameter of 25 mm (1 in).

The concentration of pyrite, the concentration of sulfide copper and the ratio of pyrite to copper sulfide minerals are shown for seven copper mine waste rock samples in Table 6.4. These examples show the wide distribution of these values among the samples. Sulfide copper is the difference between the amount of total copper, determined by assay, and the amount of acid soluble copper, determined by extraction in a standard acid leach laboratory test. However, in all samples the pyrite concentration, which was determined by micrometric analysis of heavy media concentrates, is the dominant sulfide mineral. The sulfide copper values were used to estimate the amount of chalcopyrite and the Py/Cp ratio for each of the samples.

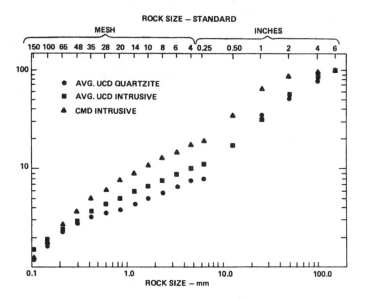

Figure 6.7. Rock fragment size distribution of three as-mined waste samples.

Table 6.4   Distribution of Sulfides in Copper Mine Waste
Rock Samples.

Sample	Open Porosity (%)	Specific Gravity (g/cm^3)	Sulfide Copper[1] (%)	Pyrite (%)	Py/Cp
1	3.67	2.64	0.137	3.55	9
2	4.99	2.68	0.114	3.13	9.5
3	3.40	2.64	0.061	2.96	16.8
4	5.78	2.73	0.354	5.01	4.9
5	5.25	2.68	0.114	2.06	6.3
6	4.95	2.74	0.180	5.04	9.7
7	6.58	2.75	0.206	6.55	11
AVG	4.95	2.69	0.167	4.04	8.4

[1]Mostly Cp, except Sample 7.

The internal open microporosity and specific gravity of these samples
were also measured and are reported in Table 6.4.

ROCK LEACHING MECHANISM AND IMPLICATIONS FOR FORECASTING PRODUCTION

As the leaching solution trickles over the wetted ore fragment, solution
penetrates the open rock pore structure as shown in Fig. 6.8. Bacteria
are attached to the rock surfaces and are bathed both by the percolating
solution trickling over the rocks and by the adjacent air. Thus, if the air is
not depleted in oxygen, bacteria readily utilize oxygen to oxidize ferrous
ions to ferric ions (including their complexes); see eqn 6.2. The cycle is
completed by coupling with the sulfide mineral grains inside the rock.
Countercurrent diffusion of ferric ions and ferrous ions accounts for most
of the transport of oxidant required for oxidation of the internal sulfide
minerals grains.

Other minor transport processes may operate in addition to the Fe^{3+}/Fe^{2+}
couple and diffusion of dissolved oxygen. Acid is generated inside the rocks
and the resulting hydrogen ion concentration gradient diffuses hydrogen
ions **out** of the rocks; this direction is opposite to the direction encountered for acid diffusion in rock micropores when leaching oxide copper
ore. Cupric and ferrous ions generated by oxidation diffuse out of the rock
through the solution filled micropores.

Cupric ions released by oxidation in the intermediate region (reaction
zone) may also diffuse further **inward** and react with chalcopyrite and

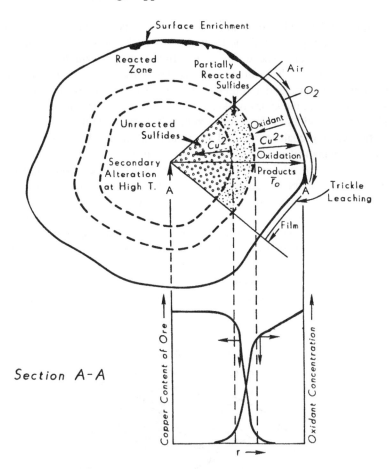

Figure 6.8. Sulfide ore fragment (cutaway rock model) showing reaction zone, shrinking unreacted core and expanding rim (reacted zone).

pyrite in the unleached core to form secondary copper minerals in a region of low oxidation potential.

Most important, the oxide copper mineral assumption of rapid chemical reaction kinetics is invalid for the slow reacting sulfide mineral grains, and the rock leaching rate depends on both diffusion and the heterogeneous sulfide mineral grain chemical reactions. This is an example of **mixed kinetics**, when two or more sequential process steps are significant in determining the overall extraction rate. Consequently, there will not necessarily be a sharp interface between the unleached shrinking core and a completely leached rim. Rather, a reaction zone, containing partially reacted sulfide grains, will be located between the fully leached rim and the unreacted core. The diffusing ion fluxes are constant for each ion across the fully leached rim but they decrease through the reaction zone as reactants are consumed and product ions produced. Some of these effects are illustrated in the cutaway sketch of the reacting ore fragment (rock) in Fig. 6.8. The width of the reaction zone will increase with increasing size of the reacting chalcopyrite and pyrite mineral grains. Two finite difference mathematical models for this leaching process will be presented in Chapter 12. For adequately crushed mixed kinetics ores, pore diffusion transport will not be rate determining in the biooxidation of sulfide ores. The simple sulfide mineral grain heterogeneous reaction rate model for quasi-spherical mineral grains is presented in Chapter 9 on refractory gold ores.

The rock leaching kinetics are complicated by changing microporosity, pH, solution concentrations of several species, and chemical weathering and **disintegration of the rocks** by the generated sulfuric acid. Fig. 6.9 illustrates the effect of weathering by dissolution of gangue constituents and penetration down existing micro-cracks in the ore fragment.

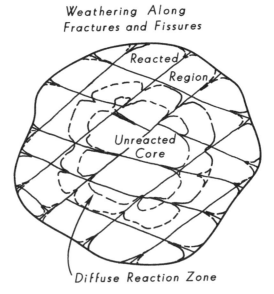

Weathering Along
Fractures and Fissures

Reacted Region

Unreacted Core

Diffuse Reaction Zone

Figure 6.9. Ore fragment after extensive chemical weathering along fissures (large micropores) due to internally generated acid from pyrite oxidation.

Continued leaching will disintegrate the rock into several smaller ore fragments. The good news is that with smaller fragments extraction of copper from initially large rocks will proceed much faster. The bad news is that the additional fine material lowers dump permeability, which is a more serious problem for air circulation, under the weak natural convection forces, than it is for solution percolation. **Poor air circulation (and oxygen depleted air within dumps) is often the biggest impediment to successful continued leaching** of mature copper waste dumps containing primary copper mineralization (chalcopyrite). Eventually, the copper yield will not justify the operating expense and maintenance of the leaching infrastructure and the dump will be abandoned.

Because disintegration of rocks from acid weathering is extremely difficult to realistically include in a computer based numerical model of the process, it has been ignored in formulating most models. The rationale for this omission is that disintegration is a late stage leaching phenomena when model forecasts of results are technically and economically less important. Braun et al. (1974), in an underground flooded copper ore leaching model, considered weathering to yield a systematic increase in rock surface area and used this to attempt to account for rock disintegration.

Quartzite is highly resistant to acid weathering, and mine wastes composed predominantly of this host rock generally do not disintegrate. Basic igneous host rocks are generally less resistant to acid weathering and disintegration than more siliceous igneous rocks.

LEACHING PRODUCTION FORECASTING

The preceding discussion indicates the extreme hazards in forecasting copper extraction yields over time from percolation leaching copper mine waste dumps and large heaps of primary sulfide copper ore. A few generalizations are worth mentioning, but they should not be relied upon. First, all ores are blends of various minerals. Even primary ore will usually contain some acid soluble copper and some secondary minerals that are relatively easily extracted and tend to give good initial leaching results. When these easily leached minerals are not a significant fraction of the total copper contained in the ore or mine waste, cumulative copper extractions rarely exceed $35 \pm 10\%$ of the original copper after many years of leaching. When secondary copper minerals predominate, copper extractions of 70% have been achieved under good leaching practices.

Although the "final" extraction will take several years for run-of-mine primary ore, a large portion of the ultimate yield, half to two-thirds, is usually extracted in two to three years after leaching begins. This is considered a short time in copper dump leaching practice, but contrast it with the few weeks that are typical for heap leaching gold ore crushed to a $-19\,mm\,(-3/4\,in)$ topsize.

Copper ores and mine wastes vary considerably in their resistance to acid disintegration. Ores that contain clay, or minerals that weather to clay, rapidly lose permeability and often produce little copper after an initial promising start.

Leaching slows markedly as the smaller ore fragments become completely oxidized and as ferric ions must diffuse farther into the larger rocks to reach sulfide mineral grains. The aggregate fractional extraction, F_t, over time, called the "extraction curve," approximates a parabola, with a leaching period measured in years. This is illustrated in Fig. 6.10.

An example of the extreme variation in copper extraction curves that can occur is shown by the data from two large scale leaching experiments conducted by Murr et al. (1982) and shown in Fig. 6.11. These extraction curves were obtained from essentially identical but separate leaching experiments using 160 tonne samples of run-of-mine ore from the Chino and Sierrita copper mines (Cathles and Murr, 1980). The tests were conducted for two years in closed vessels with aeration, periodic flushing and careful monitoring of test conditions and copper production.

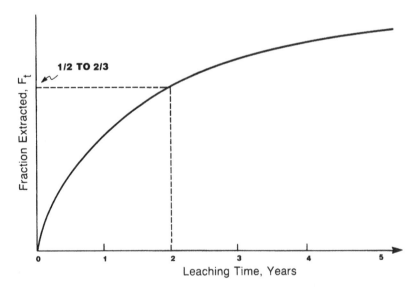

Figure 6.10.   Illustration of an extraction curve for copper mine waste rock.

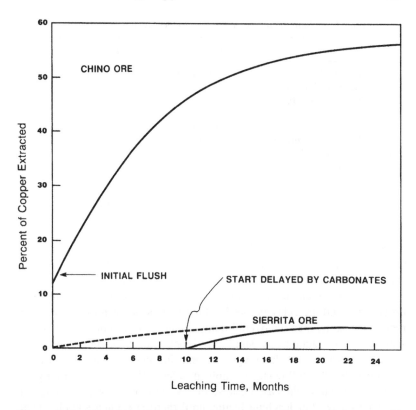

Figure 6.11. Experimental leaching extraction curves for two mine wastes with ore fragment size distribution as-mined (Murr, Schlitt and Cathles, 1982).

The ores had similar physical characteristics and heap permeabilities, which are listed in Table 6.5. Copper in the Chino sample was predominantly composed of acid soluble copper and the easily leached secondary copper sulfide mineral, chalcocite. The Sierrita sample's copper was nearly all chalcopyrite. The Sierrita sample also contained 2% carbonate in the rock, which undoubtedly delayed copper leaching while the pyrite generated acid was being consumed to neutralize carbonate.

For a new leaching operation, assuming uniform mining of waste rock of uniform composition, the aggregate copper production rate is initially small because there is an insufficient inventory of waste rock available for leaching. Production of leached copper gradually increases as the inventory

Table 6.5 Sample and Heap Characteristics for Fig. 6.11 Results.

	Chino Mine Waste	Sierrita Mine Waste
Copper Grade	0.36 wt pct	0.34 wt pct
Copper Distribution		
Acid Soluble (Oxide)	61%	9%
Chalcocite		
(secondary sulfides)	Balance	—
Chalcopyrite (primary)	less than 8%	80%+
Permeabilities (darcy)		
Dry (Before Test)	0.2 to 0.4	0.8 to 2.7
Wet	0.2 to 0.4	0.6 to 1.0
Post Leach	0.4 to 1.2	0.3 to 0.5
Carbonate	—	2 wt pct

of dump material increases. Eventually, usually after a period of one to several years, a constant (steady state) production rate of leached copper is achieved as mining and dumping continues at a constant rate.

If the waste rock dumping rate (tonnes per annum) is uniform over many years and the delay before leaching begins for each block of waste rock is the same, then the overall dump metal recovery rate should be the same as the extraction of a single ore block except for the initial delay from the beginning of dumping until leaching begins. This similarity is illustrated by the following example problem. However, uniformity of operations is rarely the case because of a variety of factors, including an extended delay after leaching begins until there is enough soluble metal inventory to justify a metal recovery plant. Consequently, the aggregate mine production rate versus time during the early period is usually an S shaped curve, as illustrated in Fig. 6.12. In the hypothetical example shown, mining is conducted for about 28 yr, but leaching is continued out to 35 yr from the initial mine start.

When mining and production of fresh ore or mine waste for leaching ceases, leach production may continue for several years, even decades, but the rate of production declines asymptotically, as illustrated by Fig. 6.12.

Table 6.6 shows the annual dumping of copper mine waste and the production of leached copper at the Bingham, Utah open pit copper mine's east side dumps over a period of 14 yr (Jackson, Schlitt and McMillan, 1979). There was no leaching of these dumps prior to 1962. Because of the very large dump accumulation of mine waste over many years before 1962, leached copper production increased rapidly once started and actually exceeded the amount of copper being dumped during the years 1967 through 1969. Then leached copper production rapidly declined toward an

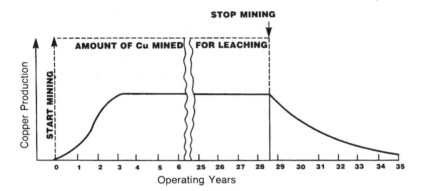

Figure 6.12. Illustration of total mine leached copper production after mine start-up and waste dumping through steady state production and termination of mining and waste dumping.

Table 6.6 Prediction Versus Measured Annual Copper Leach Production (Jackson et al., 1979).

Year	Copper Deposited in Dumps (Kilotonnes)	Dumps under Leach (%)	Predicted Copper Production (Kilotonnes)	Measured Copper Production (Kilotonnes)
1962	270.0[1]	8	0.75	—[2]
1963	32.4	8	3.7	—
1964	32.4	8	6.28	—
1965	32.4	8	8.23	—
1966	32.4	8	16.23	—
1967	32.4	70	38.63	—
1968	32.4	70	42.29	37.6
1969	32.4	70	38.18	32.0
1970	32.8	70	31.50	25.4
1971	32.8	70	28.10	23.5
1972	29.7	70	25.72	22.2
1973	26.7	70	22.16	20.5
1975	23.4	70	20.24	20.0
1976	23.4	70	18.40	20.0

[1]Deposition from 1962 and several earlier years of dumping.
[2]Dashes indicate missing data.

ultimate steady state rate that was proportional to the copper dumping rate but much less than it. By the end of the data, in 1976, a steady state production rate had not quite been achieved.

The annual production rate forecast from a computer program is also shown on Fig. 6.13. This forecasting model was a running materials

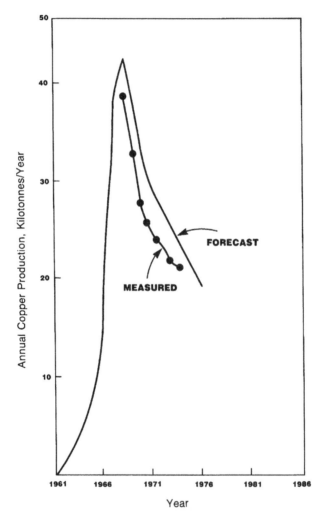

Figure 6.13. Annual leached copper production rates—actual and by computer forecast (Jackson, Schlitt and Macmillan, 1979).

balance that coupled the volumes and copper grades of mine wastes in numerous discrete blocks with an average leaching extraction curve. Note that the forecast production slightly exceeds the actual production. This indicates that the leaching extraction curve used in the computer model, which was based on laboratory experiments, was a little too optimistic.

EXAMPLE PROBLEM

Starting a new copper dump, with 27,800,000 tonne of waste rock being deposited each year, is being considered. Forecast copper metal production from this dump each year out to ten years of mining and depositing waste rock into the new dump, using the extraction curve for the Chino ore shown in Fig. 6.11 and the average ore grade of 0.36 wt pct copper shown in Table 6.5. Leaching will commence two years after dumping commences on the first year's mine production and proceed by leaching all of the rock in the dump except for rock deposited in the most recent year. This procedure is needed to provide maneuvering room for the haulage trucks on the lip of the dump.

ANSWER
(1) Determine the amount of copper dumped each year:
    27,800,000 tonne ore/yr × 0.0036 = 100,000 tonne Cu/yr.
(2) Establish annual production from a block of mine waste after leaching begins using the Chino extraction curve:

Leaching Year	Cumulative Extraction (%)	Incremental Extraction (%)	Copper Metal Recovered (tonnes)
1	48	48	48,000
2	55	7	7,000
3	56	1	1,000
4	56	0	0

(3) Array the copper production from each year's dump of waste rock in a series of vertical columns that include the delay since the beginning of mining for each year. Continue this out to ten years. The total dump copper production will be found from summing copper in the horizontal rows for each year. These data follow as Table 6.7, with copper shown in kilotonnes.

Table 6.7 Array of Copper Production for Mining Years and Leaching Years
for Example Problem.

Mining Years	Leaching Years								Total
1	0	0	0	0	0	0	0	0	0
2	0	0	0	0	0	0	0	0	0
3	48	0	0	0	0	0	0	0	48
4	7	48	0	0	0	0	0	0	55
5	1	7	48	0	0	0	0	0	56
6	0	1	7	48	0	0	0	0	56
7	0	0	1	7	48	0	0	0	56
8	0	0	0	1	7	48	0	0	56
9	0	0	0	0	1	7	48	0	56
10	0	0	0	0	0	1	7	48	56

ACID USAGE IN COPPER DUMP LEACHING

When carbonate minerals are not present, addition of sulfuric acid to the
leaching solutions rarely improves the production rate from primary sulfide
mine waste dumps and is often harmful. The additional acid reacts in the
first 5–15 m with gangue minerals. Acid attack is much faster because of
the lower pH. Consequently, rock decrepitation is accelerated in the upper
region of the dump where the fine particles are already most abundant
because of end dumping from haulage trucks and the permeability is usually marginal to begin with. The temptation to use acid either to accelerate
leaching or to discard surplus smelter acid should be resisted.

MANAGING PRIMARY COPPER SULFIDE
MINERAL LEACHING OPERATIONS

Enormous **patience** is required to effectively manage sulfide leaching
operations with waste dumps and uncrushed ores. The leaching reactions
occur over months and years rather than hours and days, which are the
customary time periods that humans are used to. The leaching production
rate is ultimately limited by oxidation of the most important sulfide minerals,
which is inherently slow; by the large fragments in the rock size distribution; and by the very slow flow of air through the mine waste dumps.

Actions taken to accelerate the leaching process and increase production
usually involve management of the circulating leach solutions and are
often counterproductive in the long run, leading to increased solution
pumping, more dilute pregnant liquors and need for an oversized copper

recovery plant (Jackson and Ream, 1980). All of these results will increase costs at the expense of a temporary increase, at best, in production of leached copper. As will be shown in the next chapter, the lowest practical solution application rates should be used to minimize rapid flow through channels in the ore mass (solution short circuiting). Adequate rest periods must be taken to allow the slow mineral oxidation process to continue long enough to provide adequate copper solution concentrations during the next solution application (wash) cycle. The length of rest periods must be extended as the dumps/heaps age.

Acid additions should not be used unless mandated by carbonates in the ore. Chemical oxidants for accelerating sulfide oxidation are used in uranium solution mining, but unfortunately they are too expensive for leaching low-grade copper mine waste dumps.

PROBLEMS

1. A planned circular open pit copper mine will be completed at a depth of 250 m and a final pit slope of 40°. The mill grade ore zone is a 200 m diameter right circular cylinder. All overburden material will be leached in mine waste dumps adjacent to the pit. Estimate the volume (m^3) of mill grade ore, the volume (m^3) of waste rock that will be leached, and the final overburden ratio. Assume the mine is located in level terrain. With an average milling ore grade of 0.6 wt pct copper, and an average mine waste grade of 0.17 wt pct copper, how much copper was excavated (a) in the mine waste rock and (b) in the mined ore, at the completion of mining? Which material, ore or waste rock, contained the most copper prior to extraction? The ore has a specific gravity of 2.6 g/cm^3.

2. Write chemical reactions describing the generation of ferric hydroxide in an aeration pond immediately following copper separation (Fig. 6.2). If iron cementation is used, what will be the relative concentrations (qualitatively) of ferric and ferrous ions in the solution entering the aeration pond?

3. From the data in Table 6.6 estimate the **cumulative** copper extracted (%) of the copper contained in **all** of the mine waste deposited through 1976.

4. The table below describes the expected **cumulative** extraction of copper from each tonne of mine waste with an average grade of 0.20 wt pct copper that will be produced from a new mine at a planned mine waste

annual generation rate of 20 million tonnes, over the next 20 yr. Using graph paper, draw a curve of annual leaching copper production out to 10 yr after mine start up. Assume a 2 yr delay before leaching starts on dumped material and continuous leaching of the waste thereafter.

Mine Waste Extraction

Year	Copper Solubilized and Extracted (%)
1	10
2	15
3	19
4	22
5	25
6	27
7	29
8	31
9	32
10	33
11	33
12	33

5. For the copper mine planning of Problem 4 it has been decided to install an SX-EW plant module with 7,000 tonne per year copper capacity followed by another module later with 7,000 tonne per year of additional copper capacity. When should the first module begin operating (years from mining start-up) so that there will be sufficient leaching production to match the capacity of the first module? when should the second module begin operating? Explain (justify) your answers.

6. Using eqn 5.39, calculate the $[Fe^{3+}]$ and $[Fe^{2+}]$ ion concentrations, in grams per liter, for each of the following two copper mine dump leaching solutions. Both solutions contain 5 g/l of total dissolved iron.
   Solution #1: $E_h = 0.69$ V
   Solution #2: $E_h = 0.78$ V

REFERENCES AND SUGGESTED FURTHER READING

Beck, J.V. (1967). The role of bacteria in copper mining operations. *Biotechnol. Bioeng.*, Vol. 9, pp. 489.

Braun, R.L., Lewis, E. and Wadsworth, M.E. (1974). In-place leaching of primary sulfide ores: laboratory leaching data and kinetics model. *Met. Trans.*, **5**, pp. 1717–1726.

Cathles, L.M. and Murr, L.E. (1980). Evaluation of an experiment involving large column leaching of low-grade copper sulfide waste: a critical test of a model of the waste leaching process. Schlitt, W.J. (Ed.), *Leaching and Recovery of Copper from As-Mined Materials*, Society of Mining Engineers of AIME, pp. 29–48.

Dutrizac, J.E. and MacDonald, R.J.C. (1974). Ferric Ion as a Leaching Medium. *Minerals Science and Engineering* **6**, *No. 2*, pp. 59–100.

Jackson, J.S. and Ream, B.P. (1980). Solution management in dump leaching. Schlitt, W.J. (Ed.), *Leaching and Recovering Copper from As-Mined Materials*, Society of Mining Engineers, pp. 79–94.

Jackson, J.S., Schlitt, W.J. and McMillan, B.B. (1979). Forecasting copper production from dump leaching. *AIME Trans.*, **266**, pp. 2009–2016.

Lewis, A.E., Braun, R.L., Sisemore, J.C. and Mallon, R.G. (1974). Nuclear solution mining—breaking and leaching considerations. Aplan, F.F. et al. (Ed.), *Solution Mining Symposium*, SME and TMS of AIME, pp. 56–75.

Malouf, E.E. and Prater, J.D. (1961). Role of bacteria in the alteration of sulfide minerals. *Journal of Metals*, Vol. 13, pp. 353.

Murr, L.E., Schlitt, W.J. and Cathles, L.M. (1982). Experimental observations of solution flow in leaching copper bearing waste. Schlitt, W.J. and Hiskey, J.B. (Eds.), *Interfacing Technologies in Solution Mining*, SME and SPE of AIME, pp. 271–290.

Nyavor, K. et al. (1996). Bacteria oxidation of sulfides during acid mine drainage formation; a mechanistic study. *EPD Congress 96*, The Minerals, Metals and Materials Society, Warrendale, PA, pp. 269–287.

Paul, B.C., Sohn, H.Y. and McCarter, M.K. (1992). Model for ferric sulfate leaching of copper ores containing a variety of sulfide minerals: part I. modeling uniform size ore fragments. *Met. Trans 23B*, pp. 537–548.

Pesic, B. and Kim, I. (1990). Electrochemistry of T. ferrooxidans interaction with pyrite. Gaskell, D.R. (Ed.), *Extraction and Processing Division Congress 90*, The Minerals, Metals and Materials Society, Warrendale, Pa, pp. 133–160.

Rossi, G. (1990). *BioHydroMetallurgy*, McGraw-Hill, New York.

Schlitt, W.J. (Ed.) (1980). Current status of copper leaching and recovery in the U.S. copper industry. *Leaching and Recovery of Copper from As-Mined Materials*, Society of Mining Engineers of AIME, Littleton, CO, Chap. 1.

Schlitt, W.J. (1992). Solution mining: surface techniques. Hartman, H.L. (Ed.), *Mining Engineering Handbook*, SME, Littleton, CO, Chap. 15.3.

Wadsworth, M.E. (1987). Solution mining systems. Szekely, J. et al. (Ed.), *Mathematical Modeling of Materials Processing Operations*, The Metallurgical Society of AIME, Warrendale, PA, pp. 311–320.

Wadsworth, M.E. (1972). Advances in the Leaching of Sulfide Minerals. *Minerals Science and Engineering* **4**, *No. 4*, pp. 36–47.

SEVEN

Solution Flow During Percolation Leaching of Ore Heaps

INTRODUCTION

This chapter and the next one are intended as brief reviews of solution and air flow, respectively, during percolation leaching. They are generally applicable to both ore heaps and mine waste dumps. Solution flow is obviously important and air flow is very important to the leaching of sulfide minerals, but these flow behaviors in ore heaps have been poorly understood. Of primary interest for sulfide ores is the combined case of downward percolation of solutions in unsaturated fragmented rock driven by gravity combined with upward flow of air, driven by a buoyancy gradient induced by changes in air temperature and composition. Saturated solution flow—flooded leaching—will be treated in a subsequent chapter. Flooding during percolation leaching obviously excludes air flow.

Water flow in unsaturated media, the vadose zone or zone of aeration in hydrology nomenclature, has been studied less than saturated (groundwater) flow. The studies that have been done were generally with soils rather than rocky ore heaps. Unsaturated flow is less quantitative and less easily modeled than saturated flow. Chemical engineering trickle bed flow correlations deal with flow in packed beds or porous media, usually of uniform particle size, whereas ore and mine waste particles are of extremely diverse particle sizes.

The present understanding of solution flow in ore heaps and mine waste dumps is derived primarily from soil mechanics, hydrogeology and chemical engineering theory coupled with experimental information from: (1) leach column flow experiments, including tracer studies, (2) water flow tracer studies in mine dumps, (3) permeability measurements in mine dumps using cased boreholes, and (4) laboratory studies of capillary rise and permeability as a function of rock particle sizes.

SOLUTION AND AIR FLOW REGIONS IN FRAGMENTED ROCK

Recall from Chapter 1 that the lixiviant solution/rock interface is one of lower surface energy than the air/rock interface, causing **solution wetting** of the rock surfaces and **capillary penetration** of the internal rock micropores. Leaching solution will also penetrate between rocks and tend to exclude air from small voids separating rocks. Leachate will drain away from large voids, which will tend to be filled with air except for the water film covering the rocks.

For purposes of this discussion, it is important to distinguish between inter-rock void space or macroporosity, whether occupied by the solution or by air, and intra-rock microporosity. The former has often been referred to as the heap or fragmented rock porosity, and in the hydrogeology literature concerning unconsolidated sediments it is called the primary porosity (Fetter, 1988). However, this leads to confusion with the internal rock porosity, which is referred to as secondary porosity in the hydrogeology literature and as microporosity in this text. Fluid flow in percolation leaching can only occur outside the rock in the inter-rock void space, or primary porosity. To avoid confusion, this text limits the term *microporosity* to the open internal rock porosity (secondary porosity) and uses *void space*, *voids* or *macroporosity* for the inter-rock space (primary porosity).

It is useful to consider the distribution of space in a heap of fragmented rock between mobile and immobile fluid phases. Capillary space may include both water-filled voids which are mobile and solution filled rock micropores which are immobile. Four phases or regions of ore heap space are defined in Table 7.1 and illustrated in Fig. 7.1. The percentages shown are from a specific ore heap column test case and are presented as an example of more or less typical results.

The largest phase is solid rock, v_s. Inter-rock void space is divided into a solution filled phase, v_l, and an air filled phase, v_g. The fourth phase, open rock microporosity, ε, exists in the rock prior to mining but can be affected

Table 7.1 Phase Regions in a Fragmented Ore Heap—Flow-Test Column (Schlitt, 1984).

Phase	Symbol	Stagnant or Mobile	Measured Vol Pct in Test Volume
Solid **rocks** including closed pores	v_s	Stagnant (dead space)	59.0%
Open **microporosity** within rocks	ε	Stagnant (dead space)	2.4%
Solution void space between rocks	v_l	Mobile (water flow)	19.0–21.5%
Air void space between rocks	v_g	Mobile (air flow) and trapped air pockets	18.1–19.6%

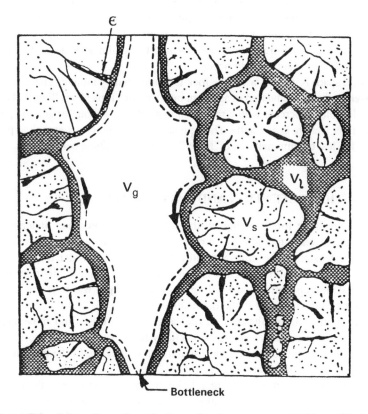

Figure 7.1. Schematic section of a heap showing space occupied by solids (v_s), rock microporosity (ε), solution void space (v_l) and air void space (v_g) (Schlitt, 1984).

by leaching. Space occupied by closed rock micropores is included in the solid rock phase.

The measured volume percentage of total space occupied by these regions in a laboratory flow-test column is also shown in Table 7.1. The column was loaded with coarse crushed copper mine waste. Several unsaturated water flow (percolation) experiments were conducted, in a manner similar to a column leaching experiment, at the Kennecott Research Center in 1975 and 1976. Some of the results have been reported by Schlitt (1984). Several experiments will be described in this chapter using this test column. The use of a common experimental basis is important to remove confusion and obtain a unified understanding of percolation flow behavior in a rubblized ore mass. The rock rubble in the column was fairly typical of a dump or heap containing a modest amount of fine material.

When larger amounts of rock fines are present, v_l will increase and v_g will decrease somewhat from the results obtained for this particular laboratory flow-test column. For most ores, open rock microporosity is about 1–6% of the rock volume or about 1–4% of the total heap space, but variations outside these limits do occur.

Void space is created when the rock is fragmented by mining. The result is "**swelling**" and a reduction in apparent bulk density. Typical swelling is about 35–40 volume percent with a reduction in swelling to about 30 volume percent after wet ore heaps settle. The equivalent swelling of broken rock to provide the test column material of Table 7.1 was 38.6 volume percent.

SOLUTION RETENTION AND CAPILLARITY

There are three forces acting on the solution in an ore heap undergoing percolation leaching: gravity, surface tension and atmospheric pressure. Surface tension is the molecular attraction that causes water to preferentially adhere to solid surfaces over air and thereby displace air from both internal microporosity and void space. **Hygroscopic water** is the water clinging to the particles in the ore heap.

As solution flows through an ore heap, forces resist fluid movement. There are shear stresses acting tangentially to the surface of the solid and shear stresses due to the internal friction in the fluid itself. The latter is the fluid's dynamic viscosity.

When application of leaching solution to an ore heap ceases, the excess water will drain away until gravity and surface tension forces are in balance, at which time solution flow will cease. The **specific retention** of the ore heap, $v_{l(0)}$, is the ratio of the volume of solution retained after drainage to the total heap volume. This water can only be removed by evaporation. In agricultural engineering, the moisture content at draindown of a soil is known as **field capacity**, and the water can be further removed by the transpiration of vegetation rooted in the soil, as well as by evaporation.

If water is introduced at the bottom of an ore heap, it will rise to reestablish a force balance. This phenomenon, known as **capillary rise**, occurs in a small bore tube, in soil, and in the void space surrounding fine particles in an ore heap.

Leaching solution and air are two immiscible fluids that compete for the available void space, but over a wide range of solution percolation rates in ore heaps below the onset of flooding, solution flow rate has only a slight effect on the void space occupied by the solution, v_l. Instead, the factors controlling v_l are the particle sizes and amounts of the finer rocks in the ore heap. This is a consequence of the limitations imposed by capillary forces. The capillary rise or retained height after draining, h_c, is related to the capillary diameter, d_c, gravity, g, the liquid density, ρ_l, the surface tension, δ_1, and the wetting contact angle, θ, between the rock and solution

$$h_c = \frac{4\delta_1 \cos\theta}{g\rho_l d_c}. \tag{7.1}$$

For a typical leach solution, $\delta_1 = 73\,\text{dyn/cm}$, $\cos\theta = 1$, and $\rho_l = 1.08\,\text{g/cc}$, the capillary rise, h_c, reduces to

$$h_c \approx 0.27 d_c. \tag{7.2}$$

An order of magnitude relationship can also be established between the effective particle diameter, d_p, and the average capillary diameter, d_c. Based on 37% void space for random packing of uniform spherical particles, the relationship

$$\frac{d_c}{d_p} \simeq \left(\frac{v_l}{v_s}\right)^{1/3} \tag{7.3}$$

provides:

$$d_c \approx 0.84 d_p. \tag{7.4}$$

Appropriate substitution then gives:

$$h_c \approx \frac{0.32}{d_p}. \tag{7.5}$$

As shown in Table 7.2, this relationship is in good agreement with experimentally observed values of capillary rise.

However, it should be noted that if particle size decreases drastically, the capillary rise will be huge and the specific retention and the total void

Table 7.2 Capillary Rise as a Function of Rock Particle Size (Schlitt, 1984).

Particle Size, Mesh	Average Particle Diameter, (mm)	Capillary Rise, (mm)	
		Measured	Theoretical
10/20	1.1	30	29
20/28	0.7	44	46
28/35	0.5	80	64
35/48	0.35	140	92
48/65	0.24	190	133

space will converge. Clays are the ultimate ultra-fine earth materials, and while they have a large void space, they tend to be saturated with water unless it is removed by evaporation.

The results of solution drain experiments for copper mine waste of different particle sizes in flow test columns are shown in Fig. 7.2. More importantly, this figure also shows the experimental correlation between rock size and the specific retention, or percentage of the void space that remains filled with solution after percolation ceases and draining is complete. Solution effectively fills all of the voids for rock sizes less than about 48 mesh (0.3 mm) with air excluded. However, for rock sizes coarser than about 10–20 mesh (1 mm), drainage will be almost complete and most of the void space will be filled with air. Similar results have been obtained by Kennedy and Stahl (1974).

The capillary rise or soil moisture tension can be measured in the field with a **tensiometer**. This device consists of a tube that is closed at the top, with a ceramic cup at the bottom to provide a permeable membrane, Fig. 7.3. Water in the tube is in contact with moisture in the soil and a suction is exerted because of the moisture tension that can be measured against atmospheric pressure with a manometer or other pressure measuring instrument.

HYDRAULIC CONDUCTIVITY VERSUS INTRINSIC PERMEABILITY

The **superficial velocity**, u_1, of solution percolating through an ore heap and the solution percolation flow rate per unit area, Q_1/A, or **specific discharge** are identical:

$$u_1[Lt^{-1}] = \frac{Q_1}{A}[L^3t^{-1}/L^2], \tag{7.6}$$

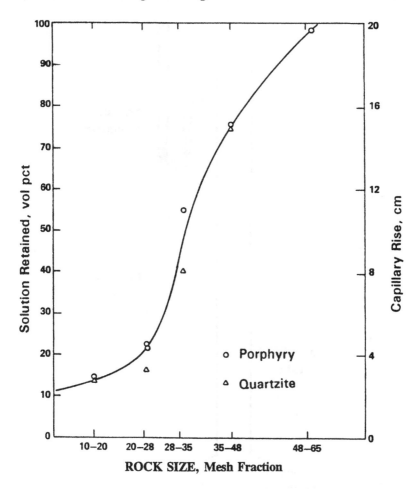

Figure 7.2. Solution retention and capillary rise as a function of rock fragment size (Schlitt, 1984).

and proportional to the **hydraulic gradient**, $h/\Delta x$, where h is the hydrostatic head, provided the flow is laminar. For the slow solution flows encountered in percolation leaching, turbulent flow is not encountered. As previously defined in Chapter 4, the proportionality coefficient is the hydraulic conductivity, K, and for a one dimensional flow geometry, the relation is

$$u_1 = \frac{Q_1}{A} = K\left(\frac{h}{\Delta x}\right). \tag{7.7}$$

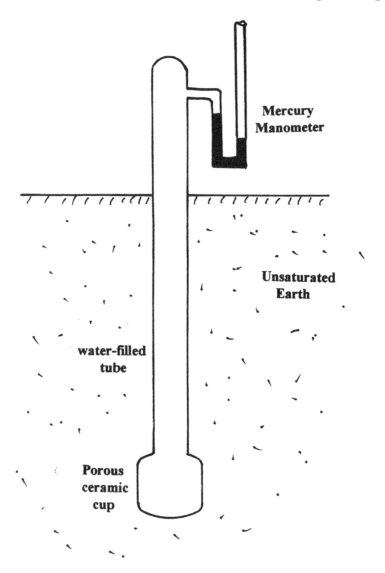

Figure 7.3.   Porous-cup tensiometer with mercury manometer to measure soil moisture tension.

The hydraulic conductivity has dimensions L/t. A variety of other names are used for K, including effective permeability, coefficient of permeability, and seepage coefficient. Since most of the ore heap space is occupied by a combination of solid ore and air, the *average* solution flow velocity in the liquid region of the heap is u_l/v_l. The hydraulic conductivity is a function of both the **intrinsic permeability** of the media through which flow is occurring, k_i, and the fluid viscosity and density,

$$K = \frac{k_i g \rho}{\mu},$$ (7.8)

where g is the gravity acceleration constant, ρ is the fluid density and μ is the dynamic viscosity of the fluid. The intrinsic permeability, with dimensions L^2, is a property of the ore media.

The superficial velocity of any fluid, gases as well a liquids, flowing through solid media at saturation can be expressed in terms of the intrinsic permeability of the media, the fluid viscosity and the fluid pressure gradient by:

$$u = \frac{Q}{A} = \frac{k_i}{\mu}\left(\frac{dp}{dx}\right).$$ (7.9)

Equation 7.9 expresses **Darcy's Law** for flow in one direction, which is sufficient for leaching solution percolation flow induced only by gravity. However, in other situations, e.g., saturated flow to wells, multi-dimensional linear coordinates or other coordinate systems are required.

A frequently encountered unit of intrinsic permeability is the **darcy**, which is defined by eqn 7.9 with the following units: viscosity, μ, in centipoise (cP), Q_l/A in cm/s, pressure, p, in atmospheres and distance, x, in cm. As a result of those units, one darcy is also equal to $0.987 \times 10^{-8}\,cm^2$, $(0.99 \times 10^{-12}\,m^2)$.

At approximately room temperature, the viscosity of air is $0.018\,cP$ and the viscosity of water is $1.0\,cP$. The temperature dependencies of air and water viscosities up to $100°C$ are shown in Fig. 7.4.

For the special case of water at $20°C$ with a pressure gradient expressed dimensionlessly [as "head" (cm)/Δx(cm)] the relation between intrinsic permeability, k_i, and hydraulic conductivity, K, in the bracketed units is

$$k_i[\text{darcy}] = 103.4K\,[\text{mm/s}],$$ (7.10a)

$$k_i[\text{m}^2] = 1.02 \times 10^{-10}K\,[\text{mm/s}],$$ (7.10b)

based on one atmosphere equals $1{,}033.26\,cm$ of water head. Other conversion units are in the appendix.

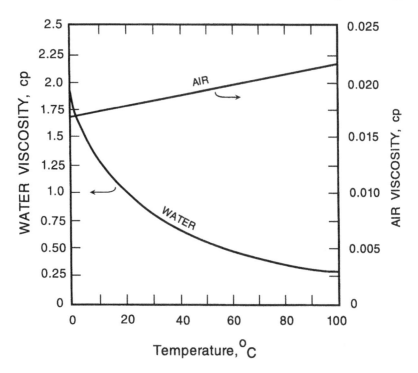

Figure 7.4. Temperature dependence of dynamic viscosities of water and air at 1 atm pressure (Bird et al., 1960).

PERCOLATION FLOW RATES AND FLOODING

Based on water percolation studies with the Kennecott laboratory flow-test column, there were three distinct solution flow regimes: (1) capillary drainage with no solution flow, (2) percolation, and (3) solution flooding. Air flow, often needed to oxidize sulfide minerals, can only occur during capillary drainage and percolation.

As solution percolates at very small application rates into a drained heap, liquid flow occurs and some additional void space will be occupied by the solution, primarily by liquid film expansion at air/solution interfaces. Experimental percolation studies with crushed ore in columns have shown that very low solution application rates cause an increase in solution void space of about 1–2% of the total space. The amount of this capillary solution swelling is small because most of the capillaries in the void space were already solution filled at draindown.

Very large vertical film flow velocities can be attained with only a small increase in solution film thickness on a vertical wetted wall. Consequently, as the solution percolation rate, Q_1/A, is progressively increased, the void space occupied by solution changes imperceptibly until the percolation flooding limit is reached.

Confirming experimental results from the Kennecott flow test column study are summarized in Table 7.3. Changes in water filled void space are shown for various solution application rates through the three flow regimes: capillary draindown, percolation and flooding. Note the negligible change in water filled void space over a wide range of percolation flow rates, between 1.2 and 3050 mm/hr. The flooding limit was reached at 4070 mm/hr for this particular example; then the entire upper column quickly flooded with solution. Thus, hydraulic conductivity of this particular heap is 4070 mm/hr, and its intrinsic permeability can be determined through eqn 7.8. Typical percolation velocities used in ore heap leaching are usually well below the ore heap's hydraulic conductivity. Ores that do not meet this criterium must be agglomerated to increase their intrinsic permeability and hydraulic conductivity.

Flooding is always initiated when a drainage bottleneck becomes pinched off or bridged with solution. This has been observed in repeated water percolation flow experiments using the transparent (plexiglass) flow-test column of crushed ore. The air void space above the initiating bottleneck rapidly fills with solution once the critical solution flow rate is achieved. Flooding always proceeds upward from a bottleneck. A bottleneck being

Table 7.3 Effect of Solution Percolation Rate on Solution-Filled Void Space in Flow-Test Column, Schlitt (1984).

Application Rate		Water-Filled Void Space (%)[1]	Comments
mm/hr	gal/ft^{-2}/hr		
0	0.0	19.0	Drained capillaries
1.2	0.03	20.6	filled
6.0	0.15	20.7	Typical range
24.0	0.60	21.1	of application
120.0	3.0	21.3	
3050.0	75.0	21.5	
4070.0	110.0	38.6	Flooded

[1]Expressed as a percent of the total space within the heap from the original unpublished Kennecott study.

pinched off by opposite water films merging to bridge the inter-void space was schematically illustrated in the lower part of Fig. 7.1.

SOLUTION PERCOLATION FLOW RATES AND WASHING EFFICIENCY

With solution confined by capillarity to the smaller voids, most of the flow will be through these capillaries if solution percolation is sufficiently slow. The velocity of flow in the voids, small enough to be fully wetted, is limited. Regardless of the solution application rate at the top of the heap, the solution flow through local fragmented rock media cannot exceed a velocity determined by the *local* hydraulic conductivity of the solution filled voids and the pressure gradient associated with the hydrostatic head at that location. And since the maximum possible hydraulic gradient is the vertical gradient vector, $h/\Delta x = 1$, then:

$$u_l \leqslant K. \tag{7.11}$$

Substituting from eqn 7.10 yields

$$u_l[\text{mm/s}] \leqslant 9.67 \times 10^{-3} k_i[\text{darcy}],$$

$$u_l[\text{m/s}] \leqslant 9.78 \times 10^3 k_i[\text{m}^2], \tag{7.12}$$

$$\frac{Q_l}{A}[\text{gallons per hr/ft}^2] \leqslant 0.86 k_i[\text{darcy}].$$

Because of the wide distribution of ore particle sizes and less than ideal ore mixing, a heap will consist of a wide range of *localized* hydraulic conductivities. This will result in a wide range of local maximum possible percolation flow velocities for each of the nearly infinite microregions of the ore heap or mine waste dump. The solution application rate, distributed to the top of the heap or dump, will be the *average* percolation flow rate.

When the average percolation velocity exceeds the local limiting velocity, local flooding will occur and excess solution must move laterally to find a path of higher hydraulic conductivity. As solution application rates are progressively increased, the additional solution is shunted to fewer, ever larger, channels because of the inability of the smaller voids to carry the additional flow. In extreme cases when the average hydraulic conductivity is low and the application rate is high, large perched water tables may form.

Large flows of leaching solution short circuiting through channels will be ineffective in washing (flushing) solubilized metal and should be avoided by limiting the solution application rates to a rate consistent with the matrix permeability of the ore heap, according to the relations expressed in eqns 7.11 and 7.12. Because metal production from sulfide ore heaps and mine waste dumps is limited over the long run to the rate of oxidation of the sulfide minerals, attempts to increase production over the short run by increasing the solution application rate will generally not improve washing, as intended, but simply increase solution short circuiting through channels and dilute the grade of pregnant liquor (Jackson and Ream, 1980). This is a very important consideration for managing leaching operations. Pouring more lixiviant on the ore heap is not always better, and often it's worse.

Unsaturated solution flow studies have been made in a pilot plant scale copper leaching column, (Murr, 1979; Murr et al., 1982), and by tracer studies in copper mine dumps (Armstrong et al., 1971; Howard, 1968). Field measurements of dump permeability (Whiting, 1977) and modeling efforts based on the results from dump leaching measurements (Harris, 1969; Roman, 1977) have been made. The results from mine dump studies have been somewhat confusing, suggesting complex phenomena. The critically important percolation flow behavior is more clearly shown by simpler experiments in a column packed with ore.

Pulse **tracer studies**, using chloride ions at different solution application rates were made with the Kennecott laboratory flow-test column described earlier in this chapter. The study results summarized below show that (1) any excess solution quickly passes through the column and (2) these short circuiting channels carry increasing proportions of the flowing water as the application rate increases.

The experimentally derived tracer curves for three application rates are shown in Fig. 7.5, where $\mathbf{C}^*$ is the dimensionless solution concentration exiting the column,

$$\mathbf{C}^* = \frac{C_{Cl^-} v_l}{M_{Cl^-}},$$
(7.13)

and θ is the dimensionless time,

$$\theta = \frac{t}{t_{avg}}.$$
(7.14)

The undefined terms in these expressions are the chloride ion effluent concentration above background, C_{Cl^-}, the mass of chloride pulse, M_{Cl^-}, the volume of water in the column, v_l, and the average time for tracer exit, t_{avg}.

Figure 7.5 shows that for the lowest application rate, 6 mm/hr, the effluent distribution is nearly symmetric about the average residence time, $\theta = 1$.

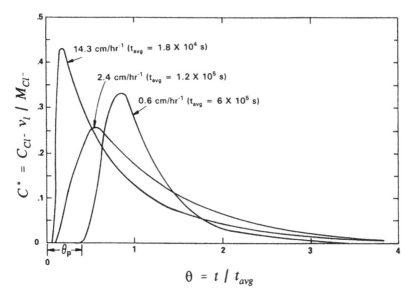

Figure 7.5. Pulse tracer results for ore column effluents at three solution application rates in dimensionless units.

As the application rate increases, with attendant short circuiting of water in channels, the effluent distribution shifts to earlier dimensionless time. Plotting in real time increases the apparent difference between these three tracer test distributions. The differences in the tracer results are more dramatically revealed in Fig. 7.6, which is a plot of tracer effluent, C_{Cl^-}/M_{Cl^-}, versus real time in days for two of the application rates. In this plot, neither the tracer effluent concentration nor time are shown as dimensionless units.

Efficient washing of the dissolved metals occurs when the solution passes uniformly through all of the ore matrix in as close to plug flow as possible. The volume percentage of the column involved in plug flow through the ore was computed from the water application rate and first appearance of the tracer at the bottom of the column. The results are summarized in Table 7.4. An application rate of 6 mm/hr provided uniform velocity and nearly plug flow, through the water-filled voids. At this application rate little of the water short circuits through channels. However, at the highest application rate, 143 mm/hr, virtually all of the water passed through channels in rapid flow. Only 12% of the water-filled void space was conducting the rapid flow, but the average superficial velocity for

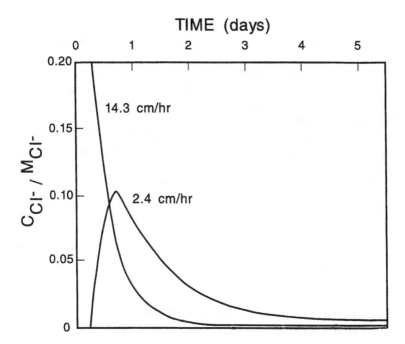

Figure 7.6. Pulse tracer effluent concentration versus time in non-dimensionless units.

Table 7.4 Summary of Tracer Results for Laboratory Flow-Test Column.

Application Rate (mm/hr)	Percent of Water-Filled Voids	Velocity (mm/hr)	Estimated Rapid Path Permeability (darcy)
6	100	0.6	0.2
24	21	92	2.8
143	12	1,000	25

these "channels" was 1000 mm/hr, indicating a channel intrinsic permeability of at least 25 darcy.

The slowest water flow velocity required 140% more water than required at the highest water flow velocity to remove one-half of the chloride tracer from the test column. Again, this results because with slower flow, there is less short circuiting and more hold up of the tracer in the fine capillaries.

This trend shows that as the water flow rate increases, higher fractions of the total flow are shunted to ever larger channels that can handle the larger water flows at ever higher velocities. The volume fraction of channels that are sufficiently large decreases as the water flow increases, because larger channels are required and there are fewer of them. Large channel fissures are often observed intersecting the surface of settled copper mine dumps.

Dimensionless $\mathbf{C}^*$ curves for solution flow in an ore heap, such as those shown in Fig. 7.5, can be described in terms of an axial dispersion coefficient, D_A, which has the same dimensions as a diffusion coefficient, L^2/t. Whereas ordinary chemical diffusion results from a statistical interplay of molecules in motion randomly "walking" about, the dispersion coefficient results from the random retardation (dispersion) of the leaching solution as it flows at different velocities through the interstices between ore particles of different size. The dimensionless group, D_A/u_1H, where H is the height of an experimental ore column or heap, actually characterizes the solution flow. This group is the inverse of the Peclet number for mass transfer, which is sometimes called the Benner number in soil hydrology literature (Iwata et al., 1995).

ESTIMATING INTRINSIC PERMEABILITY FROM ROCK SIZES

Intrinsic permeability depends on the size of the openings through which the fluid is moving. Flow resistance decreases as the square of the opening diameter increases. For monosize ore particles the opening size is proportional to the squared particle size; hence,

$$k_i \propto d_p^2. \tag{7.15}$$

This result is also consistent with the dimensions of intrinsic permeability (L^2). Ranges of intrinsic permeabilities for unconsolidated earth materials are shown in Table 7.5.

While valid only for monosized particles, the Blake–Kozeny equation for packed beds can be used to estimate a relationship between intrinsic permeability and a critical smaller rock size, d_{B-K}, as follows:

$$k_i = 6.6 \times 10^{-3} d_{B-K}^2 \left(\frac{v_l^3}{(1 - v_l)^2} \right). \tag{7.16}$$

Note that k_i will be in dimensions $[L^2]$ consistent with the dimensions of d_{B-K}. Equations 7.12 and 7.16 can be combined to estimate maximum percolation rates as a function of the critical rock size in the heap matrix for various values of v_l.

Table 7.5 Ranges of Intrinsic Permeabilities and Hydraulic
Conductivities for Unconsolidated Earth Materials.

Material	Intrinsic Permeability (darcy)	Hydraulic Conductivity (mm/s)
Clay	10^{-6}–10^{-3}	10^{-8}–10^{-5}
Silt, sandy silts, clayey sands, till	10^{-3}–10^{-1}	10^{-5}–10^{-3}
Silty sands, fine sands	10^{-2}–1	10^{-4}–10^{-2}
Well-sorted sands, glacial outwash	1–10^{2}	10^{-2}–1
Well-sorted gravel	10–10^{3}	1–10

Figure 7.7 graphs intrinsic permeability versus application rates (average percolation rates) over a range of these parameters that are applicable for ore heap and mine waste dump leaching. Flooded zones, based on eqn 7.12, are shown. The upper abscissas also show the critical particle sizes, d_{B-K}, from eqn 7.16 at $v_l = 0.1$ and $v_l = 0.2$. Figure 7.7 attempts to show various regions of heap matrix percolation behavior. The limited range of solution application rates for sprinklers given in Chapter 3 is from 7 to 11 mm/hr. This range is also superimposed on Fig. 7.7.

For a given median rock particle diameter, permeability will decrease as the standard deviation of particle size increases. The finer particles fill the voids between larger particles and dominate the flow of fluid and the permeability. An "effective particle" diameter smaller than the arithmetic mean diameter, if known, could be inserted in the Blake–Kozeny equation to estimate permeability.

One approach that has been used over the past century in soil science (Iwata et al., 1995) is to use the largest diameter of the smallest ten mass percent of the particles, d_{10}, as the effective particle diameter. A critical mean diameter, d_p^*, provided by the following equation, where N_y is the mass fraction of ore in size interval y, provides another estimate of the effective rock particle diameter that is based on the specific surface of the particle heap controlling its pressure drop (Kunii and Levenspiel, 1969):

$$d_p^* = \frac{1}{\sum_{\text{all } y}(N/d_p)_y}. \tag{7.17}$$

Using this equation, the two most extreme values of d_p^* were computed for a Gates–Gaudin–Schuhmann (GGS) distribution of rock sizes

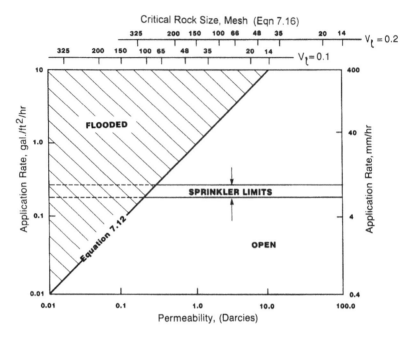

Figure 7.7. Fragmented rock critical size, permeability, and allowable solution application (percolation) flow rates.

with a top size $d_{p(max)}$, of 19 mm (3/4 in). The results are d_p^* equal to 2.0 and 1.2 mm. These values were used with eqn 7.16 to compute variations in permeability with variations in v_l. The Blake–Kozeny computed results at the two values of the critical mean ore particle diameter are plotted in Fig. 7.8. The permeability is very sensitive to the void fraction, v_l, which depends on both the particle size distribution and the fraction of total voids that are filled with water in the wet ore heap, and is generally unknown. Over the expected range of water-filled void fraction from 0.07 to 0.25 (7% to 25%), the estimated permeabilities range between about 10 and 800 darcy (10^{-11} to $10^{-9} m^2$), and they are only average values of permeability that say nothing about the wide variation in flow velocity that occurs in different parts of an ore heap that cause axial dispersion and slow washing. These very broad estimates of average permeability are poor substitutes for measured permeabilities obtained from several different crushed ore samples under heap leaching conditions, but reliable, measured ore heap permeabilities are rarely available at operating mines.

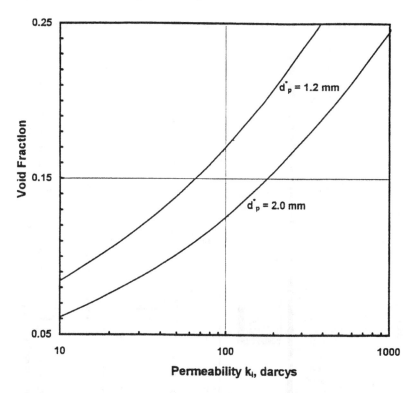

Figure 7.8. Expected range of intrinsic permeabilities in ore crushed to −19 mm (−3/4 in).

In Fig. 7.9. the relationships between hydraulic conductivity and intrinsic permeability over several orders of magnitude are shown with different fragmented rock matrix mixtures from gravel to silts and high clay soils. Commonly used mine operator solution application rate units are also shown.

Measurements of unsaturated hydraulic conductivity in rock size distributions typical of ore heaps being leached are rare, but many such measurements have been made in soils (Iwata et al., 1995). The results generally agree that the ratio of unsaturated to saturated hydraulic conductivity, K/K_{sat}, obey the following empirical relation,

$$K/K_{sat} = m^a, \tag{7.18}$$

where m is the effective saturation ratio defined as follows:

$$m = \frac{v_1 - v_{1(0)}}{1 - v_s - v_{1(0)}}. \tag{7.19}$$

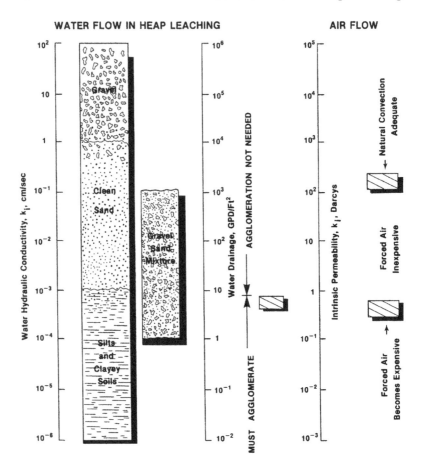

Figure 7.9. Fragmented rock matrix mixtures and permeability.

A number of these soil hydraulic conductivity studies show that the exponent, a, in eqn 7.18 varies between 3 and 4, and a graph of K/K_{sat} versus m for eqn 7.18 is shown in Fig. 7.10. However, these results with water passing through soil, which is a media of relatively uniform fine particles, do not correspond with the results obtained with the Kennecott ore column reported earlier in this chapter and summarized in Table 7.3. Because of the coarse rock particle sizes and larger open spaces in the ore column there is ample capacity for very high velocity water flow bypassing the finer material resulting in a saturation hydraulic conductivity that is very

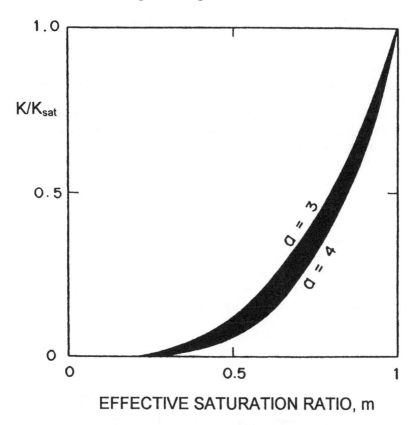

Figure 7.10. Decrease of unsaturated hydraulic conductivity with saturation ratio, *m* (Iwata et al., 1995).

large, several orders of magnitude greater than a typical soil hydraulic conductivity or the local hydraulic conductivity through the ore fines. Therefore, unsaturated water flow in soils and eqns 7.18 and 7.19 may have little relevance to solution flow in ore heaps.

In summary, at the present time there is not a good theoretical or practical method to estimate the intrinsic permeability (and hydraulic conductivity) of an ore heap. Most important, the dispersion of axial solution flow through an ore heap and washing efficiency cannot be predicted from ore heap physical properties such as particle size distribution. More research is needed in this area using rock size distributions typical of ore heaps. Dixon's recent work mentioned in Chapter 5 is promising.

MEASURING HEAP HYDRAULIC CONDUCTIVITY

Intrinsic permeability can be deduced from a measurement of hydraulic conductivity. Hydraulic conductivity at **flooding** can be measured in an ore column set up as either a constant head or a falling head **permeameter**. For a constant head permeameter, sketched in Fig. 7.11, the hydraulic conductivity is determined using eqn 7.7. The falling head permeameter, shown in Fig. 7.12, requires less water and is most suitable for a media with low hydraulic conductivity. The hydraulic conductivity is given by the following equation:

$$K = \frac{A_{\text{tank}} L}{A_{\text{heap}} t} \ln\left(\frac{h_0}{h_t}\right), \tag{7.20}$$

where h_0 is the initial head, h_t is the final head at time t, and L is the media length through which the flow is occurring. A_{heap} and A_{tank} are the cross-section areas of the media (heap) and the water reservoir (tank), respectively.

For solution percolation in ore heaps, the practical issue is to determine whether the hydraulic conductivity is adequate to convey solution through

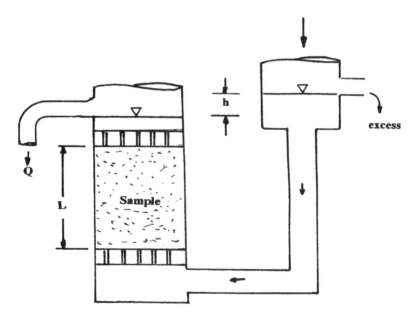

Figure 7.11. Constant head permeameter.

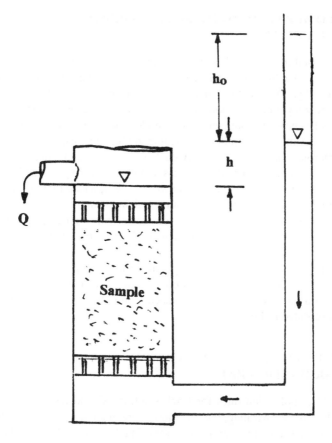

Figure 7.12. Falling head permeameter.

the heap at a velocity that matches the solution application rate, which is fixed by the sprinkler or other solution application device. Otherwise, flooding will occur. This can be accomplished in conjunction with column leaching tests by increasing the solution flow rate, either well beyond the planned application rate or until flooding begins. If this flow rate is insufficient, agglomeration of the ore is needed and flow measurements on the agglomerated ore should be done.

If air flow is needed, e.g., to oxidize sulfide minerals, then the intrinsic permeability must be much higher than needed only to convey the leaching solution.

AGGLOMERATION TO IMPROVE INTRINSIC
PERMEABILITY

As discussed in Chapter 3, agglomeration can be used to improve permeability by attaching fine particles to large particles and agglomerates and, thereby, shifting the particle size distribution to larger sizes. The Blake–Kozeny equation shows how increasing the particle diameter powerfully affects k_i. The agglomeration binders, used in heap leaching gold ores with cyanide, are limited to leaching at alkaline conditions.

Numerous binders have been evaluated for use in acidic heap leaching. The preferred binder for effectiveness and cost is mixing slaked lime and sulfuric acid into the ore mass (Southwood, 1985), where their *in situ* reaction forms calcium sulfate (gypsum). Calcium sulfate plasters used as binders often produce adequate agglomerates with the exception of their poor tolerance of strong sulfuric acid (pH $\leqslant 1$) where the agglomerates disintegrate.

The permeability below which agglomeration is required, prior to heap leaching, to obtain good percolation is about one darcy ($1 \times 10^{-12}\,\mathrm{m}^2$). This is consistent with the previous discussion in this chapter and the summary presented in Fig. 7.9.

OPTIMUM SOLUTION APPLICATION
AND IRRIGATION RATES

Desirably, application rates for leach solutions should be low enough to provide efficient washing, but all solution application systems have lower limits on their effective application rates. The lowest available application rates are obtained with drip emitters. Impact sprinklers apply fairly uniform coverage, but a high flow rate at the nozzle must be applied to obtain proper coverage. Consequently, there is a narrow limiting range on impact sprinkler application rates.

Ponding or flooded basin leaching, often employed on copper mine waste dumps, is the most inefficient mode of solution application because it promotes channel flow (short circuiting). Surface solution flowing to channel downspouts can occasionally be observed on the top of dumps during pond leaching.

If application rates are too low or too infrequently applied, the ore heap may partially dry out, which will stop bacteria action and mineral oxidation, and may plug rock micropores and interstices with precipitated salts.

With copper sulfide ores and mine waste, the rate of copper solubilization usually steadily decreases over months and years, and short solution application periods should be followed by rest periods during which aeration and oxidation of the sulfide minerals will continue. The leach solution distribution rate averaged over both application periods and rest intervals is sometimes referred to as the leaching solution "**irrigation rate**", which is defined as:

$$\text{Irrigation Rate} = \text{Application Rate} \left(\frac{\text{Leach Time}}{\text{Leach Time} + \text{Rest Time}} \right).$$

The optimum irrigation rate depends on the rate of mineral solubilization, which is usually governed by factors other than the amount of lixiviant used (Jackson and Ream, 1980).

It is psychologically difficult for a leaching supervisor to lower application rates and irrigation rates while under management pressure to increase metal production. The natural human tendency is to increase the flow of circulating leach liquor. However, the rate of leaching of copper sulfide ores in heaps is little affected by either the solution application rate or the irrigation rate. Short term washing gains are ephemeral and offset by later losses and a decrease in solution copper grade.

Excellent record keeping and running materials balances on metal extraction and inventory for each ore heap, mine waste dump or leaching section are very important in guiding the optimum rest intervals over time. An orderly management plan, coupled with the availability of several heaps or dumps for leaching, but with relatively few being actively leached at any particular time, is necessary. The duration of each solution application period can be determined from the production history (analyses of pregnant solution grade over time). Orderly planning should be based on data charted over a long period, several years for slow leaching copper mine waste dumps.

RUBBLIZATION WITH CHEMICAL EXPLOSIVES

Rubblization prior to solution mining using either percolation leaching or flooded leaching may be necessary for many ore deposits. Earlier, the use of nuclear explosives was seriously considered and the former Soviet Union conducted large scale earth construction projects with nuclear devices. Now, an environmental prohibition on nuclear mining exists. However, large scale blasting with chemical explosives has been used to rubblize smaller ore deposits near the surface prior to copper percolation leaching. Some examples were given in Chapter 5. Single blasts containing up to 2,000 tonnes of explosive have been detonated.

Explosives are emplaced either by using abandoned underground mine workings or vertical blast holed drilled from the surface. Blast holes are economically limited to a depth of about 250 m. The blast must raise the ore body slightly so that volume expansion will accompany rubblization, and provide the void space needed for solution flow. This is accomplished by an elaborate system of sequencing delays in large blasts so that the upper section of the ore body rises first followed by lifting a sequence of layers progressively downward.

The lowest cost chemical explosive is ANFO (ammonium nitrate with about six wt pct diesel oil). Usually 0.5–2 kg of ANFO are required per cubic meter of rock to be blasted. In wet holes ANFO is not effective, but slightly more expensive slurry explosives using a gelling agent can be substituted.

Safety factors that must be considered are the seismic effects that may damage surrounding property and rock missile throw. Seismic effects can often limit the size of a blast even in relatively remote locations.

PROBLEMS

1. Compute the hydraulic conductivity in cm/hr at room temperature (20°C) of an ore heap with an intrinsic permeability of $1 \times 10^{-11} \, m^2$. Next, compute the hydraulic conductivity in cm/hr for the same heap if its internal heap temperature is 60°C.

2. Will the ore heap of the preceding question likely encounter flooding problems using impact sprinklers to distribute the solution onto the heap? Explain your answer.

REFERENCES AND SUGGESTED FURTHER READING

Armstrong, F.E., Evans, G.C. and Fletcher, G.E. (1971). Tritiated water as a tracer in the dump leaching of copper, *U.S. Bureau of Mines RI 7510*.

Bird, R.B., Stewart, W.E. and Lightfoot, E.B. (1960). *Transport Phenomena*, J. Wiley, New York, NY, p. 8.

DeNevers, N. (1970). *Fluid Mechanics*, Addison-Wesley, Reading, MA, Chap. 12.

DeWiest, R.J.M. (1965). *Geohydrology*, John Wiley, New York, Chap. 4.

Fetter, C.W. (1988). *Applied Hydrogeology*, Macmillan, New York, Chaps. 4 & 5.

Harris, J.A. (1969). Development of a theoretical approach to the heap leaching of copper sulfides, *Proc. Australasian Inst. Mining and Metallurgy, No. 230*, pp. 91–92.

Howard, E.V. (1968). Chino uses radiation logging for studying dump leaching processes. *Mining Engineering, Vol. 20*, pp. 70–74.

Iwata, S., Tabuchi, T. and Warkentin, B.P. (1995). *Soil–Water Interactions*, Marcel Dekker, New York, NY, p. 382.

Jackson, J.S. and Ream, B.P. (1980). Solution management in dump leaching. Schlitt, W.J. (Ed.), *Leaching and Recovering Copper from As-Mined Materials*, Society of Mining Engineers, Littleton, CO, pp. 79–94.

Kennedy, W.A. and Stahl, J.R. (1974). Fluid retention in leach dumps by capillary action. Aplan, F.F. et al. (Eds.), *Solution Mining Symposium*, SME and TMS of AIME, New York, pp. 99–128.

Kunii, D. and Levenspiel, O. (1969). *Fluidization Engineering*, J. Wiley, New York, NY, p. 69.

Murr, L.E. (1979). Observations of solution transport, permeability, and leaching reactions in large, controlled, copper-bearing waste bodies. *Hydromet., Vol. 5*, pp. 67–93.

Murr, L.E., Schlitt, W.J. and Cathles, L.M. (1982). Experimental observations of solution flow in the leaching of copper-bearing waste. *Interfacing Technologies in Solution Mining*, Society of Mining Engineers, Littleton, CO, pp. 271–290.

Roman, R.J. (1977). Solution channeling in leach dumps. *Trans. AIME, Vol. 262*, pp. 73–74.

Schlitt, W.J. (1984). The role of solution management in heap and dump leaching practice. Hiskey, J.B. (Ed.), *Au and Ag Heap and Dump Leaching Practice*, AIME, New York, pp. 69–83.

Southwood, A.J. et al. (1985). The agglomeration of fine material for bacterial heap leaching. *Report M191*, Council for Mineral Technology, Mintek, Randburg, South Africa.

Whiting, D.L. (1977). Hydrology of dump leaching and *in-situ* solution mining. *SME-AIME Preprint #77-AS-336*.

EIGHT

Oxygen Diffusion and Air Flow During Heap Leaching

As discussed in Chapter 6, biooxidation of sulfide minerals in ore heaps and mine waste dumps is an important source of copper requiring very large amounts of oxygen. Heap biooxidation is also expected to become important for the pretreatment of sulfide refractory gold ores.

Oxygen can enter an ore heap or mine waste dump by the following mechanisms involving transport in the air-filled void space, v_g, of the heap: (1) ordinary chemical diffusion of oxygen through stagnant air caused by the depletion of oxygen reacting with the sulfide minerals, (2) flow of air induced by natural advection caused by the increase in air buoyancy as the composition and temperature of the air changes as it passes through the ore heap, and (3) forced ventilation of air through the ore heap using fans and distribution pipes within or under the heap. This chapter will discuss each of these in turn.

GASEOUS DIFFUSION OF OXYGEN IN ORE HEAPS

The following analysis of diffusion begins by ignoring biooxidation rate limitations related to intrinsic mineral oxidation kinetics and internal transport within solution-filled rock micropores. The latter will not be rate limiting for rocks below a critical size. It is generally applicable to rocks crushed to $-19\,mm$ ($-3/4\,in$) or to rocks below about that size in run-of-mine ore.

Without any air flow, fine grain, disseminated sulfide minerals in an ore heap can be oxidized in a zone adjacent to the surface of the heap by diffusion of oxygen through the air-filled inter-rock void space, v_g. The oxidized depth, or penetration distance, X_D, from the heap surface, depends on the biooxidation time. When the ore is adequately crushed and the sulfide mineral grain size is sufficiently small, gaseous oxygen diffusion will be the only rate limiting step, and its concentration gradient through this oxidized ore zone will be constant. Consequently, the oxidized depth, X_D, will be provided by the following relation (Bartlett and Prisbrey, 1995):

$$X_D = \left[\frac{2D_g v_g (c_0 - c)}{\tau (M_0/V)} \right]^{1/2} t^{1/2}, \qquad (8.1)$$

where D_g is the O_2/N_2 air diffusivity, v_g is the fractional air-filled void space between rocks in the heap, τ is an air path tortuosity factor, $(c_0 - c)$ is the difference in gaseous oxygen concentration through its gradient, and (M_0/V) is the moles of oxygen consumed per unit volume of ore heap at complete sulfide biooxidation. Because air flow is not involved, these results are independent of the heap's permeability, but obviously the effective oxidized depth does depend on the gas-filled void space, v_g.

From eqn 8.1, X_D versus t is plotted in Fig. 8.1 for a pyrite ore using standard values of the diffusivity and oxygen concentration in air, along with typical values of heap parameters; all of these are listed in Table 8.1. The bacteria are inactive when the oxygen partial pressure drops to about 3 kPa (0.03 atm), leaving a value of 8.03×10^{-6} mol/cc for the oxygen concentration difference, $(c_0 - c)$. The pyrite is completely oxidized, $F_t = 1$, down to the depth X_D, but completely unoxidized below X_D. In Fig. 8.1, values are plotted for three pyrite grades: 0.5, 1.0 and 2.0 wt pct. The results indicate that the diffused oxygen penetration distance, X_D, after two years of heap biooxidation varies from 3.38 to 1.69 m as the pyrite grade increases from 0.05 to 2.0 wt pct. During that period the aggregate fractional biooxidation of the sulfide minerals in a typical nine-meter high ore heap varies from about 19% to 38% which is commercially inadequate to provide a high yield of extracted mineral.

Figure 8.2 shows the effect of variations in air void space within the ore heap. It plots X_D versus t for two values of the air-filled void space, v_g, both at a pyrite grade of 1.0 wt pct.

The results of Figs. 8.1 and 8.2 are valid approximations only for sulfide mineral grain diameters below about 10 μm (10^{-5} m). When the sulfide mineral grains are larger, mixed kinetics must be considered, which

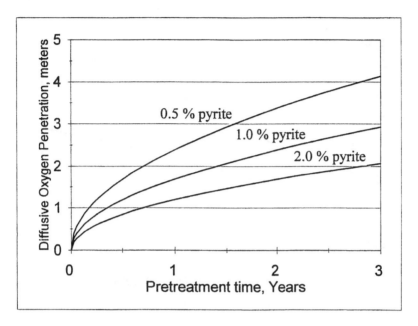

Figure 8.1. Diffusive oxygen penetration into an ore heap as a function of pyrite grade and biooxidation pretreatment time, $v_g = 0.30$ (Bartlett and Prisbrey, 1995).

Table 8.1 Constants for Calculations of eqn. 8.1 Shown in Fig. 8.1 and 8.2.

$D_g = 1.73 \times 10^4 \, \text{cm}^2/\text{d} \ (0.2 \, \text{cm}^2/\text{s})$
$\tau = 2$
$c_0 = 9.37 \times 10^{-6} \, \text{mol/cc}$
$v_g = 0.30$
$\rho_h = 1.70 \, \text{g/cc}$
$B = 3.75 \, \text{mols-O}_2/\text{mol-Py}$
$c = 1.34 \times 10^{-6} \, \text{mol/cc}$
$c_0 - c = 8.03 \times 10^{-6} \, \text{mol/cc}$
$M_0/V = 0.053 \, G_{Py}$

includes both gaseous diffusion in v_g and the intrinsic sulfide mineral oxidation rate within ore particles. These mixed kinetics were modeled for pyrite using a mineral oxidation rate expression coupled with gaseous oxygen diffusion in a finite difference simulation model (Bartlett and Prisbrey, 1996). The ore particles were crushed to small enough sizes that diffusion

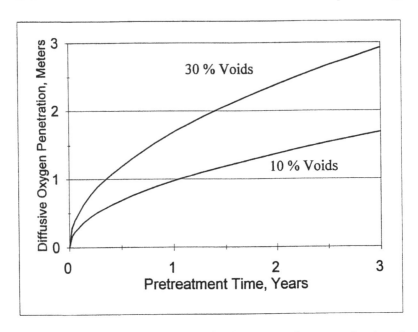

Figure 8.2. Diffusive oxygen penetration into an ore heap as a function of air-filled void space, v_g, and biooxidation pretreatment time, Pyrite $= 1.0$ wt pct (Bartlett and Prisbrey, 1995).

of ferric ions in solution-filled micropores was not affecting the oxidation rate. Oxygen concentration profiles and fractional reaction of the pyrite grains were computed as a function of depth from the ore heap's upper surface. The computed results show that a sharp interface no longer exists between fully oxidized ore and unoxidized ore. As pyrite grain size increases, oxidation penetration into the ore heap increases for the same biooxidation time, but a diffuse zone of incompletely reacted sulfide mineral grains (and incompletely oxidized rocks) exists in the ore and widens over time. This zone is parallel to the ore heap surface. Some computed results are shown in Fig. 8.3, where fractional oxidation of the local pyrite is plotted against depth into the heap, and the effects of variations in biooxidation time, pyrite grade and pyrite grain size are shown. While oxygen penetration increases somewhat as the mineral grains increase in size, aggregate biooxidation of sulfides in the heap actually decreases for the same biooxidation time. Some refractory gold ores show high cyanide extractions after incomplete sulfide biooxidation. For these ores, a larger average mineral grain size may not be a disadvantage.

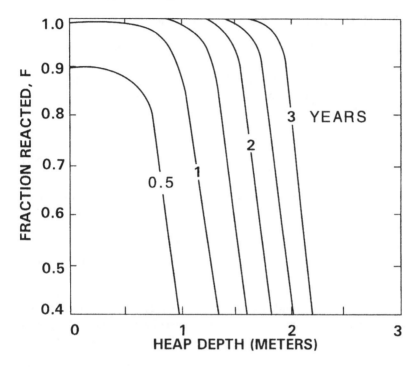

Figure 8.3. Series of fraction reacted, F, profiles computed by mixed kinetics/ diffusion at six-month intervals for heap biooxidation of an ore with 2.0 wt pct pyrite and 20 μm pyrite grains, $v_g = 0.30$ (Bartlett and Prisbrey, 1996).

In conclusion, diffusion of oxygen from the top of an ore heap limits biooxidation to depths of about one, or at the most two meters, and then only provided the treatment time is at least one year. If the heat released from the exothermic sulfide mineral oxidation is balanced by water evaporation within the ore heap, the upward water vapor flow will impede oxygen diffusion from the heap surface. Typically, oxidation depths, X_D, are cut in half when this occurs (Bartlett and Prisbrey, 1997). Oxidation depths are significantly less than the height of a typical ore heap stacked either with large wheel loaders or with radial conveyor belt stackers. Substitution of one or two-meter high bioheaps for heaps of conventional height would require very large heap pad areas, which may not be available because of the mine topography. An unacceptably high cost for construction of these larger pads may result.

VERTICAL AIR FLOW BY NATURAL AIR ADVECTION

The average superficial air velocity in the vertical direction through an ore heap or mine waste dump is related to the permeability and average pressure gradient by Darcy's law, and it can be estimated by assuming that air flows vertically through the dump. This is an oversimplification, but it is adequate for the order of magnitude calculations intended; hence eqn 7.9 becomes

$$u_g = \frac{k_i}{\mu_g}\left(\frac{dp}{dx}\right) = \frac{k_i}{\mu_g}\left(\frac{\Delta p_H}{H}\right), \tag{8.2}$$

where Δp_H is the pressure change from bottom to top of a dump or heap of height H, and μ_g is the gas (air) viscosity. This equation is also useful in interpreting air flow measurements in ore column experiments. Because of the small air pressure gradients in natural advection, air flow is laminar in heaps and dumps. The average buoyancy pressure gradient is related to the decrease in air density as it passes through the heap. The density decreases because it (1) is heated by the large thermal mass of the heap and the exothermic sulfide oxidation reactions, (2) becomes saturated with water vapor from intimate contact with the wet heap matrix, and (3) loses oxygen by chemical reaction. Consequently, the average pressure gradient, $\Delta p_H/H$, is

$$\frac{\Delta p_H}{H} = g(\rho_0 - \rho_H), \tag{8.3}$$

where g is the gravitation constant and ρ_0 and ρ_H are the gas densities at the bottom and top of the heap respectively. Substituting values for the gas densities using the perfect gas law yields

$$\frac{\Delta p_H}{H} = \frac{g p_t}{R}\left[\frac{MWG_0}{T_0} - \frac{MWG_H}{T_H}\right], \tag{8.4}$$

where p_t is the total ambient gas pressure, R is the gas constant, MWG_0 is the molecular weight (g/mole) of the ambient air (STP) and T is the absolute temperature. MWG_H is the molecular weight of the air leaving the heap at STP. MWG_H differs from MWG_0 because of an increase in water vapor concentration and a decrease in oxygen concentration in the effluent gas. Consequently, the superficial gas velocity rising up through the ore heap is given by:

$$u_g = \frac{k_i g p_t}{\mu_g R}\left[\frac{MWG_0}{T_0} - \frac{MWG_H}{T_H}\right]. \tag{8.5}$$

With p_t at one atmosphere (1.013×10^5 Pa), μ_g at 0.018 cp (1.8×10^{-5} Pa s), and the permeability, k_i, expressed in m^2, then the superficial gas velocity expressed in m/s is given by the following:

$$u_g[\text{m/s}] = 6.66 \times 10^6 k_i[\text{m}^2] \left[\frac{\text{MWG}_0}{T_0} - \frac{\text{MWG}_H}{T_H} \right]. \tag{8.6}$$

Consider the following limiting case. Oxygen is fully depleted to the level below which the bacteria are inactive (1 ppm in solution or 3% in air), the temperature rises from 25°C to 45°C, and the exiting gas is saturated with water vapor at 45°C. The molecular weight of the incoming dry air is 28.85 g/mole and the gas exits the heap at a molecular weight of 27.13 g/mole. Vertical air velocities, computed using eqn 8.6, are plotted versus intrinsic permeability in Fig. 8.4 for this limiting case. Note the extremely slow air velocities, less than 10 m/day, even at a fairly high ore heap permeability of 10^{-9} m^2, 1,000 darcy.

Large quantities of air are required to oxidize the usually prevalent pyrite, and estimates of the pyrite/chalcopyrite mole ratio (Py/Cp) in copper mine waste can be used to estimate air velocities required for specified annual copper extraction rates, on the assumption that rock kinetics are not limiting and that oxygen in air is depleted to 3%. Three examples of oxygen chemical requirements related to the expected range of Py/Cp ratios follow:

Mass Ratio (O_2/Cu)	Mole Ratio (Py/Cp)
5	1.7
15	7.4
25	13.0

The order-of-magnitude estimates, some of which are summarized in Fig. 8.5, indicate a mismatch between air velocities generated with typical dump matrix permeabilities and the requirement for air flow to supply oxygen at economically meaningful rates.

A 50-m high copper mine waste dump will contain about 200 kg of copper per square meter. Hence, Fig. 8.5 indicates that extraction in a reasonable time will require an air velocity of more than 0.3 m/hr and, according to Fig. 8.4, a dump air permeability of more than 10^{-9} m^2 (1,000 darcy). With a lower ore matrix permeability likely to occur, *significant air flow must occur primarily in channels* where local permeabilities are higher than the average permeability. This conclusion was indicated on the right side of Fig. 7.9.

Variations in heap permeability are often controlled by the angle of repose (37°) formed in casting from haulage trucks. Alternate thin layers

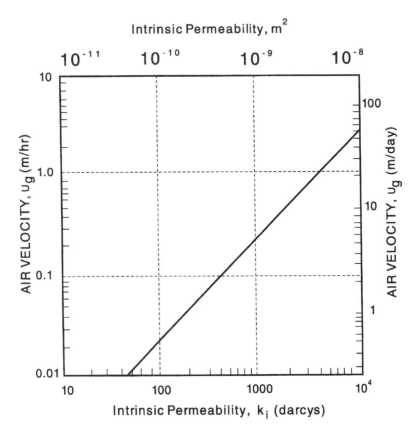

Figure 8.4. Natural advection limiting vertical air velocities and intrinsic permeability for: $T_H - T_0 = 20°C$ and 3% residual O_2 in the exhausted air.

of fine and coarse material are deposited along the direction of the dump slope, which is the bedding plane for the dumped material. Excavations through dumped and leached material, such as road cuts, often show variations of soluble copper staining stratified with the dump repose, indicating channeling parallel to it at an angle of 37°.

A perched water table, including ponding on the surface of the heap or dump, is a complete barrier to upward air flow, and the effective air permeability of a heap or dump covered with a perched water table is nil. This result often occurs when flooded basins, or ponds, are used to distribute leaching solutions for copper dump leaching, and it is another disadvantage of the ponding method of solution application. After the pond has drained,

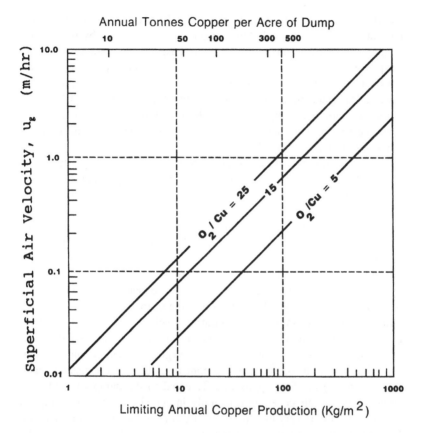

Figure 8.5. Annual copper yield per unit area of dump limited by its natural advection air velocity a 3% residual O_2 and $T_H - T_0 = 20°C$, at three arbitrary oxygen demand mass ratios (O_2/Cu).

iron salts precipitated and settled from the prior pond solution may continue to seal the dump surface and prevent air from flowing upward.

AIR FLOW BY NATURAL ADVECTION FROM THE SLOPING SIDES OF ORE HEAPS

A two-dimensional simulation model of air flow and oxygen diffusion by natural advection in ore heaps and mine waste dumps undergoing heat release from sulfide oxidation has been developed by Ritchie and coworkers (1985, 1989). This model (FIDHELM) involves both energy and materials balances, including oxygen depletion caused by the chemical reactions with sulfide minerals. It is based on measurable physical transport properties and it is particularly useful in demonstrating that air flows into the side of an ore heap, primarily near its toe, and then turns upward. The space–time configuration of this flow and the possible extent of ore oxidation are computed by FIDHELM. Given sufficient time, ore under the sloping side becomes oxidized but lateral penetration into the heap doesn't proceed much farther in reasonable times.

Figure 8.6, constructed from FIDHELM results, shows the pyrite oxidation penetration zone in the face of a 20 m high ore heap following biooxidation for periods of two years and four years. The intermediate zones represent the partially oxidized regions bounded by 90% and 10% completion of the oxidation of pyrite in the region. The results shown are for an ore heap with an intrinsic permeability of $10^{-9}\,m^2$ (1,000 darcy). The temperature was not constrained and reached a maximum of 100°C. Oxidation near the face is due to advective air flow, while at the top of the heap computed velocities were negligible and the pyrite oxidation is caused primarily by diffusion of oxygen from the top of the heap.

Interestingly, the simulation penetration results are somewhat less than indicated by measurements of oxygen concentrations in the gas-filled void space within copper mine waste dumps. Oxygen is usually depleted down to the bacteria limit of about 3% (1 ppm in solution) at a distance back from the *top of the* face approximately equal to the dump's height. The greater observed air penetration laterally in actual copper mine dumps is probably due to the very high permeability zone at the bottom of the dump that results from segregation of boulders, relatively free of fines, to the bottom of the dump following end-over truck dumping of run-of-mine rock. Apparently, rock segregation allows air to flow farther under the dump before turning upward.

The FIDHELM simulation has been extended to more realistic heap situations of biooxidation, where upper bounded temperatures caused by the bacteria limitations retard thermal convection when compared with an unbounded rise in temperature (Pantelis and Ritchie, 1993). For typical sulfide mineral concentrations in copper and refractory gold ores, the simulation results show that with the heap temperature limited to 45°C, an intrinsic permeability of $10^{-7}\,m^2$ (100,000 darcy) is required to obtain substantial oxidation of the sulfides in one year, and in general a permeability better than $10^{-8}\,m^2$ (10,000 darcy) is required for multi-year commercial operations. These are very high permeabilities, equivalent to those found in a washed coarse gravel, but not found in typical ore heaps.

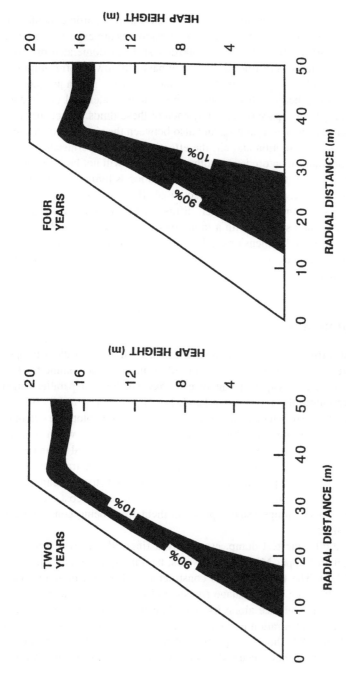

Figure 8.6. Extent of biooxidation of a truncated cone ore heap with $k_i = 10^{-9}\,\text{m}^2$, after Ritchie and Pantelis (1993).

As discussed in the previous chapter, because of fine particles contained in typical crushed ore heaps, intrinsic permeabilities are expected to range from $10^{-11}\,\text{m}^2$ to $10^{-9}\,\text{m}^2$ (10 to 1,000 darcy). Unagglomerated ore containing large amounts of fines or clay may have still lower permeabilities.

While the intrinsic permeability is a function of the media and not the fluid, in an ore heap with two immiscible fluids, water and air, it does depend on the void space occupied by each of these fluids. Hence, because of differences between v_l and v_g, and also between the sizes of voids occupied by leaching solution and air, the effective intrinsic permeabilities for air and for water in an ore heap (both unsaturated) will not be identical.

In summary, for typical ore heaps, biooxidation is limited to the sloping heap sides, caused primarily by advective air flow, and to the top surface region, caused primarily by oxygen diffusion. Furthermore, for most ore heaps, even when stacked with a radial belt stacker, heights will be much less than the lateral dimensions, allowing only a small fraction of the ore to receive adequate oxygen by both transport processes in reasonable leaching times.

FINGER DUMPS

Finger dumps, long narrow dumps, or heaps, separated by valleys of open space, were conceived by Robinson (1972) to increase the volume fraction of the ore in the heap that can receive oxygen. Several parallel finger dumps resemble the fingers of a hand. For finger dumps to be effective they must be narrow indeed, as shown in Fig. 8.7 which plots the extent of possible ore aeration as a function of the ratio of heap width to heap height. Furthermore, this computation uses the more liberal estimate that air flow into the sides of a typical copper dump can penetrate laterally a distance from the top of the face that is about equal to the heap height. Mine topography and the increased haulage distance and cost required to construct finger dumps usually prevents them from being constructed at copper mines.

An 85,000 tonne test dump, essentially a finger dump, was leached at the Bingham Copper Mine for two years, beginning in 1969 (Cathles and Apps, 1975). The test dump was constructed with run-of-mine ore using end-over dumping from haulage trucks, and it contained nine bore holes drilled from the surface along its centerline for monitoring. Temperature and oxygen concentration in the depleted air within the dump over the test period. Figure 8.8 shows the profiles of temperature and oxygen concentration in air as they occurred after four months of leaching. Air was

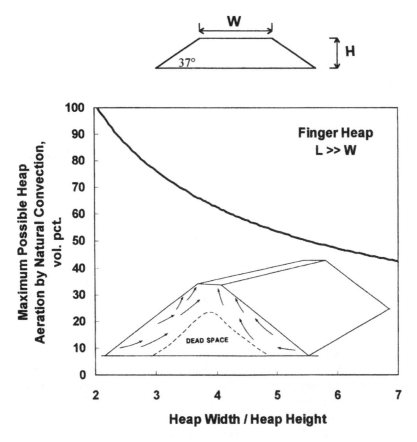

Figure 8.7. Maximum amount of copper accessible to aeration in a leaching finger dump as a function of its width to height.

drawn into the dump, primarily from its toe (right-hand side of Fig. 8.8), and gradually depleted of oxygen as it flowed through the dump (from right to left in Fig. 8.8). Temperatures are highest in the upper center of the dump.

Using test dump data, Cathles and Schlitt (1980) developed a two-dimensional model, since superseded by FIDHELM, and obtained two important insights. The buoyancy pressure difference, Δp_H, is limited to a maximum value, $\Delta p_{H(max)}$, that occurs when the internal air in the heap or mine dump is (1) fully depleted of oxygen, (2) saturated with moisture, and (3) at the maximum functioning temperature (40°C) of the bacteria.

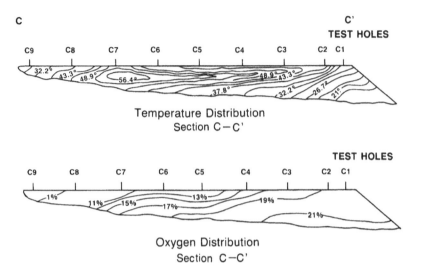

Figure 8.8. Temperature (°C) and percent oxygen in the Bingham Mine test dump after four months of leaching (Cathles and Apps, 1975).

Consequently, the maximum pressure **gradient**, $\Delta p_{H(\text{max})}/H$, and maximum air advection velocity, $u_{g(\text{max})}$, decline as heap height, H, increases. The problem is similar to that of inducing smoke to rise from a fireplace with a tall, narrow chimney constricting flow. However, higher heaps and mine dumps require proportionately greater air velocities to satisfy equal specific rates of oxygen demand and copper extraction (rates per unit mass of ore). Because the air velocity, u_g, decreases with H while the oxygen needed to maintain the same sulfide mineral oxidation rate increases with H, the specific volumetric oxygen rate of supply to the heap or dump is, at first approximation, inversely proportional to the dump height squared, H^2,

$$-R_{\text{ox}} \propto \left[\frac{1}{H^2}\right]. \tag{8.7}$$

Consequently, the leaching rate will diminish asymptotically with the dump height, H, and the probability of large sulfide ore heaps and mine dumps being starved for oxygen must be very high. Oxygen supply to the internal ore is likely to be the limiting step in copper extraction from mine dumps. However, extremely shallow dumps will not draw air, which is analogous to a fireplace with no chimney at all. Cathles and Schlitt concluded that for a typical dump ore matrix permeability, the fastest air rising velocity and mineral oxidation and leaching rates would occur at a dump height of about 20 m.

Cathles and Schlitt's second insight came from observing that the copper extraction from the test dump was highest in the late winter and early spring when the ambient air temperatures, T_0, were coldest. They concluded that, because the internal temperature of the dump, T_H, is essentially constant throughout the year, the colder ambient air temperatures in winter cause a *greater* temperature difference, $T_H - T_0$, a *greater* buoyancy driving force for advection and faster air movement, which is consistent with eqn 8.6.

Unpublished oxygen measurements from many air samples extracted from within several large dumps at the Bingham, Utah, copper mine support the simulation conclusion that air flow is minimal causing oxygen to be depleted. These samples were withdrawn through cased holes drilled vertically from the top of the dumps, and in nearly every case, the oxygen concentration in the extracted gas was less than 5%.

HORIZONTAL FLOW BY NATURAL AIR
ADVECTION UNDERNEATH A HEAP

A means of lateral transport of air underneath an ore heap is required to oxidize the minerals that are not near the sides of the heap. For commercial ore heaps, this requires a very small pressure difference over a very large horizontal distance. Consider the case of lateral air flow through the drain blanket underlying an ore heap, which normally consists of high permeability washed gravel, but above the horizon of the saturated leaching solution at the bottom of the drain blanket. The horizontal air pressure gradient, dp/dl, decreases as air flows laterally through it, because air is offloaded into the overlaying ore. Consequently,

$$\frac{dp}{dl} \propto (L-1), \tag{8.8}$$

where L is the horizontal distance to the center of the heap. The superficial lateral velocity, u_l, of air entering the heap through the upper portion of the drain blanket is proportional to the superficial vertical velocity, u_h, in the heap,

$$u_l = \frac{L}{H_{DB}} u_h, \tag{8.9}$$

where H_{DB} is the height of the drain blanket. By integrating eqn 8.8 and inserting Darcy's Law,

$$u_l = \frac{k_{DB}}{\mu_g} \frac{dp}{dl},$$ (8.10)

it can be shown (Bartlett and Prisbrey, 1996) that the lateral pressure difference, Δp_{DB}, across the drain blanket is proportional to the vertical pressure difference, Δp_H, through the heap,

$$\Delta p_{DB} = \left[\frac{k_i}{k_{DB}} \right] \frac{L^2}{2H(H_{DB})} \Delta p_H,$$ (8.11)

where k_i and k_{DB} are the intrinsic permeabilities of the ore heap and drain blanket, respectively, and H is the heap height. As a typical example, for a 5-m heap height, 0.5-m drain blanket height and a 50-m half-width, the result is:

$$\Delta p_{DB} = 500 \frac{k_h}{k_{DB}} \Delta p_H.$$ (8.12)

Hence, for the lateral pressure drop to be less than the pressure drop through the ore heap, the drain blanket permeability must be at least 500 times greater than the ore permeability. Desirably, to obtain nearly the same vertical flow velocity through the ore heap at its center as near its edge, the drain blanket pressure difference should be about one order of magnitude smaller than the pressure difference up through the ore heap. Thus, a drain blanket permeability of at least 5,000 darcy $(5 \times 10^{-9} \, m^2)$ may be needed to match required sulfide oxidation rates and provide for timely completion of oxidation, necessary for an economic process. A drain blanket with a permeability of 5,000 darcy requires very coarse gravel or cobbles that are free of any smaller rocks. While this may be approached at the bottom of some copper mine waste dumps because of rock segregation of run-of-mine ore, it is questionable whether maintaining a drain blanket with such a very high permeability is practical under operating conditions likely to be encountered at most mines using crushed ore.

An alternative to a high permeability drain blanket is to lay down rows of perforated plastic pipe, in a manner analogous with an agricultural drain field, except that air is drawn through these pipes into the heap, rather than water flowing out from under them. To be free of standing water, these air pipes must be located at an elevation slightly above the leachate drain pipes and the saturated solution layer at the bottom of the ore heap.

Even if an ore heap has adequate permeability and air can be moved laterally under it, **initiating natural convection is not a trivial problem**.

With 30 v/o air in a heap, which is about the maximum possible, there is only enough oxygen present to raise the heap temperature 0.7°C as a result of mineral oxidation. This is not enough to induce an appreciable draft. Hence, the initial convective air flow will be insignificant, and air flow may accelerate either very slowly or not at all. At the beginning, diffusion of oxygen will be the only process, and convective flow will start only after gaseous oxygen diffusion has generated a sufficient temperature increase.

Because diffusion is a slow oxygen transport and heat generating process for starting a draft, forced air ventilation may be necessary at the beginning to initiate biooxidation and raise the temperature sufficiently for natural convection to continue to supply air. This is analogous with blowing air on a campfire to start it. Once the fire becomes large enough, its natural draft will sustain it—provided it is not choked off by overlying cold wet debris reducing the permeability and flow of air. Overlying cold, wet ore has a similar damping effect.

FORCED AIR VENTILATION OF ORE HEAPS

Commercial ore heaps must necessarily be very large for economic operations, and the limitations of gaseous diffusion and natural convection for providing sufficient oxygen to the interior of large ore heaps requiring biooxidation indicates the need for forced air ventilation. With advanced planning, perforated air injection pipes, made of low-cost polymeric materials, can be installed prior to stacking ore to form the heap, as illustrated in Fig. 8.9. These air pipes are connected via a manifold to a fan. This

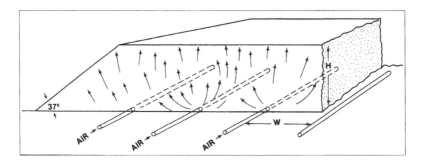

Figure 8.9. Section of ore heap with buried parallel, perforated air injection pipes (Bartlett, 1990).

approach to ventilating ore heaps was advocated for bioheap pretreatment of sulfidic refractory gold ores along with preliminary engineering and cost estimates (Bartlett, 1990). Newmont Gold Company began large scale evaluation of forced air ventilation to accelerate biooxidation of refractory gold ore heaps in 1996. Some of the air blowers involved in the Newmont test are shown in Fig. 8.10. This test involved over 500,000 tonnes of ore.

Figure 8.10. Fans blowing air under a refractory gold ore heap at the Gold Quarry Mine, Nevada (courtesy of Newmont Gold Company, 1996).

The stoichiometric air requirements to oxidize pyrite in a typical refractory gold ore are large and oxygen utilization as air flows through the heap may be incomplete, due to poor sweep efficiency. Example calculations of air requirements at different utilization efficiencies are shown in Fig. 8.11 for an ore containing varying amounts of pyrite (Bartlett, 1990).

If the parallel air distribution pipes shown in Fig. 8.9 have perforations that are closely spaced relative to the heap height, each air pipe can be considered as a line source of air flowing radially out from it. Equipotential (equal pressure) lines are hemicircles that are concentric with the air pipe in a plane perpendicular to it. As air leaves the heap, it is flowing vertically. The air flow rate per length of pipe, Q_g/L, depends on the ore heap intrinsic permeability, k_i, air pressure drop across the heap, Δp_H, air viscosity, μ_g, heap height, H, and air pipe radius, r_w:

$$Q_g/L = \frac{1}{2}\left[\frac{C_{ki}4\pi k_i\Delta p_H}{\mu_g\ln(2H/r_w)}\right]. \tag{8.13}$$

This equation results from the "method of images" (Milne-Thompson, 1956) with the air pipe a line source of air flowing to an imaginary, parallel

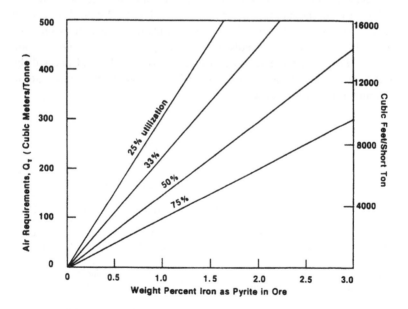

Figure 8.11. Stoichiometric air volume requirements (STP) for oxidation of pyrite in ore at different oxygen utilization efficiencies (Bartlett, 1990).

line sink located directly above the air injection pipe at a separation distance twice the height of the heap.

Definitions of the parameters used in eqn 8.13 and some consistent sets of units for use in the parameters of this equation are shown in Table 8.2. In SI units the coefficient is $C_{ki} = 1.0$. It is relatively easy, and advisable, to convert all of the given units in a problem back to SI units and then apply eqn. 8.13. Conversion tables are given in the Appendix. Note that it doesn't matter what units are used for H and r_w, provided the same unit is used for both.

A pipe line air flow rate, Q_g/L, will generally be specified by the amount of sulfide minerals that must be oxidized in a given desired bio-oxidation pretreatment time. The operational question then becomes to determine the pressure drop across the ore heap, Δp_H, using eqn 8.13. Three examples of this are shown in Fig. 8.12, for different air pipe diameters and parallel pipe separation distances, W_p. All three examples are at the same heap height, namely $H = 9$ m, and the same air flow rate, namely a superficial velocity of 0.0042 m s^{-1} (approximately 1 ft/min).

If continued for 30 days, this flow rate would provide about 690 cubic meters of air for each tonne of ore, assuming a heap dry bulk density of 1.75 tonne/m^3 (1.75 g/cc). Since bioheap pretreatment times are likely to be longer than one month, the air flow rate of Fig. 8.12 will likely exceed the rate of oxygen demanded by pyrite oxidation in a typical ore (see Fig. 8.11), even at the rapidly reacting beginning of the biooxidation process. For an adequately crushed ore, the oxygen demand rate will be controlled by the intrinsic sulfide mineral oxidation kinetics, which are in turn mineral grain size dependent.

Table 8.2 Ore Heap Air Pipe Flow Equation Parameters for eqn. 8.13. $\mu_{AIR} = 0.018 \, cP = 0.018 \, kPa \, s$.

| | Parameter | Units of Parameters | | | | |
		SI	(1)	(2)	(3)	(4)
Viscosity	μ_g	Pa s	cP	cP	cP	cP
Flow Rate	Q_g	m^{3}s^{-1}	m^{3}s^{-1}	m^{3}s^{-1}	m^{3}s^{-1}	m^{3}s^{-1}
Intrinsic Permeability	k_i	m^2	m^2	m^2	darcy	darcy
Pipe Length	L	m	m	m	m	m
Heap Pressure	Δp_h	Pa	kPa	inches-H$_2$O	inches-H$_2$O	psi
Coefficient	C_{ki}	1.0	1.0×10^6	2.49×10^5	2.45×10^{-7}	6.8×10^{-6}

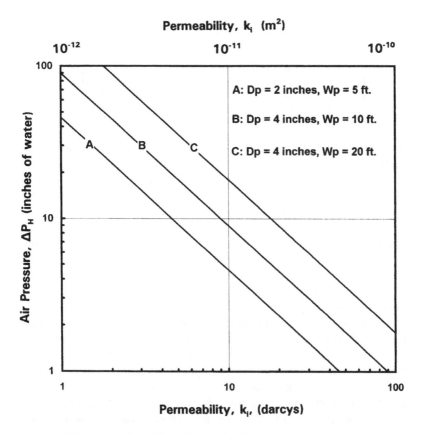

Figure 8.12. Examples of heap differential air pressure, Δp_H, versus heap intrinsic permeability, k_i, using parameters in Table 8.3.

With the superficial air flow velocity equal to the volumetric flow rate divided by the cross-sectional heap area, the air velocity is:

$$u_g = Q_g/A = Q_g/(LW_p). \tag{8.14}$$

Consequently, the air pipe flow rate is:

$$Q_g/L = W_p u_g. \tag{8.15}$$

Parameters and the resulting values of Q_g/L, determined for each example in Fig. 8.12, are shown in Table 8.3.

The significance of the results shown in Fig. 8.12, is that the required *pressure differentials necessary for forced air ventilation of ore heaps are*

Table 8.3   Parameters Used in Fig. 8.12.

Example	u_g	H	W_p	Q_g/L	D_p
A	$0.0042\,\mathrm{m\,s^{-1}}$	30 ft	5 ft	$0.0064\,\mathrm{m^2\,s^{-1}}$	2 in
B	$0.0042\,\mathrm{m\,s^{-1}}$	30 ft	10 ft	$0.0128\,\mathrm{m^2\,s^{-1}}$	4 in
C	$0.0042\,\mathrm{m\,s^{-1}}$	30 ft	20 ft	$0.0256\,\mathrm{m^2\,s^{-1}}$	4 in

modest, usually less than 5 kPa (20 in of water). This conclusion is generally valid at ore heap intrinsic permeabilities above about 10 darcy $1 \times 10^{-11}\,\mathrm{m^2}$). The total pressure required by the air movers must also include the pressure drop across the air pipe orifices and line pressure drops in both the air pipe and air manifold leading from the air mover to the air pipes. Nevertheless, the total pressure required will also be modest and not likely to exceed about 6–12 kPa (1–2 psi) in a well designed application. This is a pressure range attainable with paddle wheel centrifugal fans, which are limited to a pressure of about 15 kPa (2.5 psi). Ore heap forced air ventilation is similar to underground mine ventilation with the same range of fan pressures.

As air flows from the air pipe to the surface of the ore heap, most of the pressure change occurs near the pipe. Consequently, changing heap height does not have a major effect on the ore heap pressure drop, Δp_H, provided that Q_g/L remains constant. For example, if the heap height is doubled, without a change in ore heap retention time, the total air flow, Q_g, must double, but the spacing of air pipes, W_p, can be halved to keep Q_g/L constant. Heap height will be determined by other considerations, including pad area, pad cost and the methods of ore emplacement and removal.

Within the usual pressure operating range of centrifugal fans, the amount of electric power required is proportional to the fan pressure and the volumetric flow rate, Q_g. In designing an air ventilation system, the diameters of manifolds and air pipes must be chosen to minimize line pressure drops, so that power cost is reduced and, most importantly, so that the air is distributed nearly *uniformly* among both the many air pipes underlying an ore heap and along the length of each air pipe. To obtain uniform air sweeping flow through the ore heap, the largest pressure drop, and most of the fan pressure, should occur through the air pipe **orifices** and the ore heap.

Estimates of power consumption for a forced air ventilation system, designed with modest line pressure drops, are less than 2 kWh/tonne of ore at intrinsic permeabilities above 10 darcy $(1 \times 10^{-11}\,\mathrm{m^2})$ and pyrite grades below 5.0 wt pct in the ore (Bartlett, 1990). This range of these critical parameters will be met by most coarse crushed or run-of-mine ores.

Agglomeration of some ores to lower their permeabilities may help lower air ventilation power costs. Consequently, energy requirements and operating costs, will be modest in relation to the other costs of mining and heap leaching. **The cost of forced air ventilation is not a barrier to its application for ores needing heap biooxidation, provided they are of sufficient metal grade for mining and conventional heap processing, and it's being commercially applied on secondary sulfide copper ore in Chile and refractory gold ore in Nevada.**

Heap biooxidation is usually a batch process, and overheating can damage or sterilize bacteria. Consequently, depending on mineral availability and oxidation kinetics, the air ventilation rate may need to be limited to prevent overheating. This will likely be required early in the batch cycle. Later, as minerals are depleted by oxidation, the oxygen demand will be lower.

The temperature rise and heat capacity of the leaching solution leaving an ore heap will normally account for most of its heat loss. Heat removal by water evaporation from deep within an ore heap becomes important only at temperatures too high to sustain *T. ferrooxidans*. However, in dry, windy climates surface evaporation and radiation, especially at night, may substantially undercool the top of an ore heap down to a depth of about one meter.

FORCED AIR VENTILATION WITH NON-UNIFORM HEAP PERMEABILITY

If heaped ore is not well mixed, regions of much higher permeability than the average permeability may occur and act as high flow channels for the injected air, thereby causing poor air sweep efficiency in the heap. The air pipe pressure, p_P, is the sum of the orifice pressure drop, Δp_{Or}, and the ore heap pressure drop, Δp_H. Air channeling in the heap can be partially offset by decreasing the air pipe orifice size so that much of the pressure change occurs through the orifice. Open flow is the ultimate channel. It occurs when an ore heap is not present, i.e. $k_i = \infty$ and $p_P = \Delta p_{Or}$. Starting with a fixed air pipe pressure at open flow, loading the ore, with its less than infinite permeability onto the pipe, will decrease Q_g/L. This reduction was computed as a function of heap permeability, k_i, for three values of p_P, all not exceeding 12 kPa (2 psi), which is within the pressure capability of centrifugal fans. The results are plotted in Fig. 8.13 with the heap flow presented as a percent of open flow. Importantly, the flow in the ore heaps exceeds 75% of open flow at heap permeabilities above about 20 darcy ($2 \times 10^{-11} \, m^2$). Consequently, if a channel does occur in the ore heap, it will not have a significant effect for an average heap permeability above about 20 darcy, provided that the orifice size is selected to appropriately choke the open air flow.

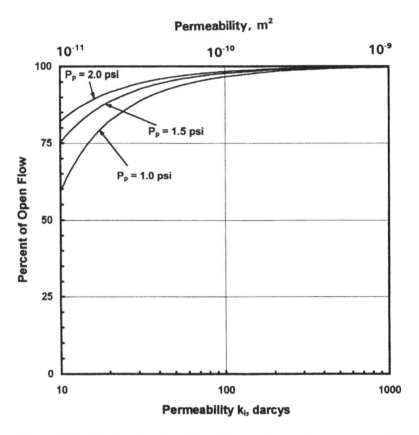

Figure 8.13. Air flow through orifices into an ore heap, at various permeabilities, as a function of the open orifice flow rate, i.e., $k_i = \infty$. Three injection pipe pressures are shown, with $p_P = \Delta p_{Or} + \Delta p_H$.

ESTIMATING INTRINSIC PERMEABILITY

Recall that intrinsic permeability is a property of the media and not the fluid. Therefore, the methods of measuring or estimating intrinsic permeability should be independent of the fluid used, e.g., either air or water. However, in two phase flow during heap leaching these fluids do not occupy the same space within the heap and their void space, v_l and v_g, are usually unequal. Furthermore, water is the wetting fluid and is pulled into the finer ore particles by capillary action, as discussed in Chapter 7 and illustrated in Fig. 7.1. Consequently the average ore particle experienced by the leaching solution is smaller than the average particle experienced by air within the heap. In principle, the Blake–Kozeny equation (eqn 7.16) could be used to estimate the intrinsic permeability for air flow in a heap by substituting in the value of v_g. However, the use of eqn 7.17 to estimate the effective particle size for insertion in eqn 7.16 will provide a lower air flow permeability than is likely to be observed.

Published information on air flow, pressure differentials and particle size distribution in wetted ore columns are virtually non-existent. When such information becomes available, useful semi-empirical correlations of air permeabilities with ore rock particle size distribution will likely result.

MEASURING INTRINSIC PERMEABILITY

Intrinsic permeability can be determined in an ore column experiment by measuring the superficial air flow velocity, u_g, or the volumetric flow rate, Q_g/A_{col}, and measuring the pressures at the top and bottom of the column to determine Δp_H. Inserting these parameters and the known viscosity of air into eqn 8.2 provides the value of k_i.

Intrinsic permeability can also be crudely measured with the set-up shown in Fig. 8.14. A pile of ore, such as one or two haultruck loads is dumped over a perforated air pipe on an impervious base. The ore pile must be leveled to provide a uniform height, H. The air pressure and flow

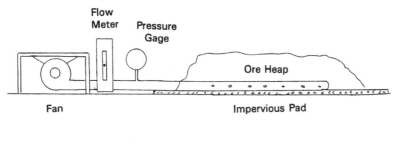

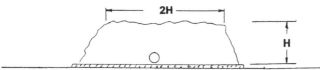

Figure 8.14. Simple air flow experiment to measure the permeability of an ore pile.

rate are monitored, and these data are coupled with the height of the ore pile to determine the permeability using eqn 8.13.

In measuring permeability, it is important to ensure that conditions expected to occur in the commercial ore heap are met. It is particularly important that the water content characteristic of two phase flow in the heap during leaching be maintained.

EXAMPLE PROBLEM I

Calculate the intrinsic permeability in darcy of a column of wet ore from the following data: The column inside diameter is 12 in and the ore height is 6 ft. At an air flow rate of 15 l/min the measured air gage pressure is 9.2 in of water. Water trickled through the ore bed at typical leaching application rates during the air flow experiment.

ANSWER
Use eqn 8.2, with the following parameters in SI units (see Appendix):

$$H = 6\,\text{ft} = 1.8288\,\text{m}$$

$$A = (12/2)^2\pi = 113.1\,\text{in}^2 = 0.07295\,\text{m}^2$$

$$\Delta p_H = 9.2\,\text{in-H}_2\text{O} = 2289\,\text{Pa}$$

$$Q = 15\,\text{l/min} = 0.00025\,\text{m}^3\,\text{s}^{-1}$$

$$u_g = Q/A = 0.0025/0.07295 = 0.003427\,\text{m}\,\text{s}^{-1}$$

$$\mu_{AIR} = 1.8 \times 10^{-5}\,\text{Pa}\,\text{s}$$

Rearranging eqn 8.2 provides:

$$k_i = (u_g)(\mu_g)(H)/\Delta p_H$$

Inserting parameters yields:

$$k_i = (0.003427\,\text{m}\,\text{s}^{-1})(1.8 \times 10^{-5}\,\text{Pa}\,\text{s})(1.8288\,\text{m})/2289\,\text{Pa}$$

$$k_i = 4.93 \times 10^{-11}\,\text{m}^2$$

Converting k_i to darcy:

$$k_i = (4.93 \times 10^{-11}\,\text{m}^2)(1.013 \times 10^{12}\,\text{darcy/m}^2)$$

$$k_i = 50\,\text{darcy}$$

EXAMPLE PROBLEM II

Biooxidation of the ore, represented in the previous problem, is planned in a large test heap that is 100 ft × 100 ft × 25 ft high, not including the side slopes. A 5,000 cubic feet per minute (5,000 CFM) fan will be used with 4 in inside diameter air pipes underneath the heap. The total perforated length of air pipes is 500 ft. Using the intrinsic permeability from the column measurements, estimate the ore heap pressure drop in inches of water.

ANSWER

The following parameters are computed and converted to SI units:

$$Q/L = 5000 \text{ CFM}/500 \text{ ft} = 10 \text{ ft}^2/\text{min}$$

$$Q/L = 0.929 \text{ m}^2/\text{min} = 0.0155 \text{ m}^2\text{s}^{-1}$$

$$k_i = 50 \text{ darcy} = 4.93 \times 10^{-11} \text{ m}^2$$

$$\mu_g = 1.8 \times 10^{-5} \text{ Pa s}$$

$$H = 25 \text{ ft} = 300 \text{ in}$$

$$r_w = 2 \text{ in}$$

$$\ln(2H/r_w) = \ln(600/2) = 5.704$$

Inverting eqn 8.13 and inserting SI units:

$$\Delta p_H = (Q/L)(\mu_g)(\ln(2H/r_w))/(2\pi(k_i))$$

$$\Delta p_H = (0.0155 \text{ m}^2\text{s}^{-1})(1.8 \times 10^{-5} \text{ Pa s})(5.704)/2\pi(4.93 \times 10^{-11} \text{ m}^2)$$

$$\Delta p_H = 5140 \text{ Pa}$$

Conversion to inches of water using Appendix A-7 yields:

$$\Delta p_H = 20.6 \text{ in-H}_2\text{O}$$

REJUVENATING DEPLETED COPPER MINE WASTE DUMPS WITH FORCED AIR VENTILATION

Over long leaching times, copper mine dumps cease to be economic and are abandoned. Nevertheless, the huge inventory of existing copper dumps contain large amounts of copper that cannot be extracted because of the limitations of natural air convection. Furthermore, with much of the original copper sulfides and pyrite still contained in mine waste dumps, they may eventually be classified by EPA as environmental hazards because of the acid mine water generation and drainage that may occur after mine closure.

In principle, it should be possible to rejuvenate old copper leach dumps using forced air ventilation. This would be a secondary recovery method, somewhat analogous with secondary petroleum reservoir extraction processes. An array of evenly spaced air injection boreholes or air wells, drilled vertically from the top of a dump would allow air injection through well casing perforations near the bottom of the dump. This concept is illustrated in Fig. 8.15 where a single injection borehole is shown penetrating the dump.

Air ventilation using this method was partially tested at a large dump during its rest cycle at the Bingham Copper Mine in 1976. Low pressure air was blown through a single borehole into the dump and injected about 50 m below the surface. Air was injected continuously for several weeks while the exhaust gas composition surrounding the injection hole at a depth of about 20 m was monitored from four shallow test wells surrounding the injection hole. The low oxygen content ($\leqslant 5\%$) of the exhaust gas indicated efficient utilization of oxygen in the ventilation air. Recall that bacteria are inactive below about 3% oxygen.

The gas compositions within several other mine waste dumps were measured at the Bingham Mine. The cumulative results indicate that the interiors of very large dumps are oxygen starved and need air ventilation. Large (high) dumps may also be the best targets for efficient utilization of ventilation because the air can be injected at considerable depth with relatively few boreholes. The efficient separation distance of an array of ventilation boreholes will be proportional to the dump height. Shallow

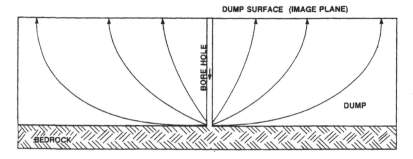

Figure 8.15. Mine waste dump section showing air ventilation through a vertical cased borehole perforated at the bottom.

dumps will require many closely spaced injection boreholes, and this may not be cost-effective.

As discussed, higher permeabilities tend to occur in the lower part of dumps because of the boulder effect resulting from edge dumping with haulage trucks. Consequently, injecting air deep into a mine waste dump will likely favor lateral air spreading and high permeabilities near the bottom of the injection borehole, where it has the greatest effect in allowing adequate air flow throughout the dump at low injection air pressures.

PROBLEMS

1. A 200,000,000 tonne mine waste dump contains 0.2 wt pct copper, all as chalcopyrite. The dump has a pyrite/chalcopyrite ratio of $Py/Cp = 10$. At a maximum oxygen utilization efficiency of 85%, calculate the volume or air required to flow through this dump to oxidize all of the Cp and Py. Express your answer in both cubic meters and cubic miles.

2. The area of the dump in Problem 1 is 3,000,000 m^2 (1,000 acres). What is the average air velocity (superficial velocity) required to permit extraction of 20% of the contained copper after one year of natural convection aeration? Using eqn 8.6, what is the minimum average intrinsic permeability needed in the dump to accommodate this flow rate, at an average temperature rise of (a) 20°C?, (b) 10°C?

3. In Example Problem I of this chapter, ore in the column was allowed to drain, without replenishing water, for two days. Then, the air flow experiment was repeated. At an air flow rate of 15 l/min the measured pressure drop through the drained ore bed was 4.7 in of water. What is the ore column permeability based on this experiment? Why is it different from the result obtained in the example problem, where the permeability was 50 darcy? Which permeability value would you use for designing the test heap, and why?

4. The test heap, of Example Problem II of this chapter, was finally operated, during which pressure measurements were made inside the air pipes. The measured values of Δp_H over the lengthy test period averaged 13 in of water with a variation of $\pm 50\%$. This average value is less than the results predicted using the column data. What is the revised ore test heap permeability in darcy? Do you consider this difference in permeabilities between the column and the test heap unusual? Explain your answer?

5. In scaling up the test heap data from Problem 4 to a commercial ore heap, quadrupling the heap height from 25 to 100 ft is being considered

in order to limit the size and cost of the heap pad. In Plan A, the value of Q_g/L will also increase by a factor of four to about $0.06\,\mathrm{m^2\,s^{-1}}$. Consequently, the pipe inside diameter will be increased from 4 to 8 in to handle the greater air flow. In Plan B, closer spaced 4-in pipes will be used so that Q_g/L remains constant. For each plan calculate the expected heap pressure, Δp_H, in psi (lbs$_f$/in) using your data from Problem 4.

REFERENCES AND SUGGESTED FURTHER READING

Bartlett, R.W. (1990). Aeration pretreatment of low grade refractory gold ores. *Minerals and Metallurgical Processing*, SME, pp. 22–29 (Feb).

Bartlett, R.W. and Prisbrey, K.A. (1995). Diffusion limited aeration during biooxidation of shallow ore heaps. *EPD Congress '95*, TMS, Warrendale, PA, pp. 365–372.

Bartlett, R.W. and Prisbrey, K.A. (1996). Convection and diffusion limited aeration during biooxidation of shallow ore heaps. *International Journal of Mineral Processing 47*, pp. 75–91.

Bartlett, R.W. and Prisbrey, K.A. (1997). Oxygen diffusion into wet ore heaps impeded by water vapor upflow. *Global Exploitation of Heap Leaching Gold Deposits*, D.M. Hausen et al., eds., TMS, Warrendale, PA, pp. 85–94.

Bennett, J.W., Harries, J.R., Pantelis, G. and Ritchie, A.I.M. (1989). Limitations on pyrite oxidation rates in dumps set by air transport mechanisms. *Biohydrometallurgy*, pp. 551–561.

Cathles, L.M. and Apps, J.A. (1975). A model of the dump leaching process that incorporates oxygen balance, heat balance, and air convection. *Met. Trans., 6B*, pp. 617–624.

Cathles, L.M. and Schlitt, W.J. (1980). A model of the dump leaching process that incorporates oxygen balance, heat balance and two dimensional air convection. Schlitt, W.J., ed., *Leaching and Recovering Copper from As-Mined Materials*, AIME, New York, pp. 9–27.

Harries, J.R. and Ritchie, A.I.M. (1985). Pore composition in waste rock dumps undergoing pyritic oxidation. *Soil Science, 140* No. 2, pp. 143–151.

Milne-Thompson, L.A. (1956). *Theoretical Hydrodynamics*, MacMillan, New York.

Pantelis, G. and Ritchie, A.I.M. (1993). Rate controls on the oxidation of heaps of pyritic material imposed by upper temperature limits on the bacterially catalyzed process. *FEMS Microbiology Reviews 11*, pp. 183–189.

Pantelis, G. and Ritchie, A.I.M. (1993). Optimizing oxidation rates in heaps of pyritic material. *Biohydrometallurgical Technologies*, A.E. Torma et al., eds., TMS, Warrendale, PA, pp. 731–738.

Ritchie, A. and Pantelis, G. (1993). Optimization of oxidation rates in dump oxidation of pyrite-gold ores. *Biomine '93*, Australian Mineral Foundation, Glenside, S.A.

Robinson, W.J. (1972). Finger dump preliminaries promise improved copper leaching at Butte. *Mining Engineer, Vol. 24*, pp. 47–49.

NINE

Biooxidation Heap Pretreatment of Sulfide Refractory Gold Ore and High-Sulfur Coal

INTRODUCTION

Depletion of direct leaching gold ores and the discovery of ores that are not amenable to direct cyanide leaching has required development of new treatment methods for these refractory ores. While there are many causes of refractory ores, the two most common and economically significant causes are encapsulation of gold by sulfide minerals (Monroy et al., 1993) and adsorption of the gold cyanide complex formed during leaching by organic carbon present in the ore (Vassilou, 1988). Both of these causes may be encountered in the same ore.

Because hydrothermal deposition of gold in the ore matrix usually occurred concurrent with sulfide mineral deposition, sulfide refractory gold ore deposits are expected to predominate, except in locations where these primary ore deposits have been naturally oxidized following deposition. A good correlation often exists between the water table and the depth of naturally oxidized ore that is amenable to cyanide leaching.

Chemical oxidation of sulfide refractory ore, as a pretreatment prior to cyanide leaching, mimics the natural oxidation process at greatly accelerated rates. Several oxidation pretreatment methods are being practiced. Among these are pressure oxidation of both acidic and basic ore slurries, biooxidation of ore slurries in stirred tanks and roasting. While pressure oxidation is sometimes not very effective on carbonaceous ore, roasting and stronger oxidizing chemicals, such as chlorine and nitric acid can be used on combined sulfide/carbonaceous ores. Froth flotation of the sulfide minerals to produce an ore concentrate, followed by chemical or biochemical oxidation pretreatment of the concentrate, is also practiced when the ore is amenable to flotation. However, the sulfide minerals encapsulating the gold in many refractory ores are often too fine-grained to permit acceptable flotation recoveries.

All of the existing oxidation pretreatment methods are too expensive for large tonnages of low-grade refractory gold mineralization. The need for a low-cost oxidation pretreatment method that will permit economic processing of these refractory sub-ores has created growing interest in biooxidation of crushed ore in large, bacteria-inoculated stockpiles, that are kept wet and aerated over the extended times needed to sufficiently oxidize the sulfide minerals (Bartlett, 1990).

REFRACTORY GOLD ORE SULFIDE MINERALOGY

Typically, pyrite is the major sulfide mineral in refractory gold ores, with arsenopyrite also often present. Gold is associated with these minerals. In gold ores, arsenic enrichment zones often occur in both pyrite and arsenopyrite grains (Monroy et al., 1993). Other sulfide minerals may also be present, but usually to a lesser extent. Usually, the sulfide minerals, ranging up to only a few weight percent at most, are fine grained, often less than $100\,\mu m$, and disseminated. Gold is found as submicron particles encapsulated in pyrite grains. Often there is a positive correlation between invisible gold and arsenic in both pyrite and arsenopyrite, as evidenced by quantitative electron microprobe analysis. It is uncertain whether this invisible gold is chemically combined in an oxidation state or present as gold atoms or clusters located in a defect crystal lattice.

Heap biooxidation of these refractory gold ores is very similar to that of copper ores and mine waste composed of primary mineralization, as discussed in Chapter 6. In both cases, pyrite is the predominant sulfide mineral, and biooxidation of sulfide minerals occurs primarily by the indirect process. *T. ferrooxidans* catalyzes the oxidation of ferrous ions to ferric ions, with the ferric ions oxidizing the sulfide minerals. In some refractory gold ores, gold extractability is proportional to the extent of sulfide biooxidation, while in other ores, incomplete sulfide oxidation will yield proportionally higher gold extractions upon cyanide leaching (Claasen et al., 1993). In the latter case, framboidal or ultrafine pyrite, rather than coarse pyrite is the likely gold host. Arsenopyrite is usually leached more rapidly than pyrite, and arsenic rich zones in both minerals are more rapidly attacked than arsenic lean zones. This selectivity may also account for ores where high gold extractions can be achieved with less than proportionate sulfide mineral oxidation.

Many sulfide refractory ores show partial gold extraction by direct cyanidation without pretreatment. While this division of the gold in an ore between "free" gold and refractory gold should be accounted for in an accurate pretreatment processing materials balance, it has often been ignored.

DEVELOPMENT OF HEAP BIOOXIDATION
OF REFRACTORY GOLD ORE

Biooxidation of ground ore slurries in stirred tanks for pretreatment of refractory gold ores began earlier than heap biooxidation of these ores. Stirred tank bioleaching is a commercial process applied to refractory gold ore concentrates through the BIOX(R) process developed by Gencor (Van Aswegan, 1993). A pilot plant operated in 1984, and several commercial plants have since been commissioned. The BIOX(R) process uses air at atmospheric pressure as the oxidant with *T. ferrooxidans*.

Biotank leaching does not have major cost advantages over pressure oxidation unless the cost of oxygen, needed for pressure leaching, is high. Long residence times, typically four days, are required in staged tanks with internal cooling coils. However, it can avoid technical problems that are encountered when pressure leaching some ore concentrates.

Adapting heap biooxidation from copper ores to sulfide refractory gold ores was a natural progression, especially for low-grade, sub-ore mineralization, where the need for a lower cost pretreatment process is mandatory. Aeration of coarse-crushed ore in wet stockpiles (very large heaps) inoculated with bacteria to oxidize the sulfide minerals was advocated by Bartlett (1990) as an inexpensive pretreatment for both lower grade refractory gold ore and small ore deposits that could not afford the high capital cost of pressure leaching or roasting. The process was described essentially as it is presently being developed, with heap biooxidation followed by cyanide leaching in a separate heap. The process flowsheet is shown in Fig. 9.1. It was concluded that successful biooxidation would require: (1) an ore with small sulfide mineral grains, (2) crushing the ore to about $-19\,mm$ ($-3/4\,in$), (3) ventilating the bacteria inoculated ore heap with air during biooxidation and (4) keeping the heap wet. Even with these restrictions, heap biooxidation times to obtain a high degree of sulfide mineral oxidation would usually be many months, which is significantly longer than the time

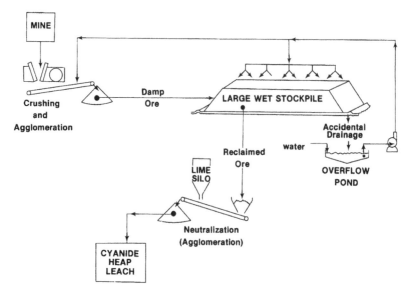

Figure 9.1. Heap biooxidation pretreatment flowsheet for low grade refractory gold ore in bacteria inoculated wet ore stockpile (Bartlett, 1990).

required for cyanide heap extraction of gold from the pretreated ore. For a continuous operation, the biooxidation heaps must be large, with fresh ore being added to the pad as pretreated ore is removed from the pad and sent to cyanide leaching. Natural convection will not likely provide adequate aeration in most cases for such large heaps, and forced air ventilation should be used to ensure an adequate supply of oxygen at all locations within the bioheap. These conclusions are still valid and pertinent.

Newmont Gold Company has been commercializing heap biooxidation of refractory gold ore in North America. An 800,000 ton demonstration bioheap was completed in 1995 at Newmont's Gold Quarry Mine (Brierley et al., 1995). Ore was crushed and agglomerated using a sulfuric acid solution containing a bacteria culture, which was blended into the ore. The ore was placed on the pad by a series of "grasshopper" conveyors ending in a radial stacker. This demonstration bioheap relied on natural convection and diffusion for supplying the needed oxygen. Newmont began a similar test in 1996 using forced air ventilation with underlying air pipes, as described in Chapter 8, to accelerate the process; see Fig. 8.10.

Combined sulfide/carbonaceous refractory gold ores are prevalent in the Carlin Mining District of Nevada and the carbon is not entirely oxidized or passivated by heap biooxidation. Consequently, heap biooxidation of the sulfide minerals to achieve excellent gold liberation will not ensure good gold recovery by cyanidation due to the adsorption of the gold cyanide complex on the carbon. Research into alternative lixiviants for extracting gold after heap biooxidation of sulfide/carbonaceous ore has brought forth a patented process using ammonium thiosulfate as a gold extractant in place of cyanide for leaching these **preg-robbing** refractory ores, but only after pretreatment by heap biooxidation (Newmont Gold Company, 1994).

ORE CRUSHING REQUIREMENTS

It is necessary to crush the ore prior to biooxidation heap pretreatment so that oxidant (ferric ions) diffusion in the rock micropores does not limit the time needed to complete biooxidation of the sulfide minerals, including those at the center of the largest rocks remaining after crushing. If adequate heap aeration is provided, crushing ore to a sufficiently small topsize will shift the biooxidation rate from a diffusion controlled rate, or mixed kinetics, to the intrinsic mineral oxidation rate.

What is a sufficiently small topsize? One approach to answering this is to compute times for nearly complete oxidation, say 95%, if diffusion was the rate controlling process for a variety of ore particle topsizes, and compare this results with the planned biooxidation pretreatment time, for example as determined based on the intrinsic mineral oxidation kinetics.

Diffusion controlled processes, with fast mineral dissolution kinetics, were described in Chapter 5. A shrinking core model of the form of eqn 5.17, but with ferric ions as the diffusant rather than acid, describes the diffusion-controlled biooxidation rate process for a refractory gold ore containing pyrite and arsenopyrite.

The heterogeneous chemical reactions for the oxidation of pyrite and arsenopyrite are similar. Outside the rocks the bacteria-catalyzed oxidation of ferrous ion to ferric ion by dissolved oxygen occurs, as previously shown in eqn 6.2, provided oxygen and nutrient are available:

$$14Fe(SO_4) + 7H_2SO_4 + \frac{7}{2}O_2 \rightarrow Fe_2(SO_4)_3 + 7H_2O. \tag{6.2}$$

While inside solution-filled micropores, pyrite is oxidized by ferric ions, as previously shown in eqn 6.3:

$$7Fe_2(SO_4)_3 + FeS_2 + 8H_2O \rightarrow 15FeSO_4 + 8H_2SO_4, \tag{6.3}$$

and arsenopyrite is similarly oxidized by ferric ions:

$$7Fe_2(SO_4)_3 + FeAsS + 8H_2O \rightarrow 15FeSO_4 + 7H_2SO_4 + H_2AsO_4. \tag{9.1}$$

Because of the identical stoichiometry of the pyrite and arsenopyrite reactions, the combined governing transport equation for the shrinking core model for ore containing both pyrite and arsenopyrite is

$$1 - \frac{2}{3}F - (1-F)^{2/3}$$

$$= 150 \frac{[Fe^{3+}]}{B} \left[\frac{1}{\rho_{ore}[(G_{FeS_2}/MW_{FeS_2}) + (G_{FeAsS}/MW_{FeAsS})]} \right] \frac{\varepsilon t}{r_0^2}. \quad (9.2)$$

Note the similarity with eqn 5.17. Equation 9.2 has been used to compute the required biooxidation time to obtain 95% oxidation ($F = 0.95$) for several cases, each with different typical values of the remaining equation parameters, listed in Table 9.1 (Bartlett, 1996).

However, as already discussed a crushed ore will have a distribution of rock sizes. When the Gates–Gaudin–Schuhmann equation (eqn 2.5) is combined with eqn 9.2, the computed time to obtain 95% conversion for a crushed ore is about 45% of the time required to obtain 95% conversion of the **largest size** in the ore. The corresponding relationships between screen opening and biooxidation time to obtain 95% biooxidation of the pyrite grains in a **crushed ore** are plotted in Fig. 9.2, for the cases of Table 9.1.

The results of Fig. 9.2 explain why refractory gold ore must be crushed, generally to about $-19\,$mm ($-3/4\,$in) and sometimes to $-13\,$mm ($-1/2\,$in), e.g., Case A of Table 9.1, before entering heap biooxidation. After adequate crushing, ferric ions easily penetrate to the center of the largest remaining rocks. While transport by diffusion will be slow at very low ferric ion concentrations, the intrinsic mineral oxidation kinetics will also be similarly retarded at low ferric ion concentrations. Therefore, with crushing to about the rock sizes indicated from Fig. 9.2, diffusion of the ferric ion oxidant in the rock micropores will not be rate limiting.

Table 9.1 Parameters Used in Equations for the Results Plotted in Fig. 9.2.

Parameter	Case A	Case B	Case C	Case D	Case E
ε, Volume Fraction	0.03	0.03	0.05	0.03	0.05
$[Fe^{3+}]$, g/l	2.8	11.2	11.2	11.2	11.2
G_{FeS_2}, Mass Fraction	0.03	0.03	0.01	0.01	0.03

These parameters are typical of refractory disseminated gold ores.

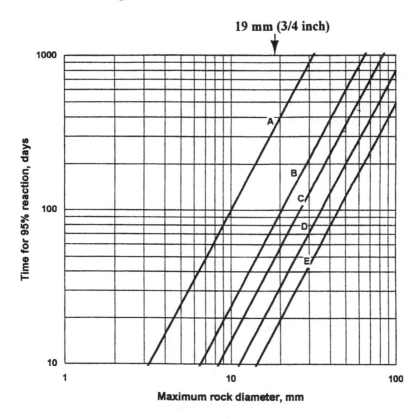

Figure 9.2. Heap biooxidation time required to obtain 95% reaction $(F=0.95)$ for a crushed ore (GGS size distribution) when diffusion of ferric ions within solution-filled rock micropores is controlling the rate of biooxidation. Conditions are given in Table 9.1.

BIOHEAP ACIDITY, BACTERIA CULTURE AND ORE AGGLOMERATION

The net sulfuric acid generated *in situ* by the oxidation of pyrite may be sufficient to prevent ferric ion hydrolysis and allow an adequate ferric ion concentration in the heap. However, acid will be quickly consumed if carbonate minerals are in the ore. The amount of acid needed, if any, to allow a high concentration of ferric ions in the leaching solution must be determined experimentally.

The potential for iron precipitation by hydrolysis within a refractory ore heap undergoing biooxidation means that the iron concentration of the leach solution leaving the heap is not a valid indicator of the extent of sulfide mineral oxidation. An accurate determination of sulfide mineral oxidation requires a total iron and sulfur inventory, including both the solution and the heap solids, before and after biooxidation leaching.

Excessive precipitation of iron salts in the ore may obscure the gold and impede gold extraction during subsequent cyanide leaching. When relatively large amounts of pyrite are present, it may be necessary to treat the bioheap drainage to partially remove iron salts outside the bioheap before the solution is recycled to it.

During the subsequent cyanide leaching, safety from HCN(g) generation and good gold recovery requires that all of the sulfuric acid everywhere in the ore must be neutralized before cyanide leaching begins. It is doubtful that this can be accomplished without reclaiming the ore from the bioheap and thoroughly blending lime or cement into it. Consequently, cyanide leaching in a separate heap is recommended, rather than trying to do so in the biooxidation heap. Ammonium thiosulfate extraction of gold from carbonaceous ores will also be conducted in a leaching heap separate from the biooxidation heap.

T. ferrooxidans adhere to solid surfaces. Because of the vast surface area of an ore heap, these bacteria are slowly mobilized within it. Consequently, merely spraying a bacteria solution onto the top of the heap may delay the penetration of bacteria and the initiation of biooxidation throughout the heap. Blending the ore with a solution containing bacteria, nutrients and acid prior to constructing the heap allows biooxidation to start more quickly. Blending the wet ore can also promote agglomeration of fines. The source of this blending solution will normally be recycled heap drainage containing ferric sulfate and bacteria, as shown in the flowsheet of Fig. 9.1. This is analogous with the "starter" in making sour dough bread. Organic polymeric binders can be included with the blending solution to strengthen the agglomerate and improve heap permeability.

BIOHEAP ENERGY BALANCE AND TEMPERATURE CONTROL

Because *T. ferrooxidans* are ineffective at temperatures above about 45°C, the heap's internal temperature must be constrained. Consider the following examples involving: (1) the heat capacity of a wet ore mass, (2) cooling the heap by passing air through it, (3) cooling by evaporation of water within the heap, and (4) cooling the heap by percolating leaching solution through it with heat transferred to the liquid solution.

For a typical crushed ore with 10 vol pct air void space and 20 vol pct water-filled void space the heap's heat capacity is 560 kcal/(m^3 °C). Based on the pyrite oxidation reaction enthalpy to produce ferrous sulfate, the heat released from the 70 vol pct ore in the heap is 46,000 kcal/m^3 for each 1.0 wt pct of pyrite contained in the ore. Further oxidation of the ferrous sulfate to ferric sulfate only increases the amount of heat generated by four percent. Hydrolysis to goethite or jarosite adds another two percent.

With a temperature rise limited to 20°C to prevent harming the bacteria, the heap's heat capacity can only absorb heat from oxidation of 0.24 wt pct of pyrite. If more pyrite is to be oxidized, then heat must be transferred out of the ore heap.

The heat capacity of air leaving the heap is about 0.312 kcal/(m^3 °C). Much larger amounts of air would be required to remove heat than to supply oxygen for pyrite oxidation. For example, a heap containing 1.0 w/o pyrite, neglecting heat stored in the heap, requires over 7000 m^3 of air for each m^3 of heap for cooling, while the air requirement to supply oxygen for stoichiometric pyrite oxidation is only 12 m^3. Hence, cooling the heap by air flowing slowly through it cannot be an important process.

Except within about one meter from the surface, heat removal by water evaporation will not be significant. The water vapor pressure at 45°C is 9.6 kPa (72 mmHg). With 12 m^3 of air passing through one cubic meter of an ore heap containing 1.0 wt pct pyrite, water evaporation will remove about one percent of the 46,000 kcal/m^3 of heat generated by complete pyrite oxidation. For the same ore at a heap temperature of 55°C water evaporation can remove about two percent of the generated heat.

Most of the heat must be removed by transferring enthalpy to the leaching solution percolating through it. Heat removal depends on the solution flow rate, heat capacity and temperature rise. Consider impact sprinklers with a flow rate constrained to 0.2 m/day and a 20°C temperature rise. At a steady heap temperature, the biooxidation rate cannot exceed the heat removal rate by the draining solution. The calculated times to complete ore biooxidation, when heat removal by percolating solution is the rate limiting process, are shown in Fig. 9.3. The limiting heat removal time is plotted against pyrite concentration in the ore (Bartlett, 1996).

Since the solution flow rate is constrained by the available solution application methods, doubling the heap height doubles the heap biooxidation time. If less than complete oxidation of the pyrite is required to liberate the gold, then shorter biooxidation times than shown would be adequate. These calculations assume that the leaching solution is cooled the required amount, 20°C in the preceding case, prior to being returned to the heap. If recycle solution cooling is less, the time needed to complete biooxidation must be extended proportionately.

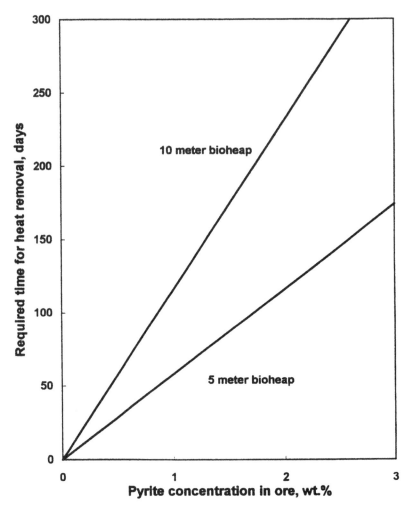

Figure 9.3. Removal of heat by percolating solution (water) from complete oxidation of pyrite; 0.2 m/day flow rate and 20°C solution temperature rise.

Though the maximum rate of heat removal from a bioheap by percolating solutions is constrained by the solution application methods, for heap heights and sulfide mineral concentrations of interest for pretreating refractory gold this is not likely to be rate limiting and cause an extension of the biooxidation pretreatment period over that required by other factors. Either limited air flow into the heap or intrinsic mineral oxidation kinetics will usually limit the rate of heap biooxidation and require leaching periods of several months. The only exception would be during the initial period in a well-aerated heap when the sulfide mineral surface area is highest and the volumetric oxidation rate is also highest. Preventing overheating and bacteria damage by constraining the rate of aeration, when using forced air ventilation, may be necessary during the early part of the biooxidation cycle. Ore heap internal temperature should be monitored and controlled by regulating air flow when using forced air ventilation.

INTRINSIC SULFIDE MINERAL OXIDATION KINETICS

A major objective in designing a heap biooxidation system for a refractory gold ore is to optimize all of the process steps *other* than the sulfide mineral oxidation kinetics so that the intrinsic mineral oxidation kinetics, which cannot be altered, become rate limiting and determine the minimum time required for bioheap pretreatment. This means that aeration must be adequate, and forced air ventilation will be required for large bioheaps, as discussed in the previous chapter. Ore crushing will be required to eliminate micropore diffusion of the reactant ferric ions from limiting the overall rate. Adequate solution percolation will be needed to remove heat and prevent excessive temperature excursions. Blending bacteria culture into the ore, and acid if the ore is basic, will provide a fast biooxidation start.

The intrinsic oxidation of pyrite has been extensively studied, and it is deemed to be an essentially electrochemical process (Zhu et al., 1993). While less attention has been paid to the oxidation of arsenopyrite, its kinetics are believed to be similar or slightly faster than pyrite under equivalent conditions. Kinetic studies of the oxidation of pyrite are not always in agreement and this may be due to different solid-state impurity concentrations affecting the electrochemical properties of pyrite, which is a semiconductor.

A. Useful Theory

For a chemically reacting spherical particle, such as an oxidizing sulfide mineral grain, that does not form a passivating film of solid reaction products, the reaction kinetics are controlled by the surface chemical reaction rate and the rate expression for the mass fraction reacted, converted, or in our case oxidized is given by the following equation (Wadsworth and Miller, 1979),

$$1 - (1 - F_t)^{1/3} = (k_s C_A^n / r_{Py})t \tag{9.3}$$

where C_A^n represents the product of reactant(s) and the reaction orders (n). However, if transport of reactants, e.g., ferric ions, is not rate limiting, then the reactant concentration should be the same, or nearly so, throughout the ore, and eqn 9.3 devolves to:

$$1-(1-F)^{1/3}=(K_s/r_{Py})t \tag{9.4}$$

where K_s includes the surface reaction rate constant and ferric ion concentration, and r_{Py} is the radius of the sulfide mineral grain, as illustrated in Fig. 9.4.

However, when a distribution of sulfide mineral grain sizes exists, as it invariably does in an ore, the smaller grains are completely reacted in earlier times than the larger grains. This causes a graph of the fraction reacted versus leaching time to become a more convex curve than it would be if

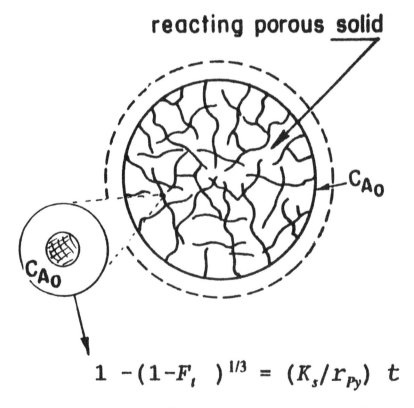

Figure 9.4. Illustration of sulfide mineral grain being oxidized inside an ore particle.

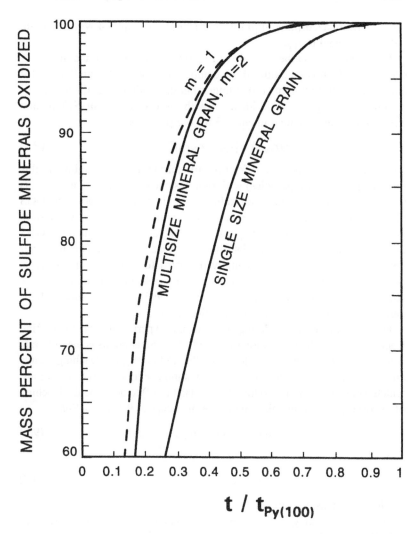

Figure 9.5. Extent of mineral oxidation versus leaching time when the reaction at the mineral surface is rate controlling, eqn 9.4—for monosize and multisize grains with a GGS size distribution.

all of the grains were of the same size. This effect is illustrated in Fig 9.5, which plots the mass percent of sulfide minerals oxidized, F_t, during the period of biooxidation, expressed as the ratio of the biooxidation time, t, to the time that would be required for complete oxidation (conversion) of the largest mineral grain, $t_{Py(100)}$. The solid curve on the right of the figure is for uniformly sized mineral grains and is a plot of eqn 9.4 with the grain radius, r_{Py}, equal to r_{Py*}.

The two curves on the left side of Fig. 9.5 are two plots of the mass percent of sulfide mineral oxidized, F_t, computed according to eqn 9.4, but for a multisize distribution of sulfide mineral grains with their grain size distributions following a relation identical with the Gates–Gaudin–Schuhmann (GGS) equation presented in eqn 2.5 for ore particles.

$$Y(r_{Py(i)}) = [r_{Py(i)}/r_{Py(*)}]^m \tag{9.5}$$

where $Y(r_{Py(i)})$ is the mass fraction of sulfide mineral grains in this instance (and not ore particles) with radii less than $r_{Py(i)}$ and where $r_{Py(*)}$ is the largest sulfide mineral grain radius in the distribution. One of the GGS distributions shown in Fig 9.5 is for $m = 1$, which is the largest value of m normally encountered with ore particles in fractured rocks. The other GGS distribution is for a narrower distribution of sizes at $m = 2$. This narrower distribution has been experimentally observed for sulfide mineral grains contained within two disseminated refractory gold ores containing pyrite with minor arsenopyrite (Harrington et al., 1993). These results will be described in the experimental section that follows.

For convenience in modeling ore leaching experiments and forecasting leaching results when oxidation of the sulfide mineral grains is the only rate controlling process, it is fortuitous that the multisize grain leaching results are nearly matched with the following simple empirical rate equations, which were first used by Taylor and Whelan (1942) to describe the leaching of exposed copper minerals from pyrite ore heaps at Rio Tinto, Spain.

$$\ln(1 - F_t) = -K_m t, \tag{9.6}$$

and

$$F_t = 1 - e^{-K_m t}. \tag{9.7}$$

These expressions are mathematically identical with first order rate equations in homogeneous chemical kinetics, but the heterogeneous sulfide mineral leaching mechanism is entirely different.

Figure 9.6 repeats the correct monosize grain oxidation curve and the correct multisize grain oxidation curve, with the GGS exponent $m = 2$, and adds dashed lines that are plots of the simple expression given by eqn 9.7 arbitrarily matched to the correct expressions at the time, $t/t_{Py(100)} = 0.6$. While a good match cannot be made for the monosize grain curve, the multisize grain dashed curve matches fairly well with the correct expression, certainly within the error of experimental measurements involving leaching in ore columns and ore heaps.

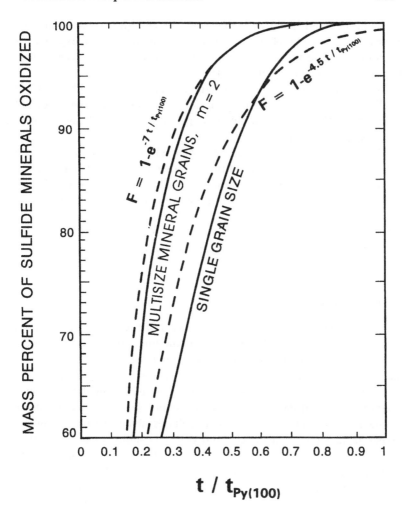

Figure 9.6. Matching the logarithmic rate expression of eqn 9.7 to a surface controlled mineral oxidation reaction for spherical particles with (1) a single mineral grain size and (2) a Gates–Gaudin–Schuhmann multisize distribution of mineral grains.

Generally, during ore testing and evaluation for heap leaching, the sulfide mineral grain size distribution and other parametric information needed for a deterministic leaching simulation will not be available. Also, a modest computer mass balancing program is required to accurately utilize eqns 9.4 and 9.5. When these are not available, then eqn 9.7, is recommended and likely to be adequate. An empirical value of the rate constant, K_m, must be obtained, either from column leaching experiments or from operating results with prior ore heaps to obtain F_t versus t data that can be used with eqn 9.7 to extract K_m.

Once the rate constant, K_m, is determined it can be used with eqn 9.7 to predict results at longer leaching times by extrapolation. Sometimes full gold extractability can occur with only partial mineral oxidation. In these cases, $K_{Au} > K_m$ in eqn 9.7.

The preceding discussion of heap biooxidation, when the only rate determining mechanism is the intrinsic sulfide mineral oxidation kinetics, has focused on refractory gold ores. Nevertheless, it should be clear that it also pertains to heap biooxidation of copper and other sulfide ores, especially where primary sulfides are dominant, provided the ore is adequately crushed and the heap is well aerated. *Equations 9.6 and 9.7 can be employed as useful approximations for design scale-up in such cases.*

B. Experimental

Experimental biooxidation rate measurements of two refractory gold ores, from mines near Carlin, Nevada and Yellow Pine (Stibnite), Idaho, were made under conditions where the biooxidation rates were controlled by the intrinsic pyrite mineral oxidation rate (Harrington et al., 1993). Both ores were sulfidic but not carbonaceous. The experiments involved thorough characterization of the ore and multiple column leaching experiments on crushed ore, inoculated with bacteria cultures obtained from the mines. Experiments were carried over a full year, and sulfide mineral oxidation rate data were correlated with subsequent gold cyanide extractions. The sulfide mineral grain size distribution for both ores was determined by photomicroscopy yielding $d_{80} = 40 \, \mu m$ for the Carlin ore sample, which contained 3.8 wt pct pyrite, and $d_{80} = 100 \, \mu m$ for the Stibnite ore sample, which contained 3.0 wt pct pyrite. Arsenopyrite was also accounted for, though much less was present in both ores.

Many replicate samples were experimentally evaluated for each well-blended ore. Drained leachate was recycled daily, and Eh, pH, $[Fe^{2+}]$ and $[Fe^{3+}]$ were monitored. At periodic intervals, an entire ore column was emptied for residue analysis followed by cyanide leaching to determine gold extractions. The extent of sulfide mineral oxidation was determined from a materials balance on both the solution and the solids. Results may be inaccurate when relying only on solution assays because variable amounts of released iron precipitate and remain in the column residue.

The initial concentrations of both sulfide iron and oxidized iron in the ore samples were determined.

Because not all of the gold in each ore was refractory, the refractory sulfide gold extractability was determined by subtracting the previously measured free gold cyanide extractability from the total measured extractability, obtained by cyanide leaching ground column residue samples after various periods of biooxidation.

Because both ores were predominantly pyritic, their experimentally determined kinetics were compared with a simulation model developed for pyrite oxidation. Both ores gave excellent correlation over time with the extent of pyrite oxidation, refractory gold extraction and the simulation model, which was based on the surface chemical reaction between ferric ions and the shrinking pyrite grains using the intrinsic pyrite oxidation kinetics independently determined earlier (Zheng et al., 1986). For details of the simulation model, see Harrington et al. (1993).

Results from the Carlin, Nevada ore are plotted in Fig. 9.7. The two outer lines on the plot represent the extraction predicted by the simulation for the largest and smallest pyrite grains in the ore. The central, bolder line is the mass-weighted simulation for all of the pyrite grain sizes; hence it predicts the aggregate conversion of sulfide minerals by biooxidation for

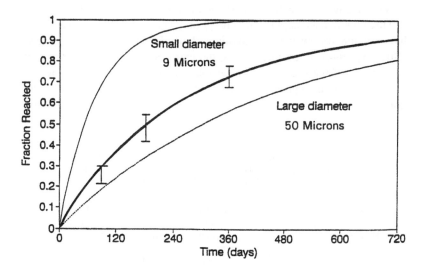

Figure 9.7. Aggregate sulfide oxidation (middle line) and cyanide extraction of refractory gold (bars) for the Carlin ore.

this Nevada ore sample at the times shown. The aggregate biooxidation curve was determined by the simulation model using only the measured ore characteristics and the published pyrite oxidation rate data, i.e., there were no adjustable parameters. Experimental sulfide oxidation, based on residue analysis at 90, 180 and 365 days, was in good agreement with the aggregate simulation curve of Fig. 9.7.

Six cyanide gold extraction tests were conducted for each of three aggregated ore residues that were removed from columns after 90, 180 and 365 days of biooxidation. The results, after subtraction of the free gold, are also displayed on Fig. 9.7 as three error bars of ± 1 standard deviation, for the six experimental refractory gold fractional extractions at each biooxidation period. The mineral oxidation rate predicted by the simulation model agrees with both the experimental results for pyrite biooxidation and the refractory gold extractions.

Two pilot heap experiments were conducted by Newmont Gold Company on noncarbonaceous sulfidic refractory from the Carlin district (Brierley, 1994). Conditions are summarized in Table 9.2. The resulting cyanide extractions of refractory gold, obtained after deducting the free gold or direct cyanide extractability, correlate very well with the aggregate sulfide oxidation and gold extraction laboratory experimental/simulation results of Harrington et al. (1993). This correlation is shown in Fig. 9.8. These pilot heaps were apparently small enough for natural convection to provide adequate oxygen transport and biooxidation throughout the heap.

Three carbonaceous/sulfidic ore pilot heaps were also subjected to biooxidation followed by ammonium thiosulfate extraction of the gold (Brierley, 1994). In all three cases, the extractions of the refractory gold component of the ores were greater than the aggregate curve of Fig. 9.7 but less than the upper curve predicting extraction from small pyrite grains. A possible explanation of these higher extractions, in addition to a

Table 9.2 Summary of Non-Carbonaceous Refractory Gold Ore Pilot Heap Biooxidation Tests Performed by Newmont Gold Company (Brierley, 1994).

Heap Size Tonnes	Leach Days	Lixiviant	Untreated Gold Extraction	Total Gold Extraction	Refractory Gold Extraction*
432	52	CN	14%	30%	19%
635	185	CN	4%	50%	48%

*Computed by R.W. Bartlett.

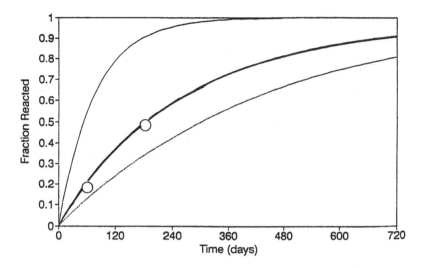

Figure 9.8. Cyanide gold extractions from two Carlin ore pilot heaps compared with laboratory experiment/simulation results shown in Fig. 9.4.

smaller average mineral grain size, is that while ammonium thiosulfate leaches gold, it also continues to oxidize pyrite and thereby liberates more gold for extraction (Logan, 1996).

ECONOMIC CONSIDERATIONS

With low bulk mining costs, favorable economies of scale and good ultimate gold recovery, sulfide refractory gold ores containing 1.0 g/tonne of gold, or less, can be profitably pretreated by biooxidation in ore heaps followed by cyanide leaching.

At 1.0 g/tonne and a gold price of $330/oz, the metal value in the ore is over $10/tonne. Contrast this with the metal value in typical copper mine waste at 0.2% and $1.00 per lb of copper, which is only $4/tonne. While the oxidation of pyrite is the dominant oxygen consuming process in both ores, the low copper waste value precludes additional ore preparation and treatment steps, including ore crushing, that will accelerate the leaching process and improve metal recovery. Ore crushing in three stages *with forced air ventilation* will be used in biooxidation heap pretreatment of approximately 10,000,000 tonnes per year of ore at the Gold Quarry Mine near Carlin, Nevada (Newmont Gold Company).

A. Optimum Bioheap Pretreatment Time

When heap biooxidation pretreatment of a refractory ore is being planned, the biooxidation time must be determined. Revenues are increased by longer biooxidation pretreatment of the ore, but revenues from *all* of the extracted gold are delayed, which has an associated cost. While the incremental operating cost is negligibly small, the capital cost increases because a larger pad and metal recovery plant are needed as biooxidation time is extended. Biooxidation pretreatment should not be continued longer than the break even time, t_{max}, where incremental gold revenues equal the incremental costs.

F_{max}, the biooxidation fractional conversion of sulfides in the ore at t_{max}, is related to the following revenue and cost factors that are typically expressed in dollars per tonne of ore per day;

Revenue generation factor:

$$C_R = G_{Au}^r P_{Au} K_m.$$

Cost of delaying revenue from additional refractory gold:

$$C_{lr} = G_{Au}^r P_{Au} \left[\frac{i}{365} \right].$$

Cost of delaying revenue from oxidized gold that may be present in the ore and directly CN leachable:

$$C_{lo} = G_{Au}^{CN} P_{Au} \left[\frac{i}{365} \right].$$

Daily interest cost on the bioheap capital investment:

$$C_c = a(T_H)^{-b} \left[\frac{i}{365} \right],$$

where G_{Au}^r and G_{Au}^{CN} are the accessible refractory gold ore grade and oxidized (readily CN leachable) gold ore grade, respectively. Both grades must be determined experimentally and not from total assayed gold. P_{Au} is the gold price, i is the firm's annual interest cost expressed as a fraction, and $a(T_H)^{-b}$ is the bioheap capital cost per annual tonne of ore.

When costs and revenues are equal, the following algebraic relation is met (Bartlett, 1995):

$$C_R(1 - F_{max}) - C_{lr}F_{max} - C_{lo} - C_c = 0. \qquad (9.8)$$

Solving for F_{max} from eqn 9.8 yields:

$$F_{max} = \frac{C_R - C_{lo} - C_c}{C_R + C_{lr}}, \tag{9.9}$$

and the break even time, t_{max}, is given by substituting F_{max} into eqn 9.6, which yields:

$$t_{max} = \frac{-\ln(1 - F_{max})}{K_m}. \tag{9.10}$$

B. When to Bioheap Pretreat Mixed Oxide/Sulfide Gold Ores

A block of mine ore will often contain both free gold, which can be recovered directly by cyanide leaching, and sulfide refractory gold. Even when a bioheap pretreatment facility exists at the mine, it may not be worth the expense and delay in obtaining revenue from gold sales to bioheap pretreat a particular block of mixed oxide/sulfide ore. Based on drill hole samples and assays, each ore block should be classified into one of the following three mine products: (1) mine waste to be discarded, (2) refractory ore requiring bioheap pretreatment, or (3) direct cyanide leaching ore. The option that maximizes the net present value of the ore when it is provided by the mine should be chosen. While a three product selection algorithm is a little more complicated than the usual waste/ore decision, based on a cut-off metal grade in the ore, a useful algebraic formula has been developed (Bartlett, 1997). Pretreatment should be bypassed if the resulting grade of gold extractable by direct cyanide leaching, G_{Au}^{CN}, exceeds a critical gold grade, G_{Au}^*; otherwise, the ore should be sent to heap biooxidation pretreatment provided that the grade exceeds the waste cutoff grade, G_{Au}^o. The critical gold grade, G_{Au}^*, is defined by the following relationship,

$$G_{Au}^* = \frac{\left[Y G_{Au}^T - G_{Au}^o - \dfrac{C_{biox}}{P_{Au}} \right]}{(1+i)^n} + G_{Au}^o \tag{9.11}$$

where Y is the combined yield of gold from both refractory ore and oxide ore after biooxidation and cyanide leaching. G_{Au}^T is the total gold grade in the ore and C_{biox} is the biooxidation pretreatment operating cost. The financial discount rate, i, and the time of pretreatment, n, which must be expressed in consistent units generally per year, are included in the derivation. All of these parameters will be known or are readily measurable from a representative sample of the ore block. G_{Au}^{CN} can be determined from a short duration laboratory test.

While the additional extractable gold resulting from biooxidation pretreatment followed by cyanide leaching takes months to determine, e.g., in a lengthy column ore test, the yield of gold, Y, will usually be fairly predictable from previous ore column leaching tests or from operating experience at the mine.

HEAP BIOOXIDATION PROCESSING OF
OTHER METAL SULFIDE ORES

While not yet of significant industrial application, ores containing several sulfide minerals other than those of copper and iron are amenable to biooxidation in ore heaps. These include the sulfides of lead, zinc, cobalt, nickel and molybdenum. Sulfide minerals are the most important source of each of these metals. At the present time these minerals are concentrated from their ores by flotation, and the concentrates are processed further to extract the metals by a variety of hydrometallurgical and pyrometallurgical methods. Laboratory biooxidation leaching studies of each of these metal sulfides have been made, and Rossi (1990) has provided an extensive review of this biooxidation leaching literature.

If, and when developed, heap biooxidation of ores containing the sulfide minerals of these other metals is expected to be very similar to that being developed for heap biooxidation pretreatment of refractory gold ores. Adequate ore preparation, including crushing and agglomeration with bacteria inoculated lixiviant, and forced air ventilation of large ore heaps over an extended period of time will be required. In some cases biooxidation and leach extraction of the valued metal will occur concurrently, while in other cases heap biooxidation will be a pretreatment to biochemically liberate the metal, similar to what is being done with refractory gold ore. Arsenopyrite and stibnite, though of little commercial value, will also be biooxidized if present in these ores and the solubilization of arsenic and antimony presents environmental problems.

DESULFURIZATION OF COAL

Coals invariably contain some sulfur, but the range of sulfur concentration is wide, from less than one wt pct to greater than five wt pct. When burned, this sulfur is oxidized to SO_2 and SO_3, which are emitted into the atmosphere. Electric power plants burning coal are the largest source of SO_x

air pollution in the United States and in several other industrialized countries. Reduction in SO_x emissions under the Clean Air Act is required in the United States. The current *new* source standard sets a sulfur emissions limit of $0.80\,mg/m^3$, with a limit of $0.365\,mg/m^3$ averaged over any 24 hr period, which means that scrubbers must be included in the power plant. The current requirements are less stringent for existing coal burning power plants, but they are equivalent to a coal limit of about 0.8 wt pct sulfur. This limit has placed a premium on burning oil and natural gas and switching to low sulfur coal (less than 1 wt pct sulfur). In the United States, low sulfur coals are located primarily in the western states with relatively sparse population and low demand for thermoelectric power. The coal mined in Ohio, Illinois, Pennsylvania and West Virginia is nearer to the largest concentration of population and demand for electricity, but very little of this coal contains less than 1 wt pct sulfur. Consequently, research on processes to remove sulfur from this more accessible high-sulfur coal has been emphasized.

Sulfur occurs in coal principally in two forms, both of which are usually present in significant amounts. Organic sulfur consists of sulfur atoms covalently bonded within the coal matrix. While the amount may vary somewhat in different macerals of a given coal, organic sulfur is highly disseminated and it cannot be extracted from the coal by flotation or any other physical separation method normally used in mineral processing. Furthermore, organic sulfur is not readily accessible to chemical or biochemical destruction without a treatment so severe that it amounts to coal liquefaction.

Sulfur is also present in coal as sulfide mineral grains, principally pyrite. Other iron sulfide minerals such as maracsite and pyrrhotite are also found. Both forms of sulfur in coal are illustrated in a schematic drawing of a cross-section of a single coal particle in Fig. 9.9.

HEAP BIOOXIDATION OF PYRITE IN COAL

Biooxidation of pyrite and other sulfide minerals in coal is very similar to biooxidation of pyrite in ores, and the factors discussed previously, beginning with Chapter 6, remain equally important. Pyrite oxidation generates sulfuric acid and ferrous sulfate. The ferrous sulfate is biocatalytically oxidized by oxygen in air to ferric sulfate. Availability of air, coal microporosity and diffusion of ferric ions into micropores to oxidize the sulfide minerals are important to biooxidation. Just as with ore, crushing coal to obtain an adequately small coal particle size will be necessary to prevent diffusion of ferric ions into coal micropores from being rate limiting.

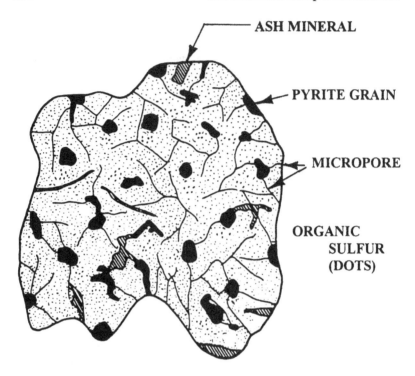

ASH MINERAL

PYRITE GRAIN

MICROPORE

ORGANIC
    SULFUR
    (DOTS)

Figure 9.9. Schematic drawing of a coal particle, after Rossi (1990).

The pyrite grain size is also important and can vary over a wide range. Depending on the particular coal, pyrite may occur as frambroids that are smaller than 1 μm, moderately sized discrete crystalline grains, long veinlets of width varying from a few microns to many millimeters, and all of the above. Framboidal pyrite and smaller pyrite grains can be biooxidized in acceptable time periods, if accessible to the ferric sulfate solution, but larger veinlets cannot. This is one of the major differences between some coals and disseminated gold and copper ores where the pyrite grain size is rarely very large. It also explains why some coals are amenable to a high fractional extraction of pyrite by biooxidation, greater than 90%, while others yield less than 10% biooxidation of the pyrite.

T. ferrooxidans was first discovered in conjunction with a study of natural (bio)oxidation of coal in a Pennsylvania mine. Biooxiation as a method to desulfur coal has been extensively studied and there are several reviews, for example Rossi (1990). Process engineering studies have been aimed at continuous processing of coal slurries. These studies have focused on agitation leaching in stirred tank reactors or in Pachuca tanks, which are reactors that are air sparged to keep the solids in suspension. The coal slurry flows through a stirred tank reactor followed by a plug-flow reactor. This sequence provides biomass propagation to the fresh coals in the first reactor with an adequate residence time in the second reactor (Huber, 1983).

Just as with ores, biooxidation of crushed coal in large heaps should provide lower per ton unit processing costs than slurry biooxidation, but this approach has received little attention. Furthermore, because of the essential needs for an uninterrupted supply of electricity by industry, commerce and homes, large stockpiles of heaped coal are usually kept at power plants and coal shipping terminals, with an average storage residence time of many months. Biooxidation of large coal heaps would be relatively easy to do utilizing these coal stockpiles.

The need for cost-effective coal cleaning processes is critical, and a very large research effort has been underway for several years, funded by the Federal government. Generally, results with methods beyond coal washing have been disappointing and newer processes including biooxidation have not been commercially adopted. There are at least three important causes preventing adoption. First, many high sulfur coals contain appreciable organic sulfur and removal of all of the mineral sulfur will often not reduce the sulfur sufficiently below 1 wt pct. However, acceptable environmental performance can be obtained by blending partially desulfurized coal with coal from low sulfur coal mines. Second, in the United States thus far it has often been less expensive to substitute low sulfur coal purchased at a higher price, rather than pay to clean high sulfur coal. Third, the substantially lower cost of heap biooxidation of large coal stockpiles, compared with coal slurry biooxidation, has not been recognized. Heap biooxidation of refractory gold ore is being commercialized. Perhaps heap biooxidation to desulfurize pyritic coal will be next.

PROBLEMS

1. Crushed ore ($-1/2$ in) was bioleached in a column for 90 days with an analysis of the leached residue indicating that 30 wt pct of the sulfide minerals were oxidized. Using eqn 9.6, calculate the empirical rate constant, K_m, in units of day^{-1}. Estimate the mass fraction, F_t, of the sulfide mineral grain oxidized in (1) six months and (2) in one year. How long must this be biooxidized to obtain 85 wt pct oxidation of the sulfide minerals?

2. Compute the maximum economic fractional extraction, F_{max}, and bioheap pretreatment time, t_{max}, for a refractory gold ore at a gold price of \$375/oz.

This gold ore has the following heap biooxidation processing parameters. The company interest rate is 12% ($i = 0.12$) and the capital cost of the bio-heap is \$4.00 per annual tonne of ore. G_{Au}^r is 0.07 oz/tonne of ore and G_{Au}^{CN} is 0.01 oz/tonne of ore. The empirical rate constant, K_m, is 0.005 d^{-1}.

3. Using parameters and results from the previous problem estimate the combined gold yield, Y, that will likely be obtained if the ore is given a heap biooxidation pretreatment for one year. The total grade of gold in the ore is expected to average 0.08 oz/tonne and the economic cutoff grade, G_{Au}^o, is 0.01 oz/tonne.

Using your value of Y, derive a simple linear expression (with only two terms) to show the variation in critical gold grade, G_{Au}^*, in terms of the total gold grade, G_{Au}^T, such that biooxidation can be bypassed if the directly leachable gold extraction, G_{Au}^{CN}, exceeds this critical gold grade. The heap biooxidation operating cost is \$6.00/tonne of ore. Next, draw and present a plot of G_{Au}^* versus G_{Au}^T on graph paper over the G_{Au}^T range from 0 to 0.10 oz/tonne.

REFERENCES AND SUGGESTED FURTHER READING

Bartlett, R.W. (1997). Economic criteria for choosing bioheap pretreatment of mixed oxide/refractory gold ore. *Global Exploitation of Heap Leaching Gold Ores*, Hausen et al., Eds., TMS, Warrendale, PA, pp. 95–103.

Bartlett, R.W. (1990). Aeration pretreatment of low grade refractory gold ores. *Minerals and Metallurgical Processing*, SME, Littleton, CO, pp. 22–29.

Bartlett, R.W. (1995). Optimum duration for bioheap pretreatment of refractory sulfidic gold ores. *Proceedings, XIX Int'l Mineral Processing Congress Vol. 2*, SME, Littleton, CO, pp. 197–201.

Bartlett, R.W. (1996). Biooxidation pretreatment of sulfide refractory ore. *Min. Proc. Ext. Met. Rev.*, 16, No. 2, pp. 89–124.

Brierley, J.A., Wan, R.Y., Hill, D.L. and Logan, T.C. (1995). Biooxidation heap pretreatment technology for processing lower grade refractory gold ores. *International Biohydrometallurgy Symposium Proceedings*, Chile.

Brierley, J.A. (1994). Biooxidation heap technology for pretreatment of refractory sulfidic gold ore. Australian Mineral Foundation *Biomine '94*, Perth, Western Australia, pp. 8.1–8.7.

Classen, R., Logan, C.T. and Synman, C.P. (1993). Biooxidation of refractory gold-bearing arsenopyrite ores. *Biohydrometallurgical Technologies*, TMS, Warrendale, PA, pp. 479–488.

Harrington, J.G., Bartlett, R.W. and Prisbrey, K.A. (1993). Kinetics of biooxidation of coarse refractory gold ores. *Hydrometallurgy Fundamentals, Technology and Innovation*, SME, Littleton, CO, pp. 691–708.

Huber, T.F., Kossen, N.W.F., Bos, P. and Kuenen, J.G. (1983). Modelling, design, and scale-up of a reactor for microbial desulphurization of coal. *Progress in Biohydrometallurgy*, Rossi, G. et al., Eds., p. 279.

Logan, T.C., Newmont Gold Company (1996). Private communication.

Monroy, M., Marion, P., Berthelin and Videau, G. (1993). Heap-bioleaching of simulated refractory sulfide gold ores by *Thiobacillus ferrooxidans*: a laboratory approach on the influence of mineralogy. *Biohydrometallurgical technologies*, TMS, Warrendale, PA, pp. 489–498.

Newmont Gold Company (1994). U.S. Patent 5, 332, 559.

Rossi, G. (1990). *Biohydrometallurgy*, McGraw-Hill, New York, NY, Chap. 5.

Taylor, J.H. and Whelan, P.F. (1942). The leaching of cuprous pyrites and precipitation of copper at Rio Tinto, Spain. *Bull. of Inst. of Mining & Metallurgy*, November, No. 457, pp. 9–11.

van Aswegen, P.C. (1993). Commissioning and operation of biooxidation plants for the treatment of refractory gold ores. *Hydrometallurgy Fundamentals Technology and Innovation*, SME, Littleton, CO, pp. 707–725.

Vassilou, A.H. (1998). Gold in disseminated carbonaceous ores. *Journal of Metals*, 40, pp. 26–28.

Wadsworth, M.E. and Miller J.D. (1979). Hydrometallurgical processes. Sohn, H.Y. and Wadsworth, M.E. Eds., *Rate Processes of Extractive Metallurgy*, Plenum Press, New York, pp. 133–244.

Zheng, C.Q., Allen, C.C. and Bautista, R.G. (1986). Kinetic study of the oxidation of pyrite in aqueous ferric sulfate. *Industrial Engineering Chemistry Process Design Development*, 25, pp. 308–331.

Zhu, X., Li, J. and Wadsworth, M.E. (1993). Kinetics of the transpassive oxidation of pyrite. *EPD CONGRESS 93*, J.P. Hager, Ed., TMS, Warrendale, PA. pp. 355–368.

TEN

In Situ *Flooded Leaching*

ORE DEPOSIT TYPES

In situ flooded leaching is used to extract nonferrous metals, particularly uranium and copper; extract soluble salts including halite, trona, potash, boron and magnesium minerals; extract molten sulfur by the Frasch process; and construct underground cavities in salt deposits. These cavities are used for petroleum reserve storage, nuclear waste storage, and other underground purposes needing a water-tight environment. Most of the geological targets are either massive deposits, such as porphyry copper deposits and salt domes, or tabular deposits. Important examples of the latter are uranium roll front deposits and evaporites. Most commercially leached evaporites are nearly horizontal.

In situ solution mining has potential for use in extracting several other nonferrous metals including manganese, molybdenum, zinc, cobalt, nickel and vanadium. This chapter is oriented toward *in situ* leaching of metalliferous ore deposits. Brines and evaporites will be treated in Chapter 13. *In situ* solution mining is reviewed by Richner et al. (1992), and it has been covered in several symposium proceedings: Aplan et al. (1974); Schlitt and Hiskey (1981); Davis and Mays (1978) and Schlitt and Shock (1979).

Figure 10.1 illustrates three general conditions encountered in shallow deposits below the water table. The first is an artificially depressed water table produced by pumping, with or without using grouting to deter water inflow. This deposit, if sufficiently rubblized, may be percolation leached. The second case has been rubblized by partial mining or fracturing by using

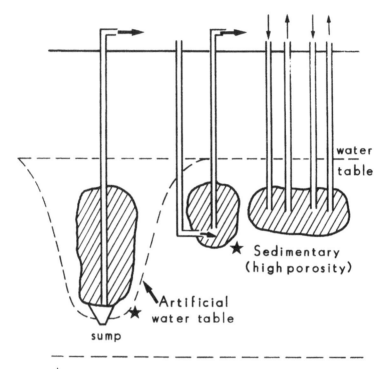

water
table

Sedimentary
(high porosity)

Artificial
water table

sump

*Rubblized (Explosives or Mining) Hydrofracted,
or Chemically Induced Porosity*

Figure 10.1.   Examples of leaching shallow ore deposits below the water table.

either explosives, hydraulic pressure (hydrofracturing), or chemicals. Leaching solutions are being injected through a well into the bottom of the deposit and rise upward. The third example has sufficient natural microporosity to permit solution flow horizontally from injection wells to producer wells.

Wells drilled from the surface are usually laid out in a plan grid pattern of alternate injection and production wells, as illustrated in Fig. 10.2, for uranium recovery from a permeable horizontal strata. This method is well suited to permeable deposits bounded by impermeable ground above and below the deposit that confine the solutions in it. For sedimentary deposits, including evaporites, the confining beds are often impermeable. Well casings are perforated in the ore bearing region so that flow is approximately parallel to the strata. Using the most common well pattern, solution flow radiates from injection wells and is radially received at production wells.

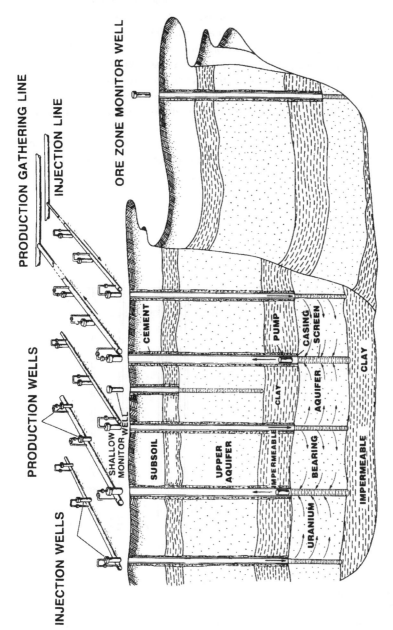

Figure 10.2. Wellfield for uranium solution mining from a confined horizontal ore deposit.

Stratabound permeable ore deposits have much in common with confined aquifers and petroleum reservoirs. The latter often exist because the
confining overlying stratum traps oil and gas. Hence, much of the solution
mining technology for these strataform deposits and evaporites has come
from hydrology and, particularly, petroleum reservoir engineering. Also,
the energy companies have been active as uranium producers, especially in
south Texas where much of the solution mined uranium has been produced. These Texas uranium deposits were discovered by logging oil and
gas exploration drill holes.

Deep deposits, well below the water table, may not be accessible by
conventional mining. Solution mining a deep primary (hypogene) sulfide
ore deposit is illustrated in Fig. 10.3. Investigators at Lawrence Livermore

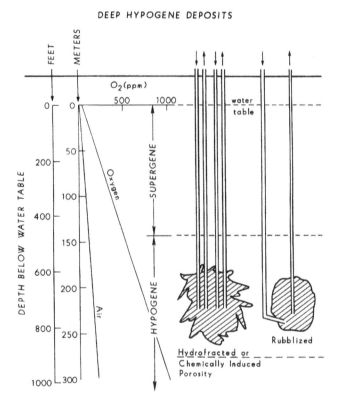

Figure 10.3. Solution mining of deep deposits showing increased oxygen solubility with depth below the water table.

Laboratory proposed the use of nuclear devices for springing and block caving deep primary copper deposits to obtain rubblization and adequate solution permeability, Aplan et al. (1974). A major advantage of flooded leaching of deep sulfide deposits is the greatly increased solubility of oxygen caused by the large hydrostatic pressure. At atmospheric pressure, water in equilibrium with air contains about 7 ppm, and for equilibrium with commercially pure oxygen the solubility is about 35 ppm. Because of increased pressure, the amount of dissolved oxygen increases rapidly with depth below the water table as shown in Fig. 10.3.

MODIFIED *IN SITU* (FLOODED) LEACHING OF FRAGMENTED COPPER DEPOSITS

Because the rate of leaching copper sulfide minerals is proportional to the dissolved oxygen concentration, leaching can proceed rapidly at significant depths below the water table with oxygen as the major reactant, which differs distinctly from mine waste dump leaching. Because the surrounding earth is a good thermal insulator, the leaching process behaves nearly adiabatically. Temperature rises gradually in the leaching zone after mineral oxidation begins, often to above 100°C. The increased temperature also accelerates the leaching rate, which can become very rapid compared with that encountered in mine waste leaching.

These general conclusions were verified for copper porphyry ore (protore containing 0.7% copper mostly as chalcopyrite with $Py/Cp = 2$) in a large scale experiment conducted at Lawrence Livermore Laboratory, Braun et al. (1974). A 5.8 tonne sample of ore was leached in a large flooded pressure vessel for nearly two years. The rock fragment size ranged from 0.1 to 160 mm in diameter, representing ore located above an underground nuclear explosion, with subsequent caving into the nuclear cavity to produce a "chimney" of fragmented rock. The fragmentation is very similar to that formed by block caving in conventional underground mines. Leaching was conducted by direct sparging of oxygen into the bottom of the pressure vessel at 2.76 MPa (400 psi) with the temperature maintained at 90°C.

The resulting extraction curve obtained over a leaching period of nearly two years is shown in Fig. 10.4. At the elevated pressures, the dissolved oxygen concentration greatly exceeded the ferric ion concentration and direct sulfide mineral oxidation by dissolved oxygen was the dominant chemical process. The linear portion of the extraction curve after 200 days was attributed to weathering of the larger ore rocks over time, giving them a progressively decreasing effective diameter.

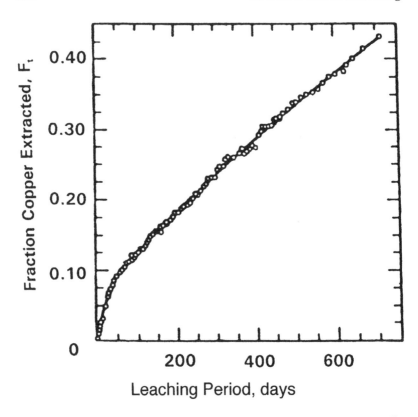

Figure 10.4. Extraction of copper from coarse sized primary sulfide ore at 90°C and 2.76 MPa oxygen pressure (Braun et al., 1974).

Although sulfuric acid was produced from the oxidation of pyrite, the solution pH never dropped below pH = 1.5, and because of the elevated temperature, iron hydrolyzed as hematite, rather than jarosite. Other reaction products obtained in the Lawrence Livermore Laboratory experiment included elemental sulfur and basic iron sulfate. Oxygen consumption was 7.9 moles for each mole of copper dissolved (4 kg-O_2/kg-Cu). An oxygen materials balance indicates that approximately one-half of the pyrite was oxidized and hydrolyzed. This is slightly more than one mole of pyrite per mole of copper. Nevertheless, the reaction zone morphology is very similar to that observed for percolation leaching of sulfide copper ores in dumps, and a similar rock leaching morphological model (shown in Fig. 10.5), is used.

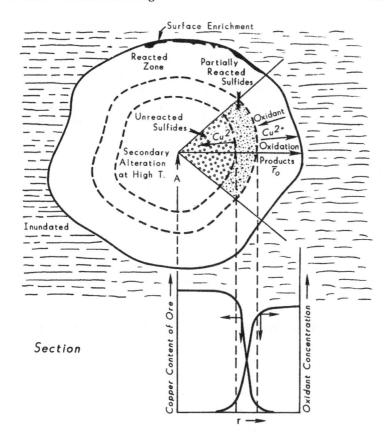

Figure 10.5. Model for flooded leaching of sulfide ore rock.

TRUE *IN SITU* (FLOODED) LEACHING OF
COPPER ORE DEPOSITS

A few true *in situ* leaching copper projects conducted below the water table have been field evaluated. These projects, conducted without fragmentation of the rocks, have depended on natural fractures in the ore deposit for both permeability and adequate sweep efficiency. A major program was conducted by Kennecott on a deep primary sulfide deposit near Safford, Arizona in the 1970s. An ammonia/ammonium sulfate lixiviant with emulsified oxygen was selected to leach chalcopyrite. Very high solution grades were needed for project profitability. Because of the low permeability of the ore deposit the amount of solution that could be transported through the deposit from injection wells to production wells was limited.

Many difficulties were encountered, including the capture of copper by ion exchange with zeolites and clays in the ore. Although only a few percent of zeolite is present in the ore mass, it was a very large chemical sink that in effect had to be charged with dissolved copper before metal could be recovered from a production well. An alternative explanation for the missing copper is secondary enrichment of chalcopyrite, caused by reaction with 'dissolved copper as the solution became depleted in oxygen while traveling from the injection well to the production well. A substantial amount of copper oxidation and secondary enrichment may be required before much soluble copper appears at the production well.

Another major *in situ* leaching project, being conducted at the Santa Cruz site west of Casa Grande, Arizona, involves an attempt to extract copper from a copper chloride salt horizon (the mineral atacamite, $CuCl_2 \cdot 3Cu(OH)_2$), and a deeper, predominantly chrysocolla horizon. These minerals are acid soluble. The deposit is located 365–915 m below the surface, and at a considerable depth below the 180 m deep water table (O'Neil, 1991). The Santa Cruz project has been directed and funded primarily by the former Bureau of Mines, but industry is also participating. Pumping tests on wells drilled into the deposit have been conducted along with a tracer test using a sodium chloride/sodium bromide solution injected into the center well of a five spot well test configuration. The planned development steps for the Santa Cruz project were:

(1) Characterize geology and hydrogeology with three test wells and one monitor well; water pumping tests.
(2) Install two additional wells to complete the five spot pattern; drill three monitor wells; tracer test; apply for aquifer protection permit.
(3) Obtain permits and construct SX/EW plant.
(4) Conduct acid leach test in the five spot well pattern.
(5) Decommission the site; interpret and publish test results.

Step 4 began in 1996.

True *in situ* copper leaching experiments were performed on the deep Van Dyke oxide copper ore deposit within the town of Miami, Arizona by Occidental Minerals. In a pilot test, two holes were drilled 23 m apart to a depth of about 300 m. The unique aspect of the approach was to use horizontal hydraulic fracturing in the production and injection wells but at different elevations within the ore zone. Hence, the sulfuric acid leaching solution entered one set of fractures and swept through the ore vertically to the other fracture horizon. Another firm has continued copper production by *in situ* solution mining the Van Dyke ore deposit on a small scale.

The Cyprus Casa Grande Mine (Arizona) is a commercial operation involving the leaching of both underground caved ore and undisturbed ore. Solution is pressure injected from a wellfield drilled from the surface. Pregnant liquor is collected below the ore in accessible underground workings and pumped to the surface. A similar operation occurs at the San Manuel Mine (Arizona).

Many major problems with solution mining deep copper deposits require more research. Among these are controlling solution migration and recovery, and obtaining adequate permeability. Deep *in situ* systems can be operated at elevated temperature and at increased chemical oxidant concentrations compared with mine waste leaching in dumps. For most deep *in situ* systems, rates of sulfide leaching reactions will depend upon the direct chemical attack by oxygen or other chemical oxidants, rather than by ferric ions. Both sparged oxygen and stable emulsions of oxygen in the leaching solution are possible. The autotrophic bacteria important to dump and heap leaching cannot survive under the harsher conditions but, fortunately, they are not needed at the higher temperatures and oxygen solubilities that can be encountered at considerable ore deposit depth.

URANIUM ORE DEPOSIT GEOLOGY

Uranium deposits in the United States are mainly of the sandstone type. They are located in the Colorado Plateau, in Wyoming basins, and in the Texas Gulf coastal plain. In most cases they are believed to have been formed from the migration of oxidizing ground waters carrying soluble uranium down the dip of the host formation. In Wyoming, large granite intrusives are located near the ore deposits at higher elevations. These granites are believed to be the source of the uranium, which was released by groundwater leaching. These migrating solutions precipitate uranium when reducing conditions, often associated with organic detrus, are encountered. Some petrified trees in the Colorado Plateau contain very large concentrations of uranium.

Often a chemical zone of deposited uranium minerals develops in the reducing environment immediately in front of the oxidizing environment. This is called a "roll front" and it is illustrated in Fig. 10.6. Migration may be progressive over long periods as oxidizing solutions enter, mobilize, and move uranium a short distance further on in a manner similar to an ion exchange chromatographic column. Over extended periods, considerable enrichment of the deposit at the roll front can occur from the very dilute migrating solutions. The actual shape and grade of the deposit will vary considerably depending on the local permeability of the host rock and the organic material distribution. This is illustrated in Fig. 10.7. Average mine ore grades also vary considerably in these deposits. In the United States,

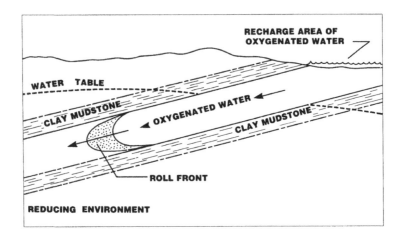

Figure 10.6. Cross-section of a uranium roll front showing formation and migration down dip.

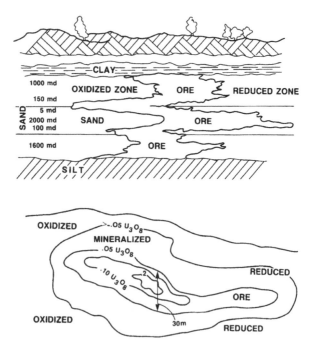

Figure 10.7. Typical section and plan views of uranium roll front ore deposit (Shock and Conley, 1974).

ore grades vary from less than 0.05% to 0.5% uranium, and commonly averaging 0.2–0.3% uranium. Much higher grade ore deposits exist in Canada and Australia and have come to dominate the international uranium mining industry.

URANIUM LEACHING CHEMISTRY

Uranium occurs in reduced ores in the tetravalent form in oxide minerals such as uraninite and pitchblende. Uranium is often associated with vanadium, and in mixed chemical compound minerals such as carnotite, $K_2(UO_2)_2(VO_4)_2(1-3H_2O)$. Uranium is also found in phosphates.

The leaching process consists of oxidation to the hexavalent uranium, UO_3, or the uranyl ion, UO_2^{2+}, which is soluble when complexed with either carbonate ion or sulfate ion, depending on the lixiviant used.

The preferred lixiviant in surface plants, leaching milled uranium ore, is usually sulfuric acid with air sparging. Ferric ion is the intermediate mineral oxidant. When considerable carbonate rock or other minerals in the ore consume large amounts of acid, the lixiviant choice may switch to a solution of either sodium carbonate or ammonium carbonate.

During uranium *in situ* solution mining, sulfuric acid may generate precipitates, such as gypsum, that hinder ore deposit permeability. Sulfuric acid mobilizes other species that are often in the ore deposit aquifer and environmentally deleterious (e.g., molybdenum and selenium). Consequently, ammonium carbonate leaching has usually been the preferred lixiviant for solution mining operations.

Ammonium carbonate leaching is highly selective for uranium (Merritt, 1971) and does not generate precipitates. It produces a pregnant liquor by the following reactions,

$$2UO_2 + 2H_2O_2 = 2UO_3 + 2H_2O, \tag{10.1}$$

$$UO_3 + (NH_4)_2CO_3 + 2(NH_4)HCO_3 = (NH_4)_4UO_2(CO_3)_3 + H_2O. \tag{10.2}$$

This dilute alkaline leach liquor can be sent to a solvent extraction or ion exchange step, and the resulting organic solvent or ion exchange resin can be stripped with a strong sodium carbonate solution to produce a solution (eluant) more concentrated in uranium. The eluant can be treated with sodium hydroxide to yield a marketable hydrolysis product, called "yellow cake," that can be refined to UO_2:

$$Na_4UO_2(CO_3)_3 + 4NaOH = Na_2UO_4(s) + 3Na_2CO_3 + 2H_2O, \tag{10.3}$$

$$2Na_4UO_2(CO_3)_3 + 6NaOH = Na_2U_2O_7(s) + 6Na_2CO_3 + 3H_2O. \tag{10.4}$$

The caustic precipitation step is highly selective, which is beneficial for purification.

Figure 10.8 shows the E_h–pH relationship for the uranium–carbonate–water system. The mineral uraninite is stable under reducing potentials, 0.0 to -0.4 V at pH 9, the usual pH for ammonium carbonate leaching, but it is dissolved at higher oxidation potentials. The double hatched region shows the range of E_h and pH employed for *in situ* leaching with ammonium carbonate and the stable complex, $UO_2(CO_3)_3^{4-}$.

As can be seen from Fig. 10.8, oxidants with an oxidation potential, E_h, higher than about -0.06 V should be effective for ammonium carbonate leaching of uranium. The list of acceptable oxidants includes air, O_2, H_2O_2, ClO_3^-, ClO^-, Ce^{4+}, and MnO_2. However, most of these are either too expensive, unavailable in North America (e.g., MnO_2), or have

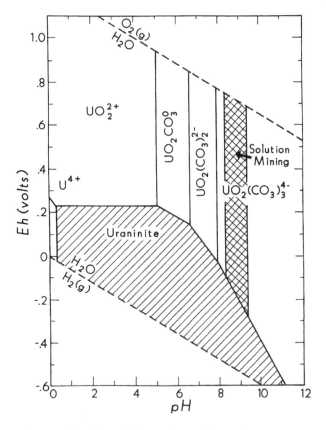

Figure 10.8. E_h–pH diagram for uranium–carbonate–water.

solubilities that are too low (e.g., O_2) to have adequate uranium transport capability. Uranium ore often contains other oxidant consumers such as kerogen (organic matter), pyrite, vanadium minerals and MoS_2.

Although air is the least expensive oxidant, it is usually not used in solution mining uranium because of the low oxygen solubility in the leach solution. Hydrogen peroxide is often chosen for solution mining of uranium in North America, and it was the oxidant used in eqn 10.1.

FLOODED LEACHING WITH WELLS

Leaching solutions must be injected as a barren liquor and recovered as a pregnant liquor. The most common method of doing this is with an overlaying pattern of vertical injection wells and production wells, Fig. 10.2. A flooded leaching plan must be based on an understanding of the hydrologic characteristics of the ore deposit and its confining rock. The degree to which solution flow can be predicted and controlled determines to a large extent the efficiency of ore recovery. The principle hydrologic characteristics influencing solution flow are:

- void space (porosity)
- permeability
- thickness (height of ore zone being swept)
- transmissivity
- storage coefficient
- water table or piezometric surface
- hydraulic gradient

The permeability ideally would be adequate and uniform, but this is rarely achieved. Large variations in permeability will cause uneven flow with "short circuiting" of solutions through the high permeability paths— poor "**sweep efficiency**." Low permeability regions will not be adequately swept by the leach solutions and the yield of leached solute will be poor. Poor sweep efficiency is often a serious problem in *in situ* solution mining and reservoir engineering generally. "**Transmissivity**" is the product of permeability (uniform or average) times the deposit thickness. "**Storage coefficient**" is the volume of solution released from the reservoir (ore deposit volume being leached) per unit decrease in head.

PREGNANT LIQUOR RECOVERY USING WELLS

In addition to permeability and permeability variations, three important factors affect pregnant liquor recovery and grade: (1) mineral leaching (solubilization) **kinetics** of the ore formation (similar to phenomena discussed in previous chapters), (2) **injectivity response**, and (3) the **solution displacement** mechanism that will optimize the concentration of valuable solute in the produced solution. It is usually cost-effective to produce the maximum concentration of solute in the pregnant liquor with a minimum injection of solution.

Leaching Kinetics

Leaching experiments using representative drill core taken from the ore deposit and flooded with the leach solution are employed to measure permeability and obtain empirical extractions rates per unit volume of core. Also, problems encountered with precipitation and plugging and improved void space benefits, if any, can often be identified from drill core leaching experiments. Variations in oxidation, oxidant consumption by gangue minerals, lixiviant ion exchange (e.g., with zeolites), sorption and other changes to the rock, should be carefully evaluated from laboratory tests with drill core samples.

All of the mineral to be extracted may not be exposed in open flow paths. Most may be in dead end micropores, and these can only be extracted by a countercurrent diffusion chemical process similar to the extraction of mineral **within** micropores of individual rocks during percolation leaching. Actual flow fissures may be few and far apart.

Obtaining core that has not been further fractured or damaged during drilling is difficult. Altered core can give optimistic but false leach extractions and permeability results.

Drill core leaching experiments are conducted in pressure vessels with seals so that the solutions must flow through the sample in a defined path, usually "plug flow", without bypassing. Lixiviant composition, pressure, temperature and other conditions expected in the solution mining operation must be represented as closely as possible.

A typical uranium extraction curve from a core sample using sulfuric acid, taken from Shock and Conley (1974), is shown in Fig. 10.9. The solubilized mineral concentrations and pH are shown in relation to the volume of solution flowing through the core, expressed in number of pore volumes. At a constant lixiviant flow rate, pore volumes are proportional to leaching time. Note that in this particular leaching case, uranium extraction is favored over vanadium and iron extraction. Uranium recovery was essentially complete after a half pore volume of leach solution had passed through the core sample in plug flow.

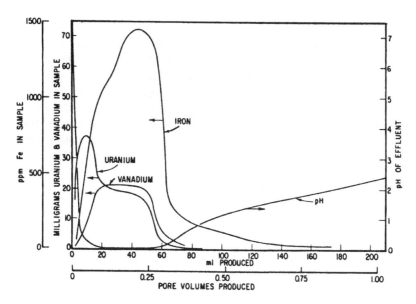

Figure 10.9. Extraction curves for an acid leached uranium bearing sandstone core sample (Shock and Conley, 1974).

The concentrations in Fig. 10.9 are similar to what would be expected from a "line drive" of injection wells in the ore deposit reservoir, which is a nearly plane source of injected solution, similar to that in a plug flow experiment. In a line drive the flow lines are parallel to each other and perpendicular to hydraulic gradient equipotential lines, which, of course, are also parallel to each other. Thus, the concentrations in Fig. 10.9 do not represent the uranium extraction concentrations expected from a radial displacement of lixiviant from a pattern of vertical injection wells.

Over a much longer distance between lixiviant injection and pregnant liquor production wells, solubility limits may prevent further extraction, acid may be consumed by carbonates, uranium may be reprecipitated, double salts may form, plugging may occur, etc. In other words, the core kinetic leaching tests provide information under optimum lixiviant displacement conditions and may lead to excessive optimism. Cores are extremely small samples and thus of dubious value in representing an entire ore body. Consequently, it is usually advisable to conduct several core experiments from different regions of the ore deposit and conduct **stream tube leaching experiments**.

Injectivity Response

The compatibility of the injected solution and the ore matrix must be determined. Bad injectivity response can result because of (1) formation plugging due to precipitation, (2) pregnant liquor robbing, including secondary enrichment reactions, (3) ion exchange, (4) swelling of clays, (5) gangue mineral decrepitation, etc. Sometimes chemical additives or other changes to the solution can offset these negative characteristics. But the injectivity response should be determined first in the laboratory with core samples and later in early field trials.

Displacement Studies (Dilution)

With radial displacement from an injection well, pore volumes increase rapidly with distance from the well (flow velocity decreases). At 1 m from the injection well at 30% void volume, which is not untypical of a sandstone, the pore volume is about $1\,m^3$ per meter of height (thickness of the formation). But at 50 m distance the pore volume is nearly $100\,m^3$ per meter of height.

If the wells are in an ore deposit that is confined at the roof and floor, but unconfined laterally, as is usually the case for uranium and other sedimentary ore deposits and evaporites, the problem becomes one of choosing the optimum well spacing for injection and production wells to optimize solution recovery. Not all of the solution is recovered.

For the usual five spot flood pattern (a central injection well surrounded by four equidistant production wells) in a laterally unconfined reservoir, the displacement liquid breaks through quite rapidly. Then the four production wells generate a solution which decreases rapidly in uranium concentration, even assuming that all uranium is accessible to the solution and dissolves rapidly. The injection sweep drives liquid with dissolved uranium beyond the producing wells, and some of the soluble uranium is permanently stored, in more diluted form, within the extended reservoir. This is analogous with the spaceship mission to the planet Jupiter that misses and continues out into space forever.

Recovery for a five spot well configuration computed by Shock and Conley (1974), is shown in Fig. 10.10, matched to the extraction experimental results in Fig. 10.9. The peak concentration of dissolved uranium after injection of 1.2 pore volumes is $8\,mg/l$ of U_3O_8, which is considerably less than the peak of $37\,mg/l$ in the core test. Most importantly, the recovery is only 55% after leaching out to 2 pore volumes. Recall from Fig. 10.9 that all of the uranium was extracted in 0.5 pore volume in the core test.

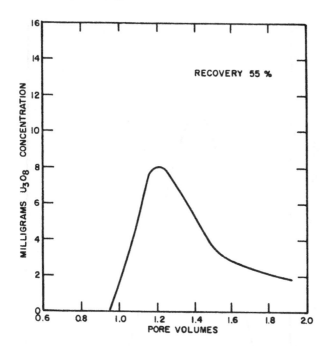

Figure 10.10. Computed recovery from a **laterally unconfined** five spot well array (Shock and Conley, 1974).

Fortunately, recovery increases as the wellfield becomes larger and more complex and the number of wells increase, as is usually the case in a commercial application of solution mining. However, the wellfield will always be inefficient at its edge if the deposit is laterally unconfined. One approach to mitigating this problem is to produce more solution than is injected, coupled with using the outer ring of wells as solution production wells. Another approach is to drill a row of water injection wells around the periphery of the solution mining field. A slight injection of water at each peripheral well will provide a hydraulic barrier to flow. These **guard wells** fully confine the system and can, in principle, provide 100% solution recovery. However, the solution will be diluted and slightly more solution will be produced than injected.

The production curve for a laterally confined five spot is shown in Fig. 10.11, and can be compared with the unconfined five spot in Fig. 10.10. In both of these cases, solution flow is horizontally confined by impermeable strata above and below the uranium bearing aquifer.

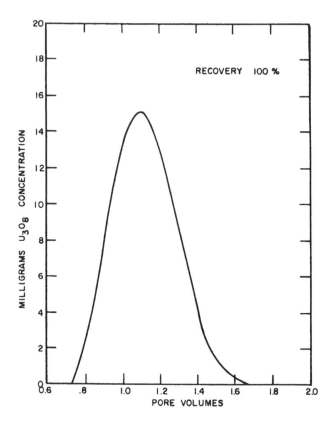

Figure 10.11. Computed recovery from a **fully confined** five spot well array (Shock and Conley, 1974).

STREAM TUBE LEACHING EXPERIMENTS

During development of the south Texas *in situ* uranium mines, Murphy et al. (1984) developed the stream tube leach test. It is similar to a column leach test except that it is saturated, velocities are usually lower and several tubes in series are used to obtain a very long flow path similar in length to that along a streamline from an injection well to a production well, which may be many meters. The system is flooded with lixiviant, often under pressure, which flows through the stream tube at a slow constant rate and is not recirculated. The concentration of metal solute in the final effluent is monitored over time (number of pore volumes), and small samples of solution may be collected at intermediate points along the flow path.

Figure 10.12 shows a soluble gold concentration profile in the effluent from leaching a gold ore in a stream tube (Jacobson and Murphy, 1988). The **breakout**, when dissolved metal first arrives in the effluent, is a function of the ore, stream tube length and lixiviant flow velocity. In Fig. 10.12, breakout occurred after 3.5 pore volumes had been introduced at a constant velocity over a period of 250 hr. Following breakout, the concentration rises to a peak value quickly (14.5 hr) and then declines.

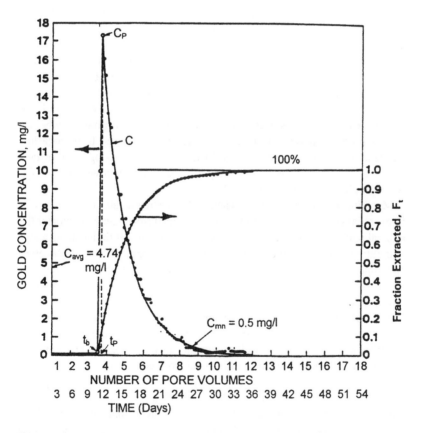

Figure 10.12. An actual plot of gold effluent concentration and fractional extraction from a stream tube test (Jacobson and Murphy, 1988).

LOGARITHMIC DECLINE OF WELL PRODUCTION
DURING *IN SITU* LEACHING

In Chapter 9 it was noted that leaching ore with a distribution of mineral grain sizes that is not too wide will devolve to an approximately logarithmic relation in the fraction reacted:

$$F_t = 1 - e^{-K_m t}. \tag{9.7}$$

And, in differential form this is a quasi-first order reaction in the remaining mass of mineral, $1 - F_t$; hence:

$$dF/dt = K_m(1 - F_t). \tag{10.5}$$

The overall *in situ* leaching extraction rate from a production well, or a group of wells started simultaneously, is complicated by the usual factors of mineral leaching chemical kinetics and lixiviant/solute transport by diffusion that were discussed in previous chapters. However, it is further complicated by the complex flow patterns, variable permeabilities and different distances of the streamlines connecting injection and production wells. Even at a steady fluid flow rate, production from a well invariably declines over time. This decline in dissolved metal concentration, whether from a production well or stream tube, is often observed to be logarithmic (Jacobson and Murphy, 1988). In these cases, semi-log plots of c/c_P and of $1 - F_t$ versus time will be straight lines with a negative slope, $-K_m$. An example is shown in Fig. 10.13.

Matching experimental data, whether from stream tube experiments or from actual well production results, to an assumed first order extraction of the mineral being leached will often provide a good empirical fit, including a logarithmic decline over time in the dissolved metal concentration in the pregnant liquor from the production well. In the absence of better information for a specific *in situ* leaching situation, the logarithmic well decline (first order) model is recommended for use with well and stream tube production data.

The following equations have been derived by Murphy et al. (1984). With the logarithmic production decline model, in a stream tube leaching experiment or a production well at constant flow rate, Q, the effluent concentration, c, is proportional to dF/dt, and from eqn 10.5,

$$c = (W_0/Q)K_m(1 - F_t) \tag{10.6}$$

where, W_0 is the total amount of accessible mineral that can be dissolved (at $F_t = 1$). Consequently, if the leaching rate follows the logarithmic relation then,

$$c = (W_0 K_m/Q)e^{-K_m t}. \tag{10.7}$$

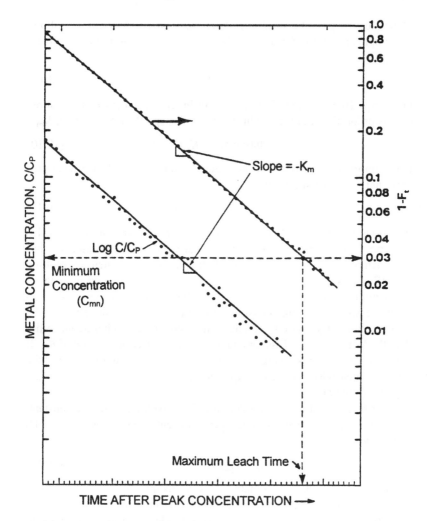

Figure 10.13. Semi-log plots of c/c_p and $1-F_t$ versus time (Jacobson and Murphy, 1988).

Beginning approximately at breakout, with $t=0$ defined to occur when the peak concentration, c_p, is reached, eqn 10.7 can be used to define c_p; namely, at $t=0$:

$$c_p = W_0 K_m/Q. \qquad (10.8)$$

This value of c_p can be substituted back into eqn 10.7 to yield the following simple expression for the dissolved metal concentration in the effluent at any time

$$c = c_p e^{-K_m t}. \tag{10.9}$$

From eqn 10.9, the logarithmic decline in dimensionless concentration, c/c_p, is proportional to the leaching time, t, through the rate constant, K_m:

$$\ln(c/c_p) = -K_m t. \tag{10.10}$$

Furthermore, the logarithm of the mineral fraction unleached, is also proportional to the leaching time.

$$\ln(1 - F_t) = -K_m t. \tag{10.11}$$

Equations 10.10 and 10.11 are the two semi-log plots shown in Fig. 10.13.

In principle, once the value of the logarithmic rate constant, K_m, has been determined from an appropriate stream tube experiment by plotting the effluent concentration data on semi-log paper, it can be used in an extension to a production well to predict the decline in metal production from the well following breakout. However, the ore in the stream tube experiment often is more fragmented and with a much higher permeability than the formation ore. This may yield an erroneously false value of K_m. It is partially offset by keeping the solution flow velocity in the stream tube low to match what would be expected at operating pressures in the ore deposit formation.

After breakout at a new production well, or wellfield, the rate constant, K_m, can be approximated from the peak concentration, c_p, using eqn 10.8 as follows:

$$K_m = c_p Q / W_0. \tag{10.12}$$

While this is a crude estimate, it is a distinct planning advantage. Production can be forecast early and decisions can be made about the optimum time to start new wells that are needed to keep a nearly constant supply of metal in the aggregated pregnant liquor feeding the metal recovery plant.

EVALUATING ORE DEPOSIT HOST FORMATIONS FOR *IN SITU* (FLOODED) LEACHING

Permeability tests in the field involve forcing a measured quantity of water into a formation through a well at a known pressure. Tests generally fall into three categories: constant head, variable head, and pumping tests. The constant head and variable head tests are similar to permeameter tests described in Chapter 7. The test set-up for each is similar (shown in Fig. 10.14, Chamberlain, 1979). A test interval is confined by **stradle packers** located in the well above and below it. The packers prevent fluid from being injected or pumped except within the test interval.

In the constant head test, the volume of water injected is measured over a fixed time at a constant pressure above the static pressure. In the variable head test, a fixed amount of fluid is added or withdrawn from the well and the rate at which the fluid level returns to its original level is used to

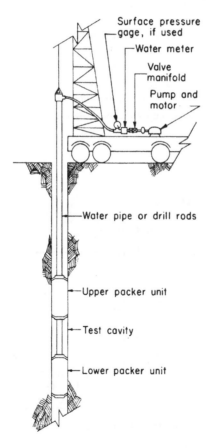

Figure 10.14. Permeability measurement in a well, using straddle packers to isolate the test zone (Chamberlain, 1979).

measure permeability. Pumping tests are performed with a pumping well and two or more observation wells drilled at various distances away from the pumped well (shown in Fig. 10.15, Chamberlain, 1979). As water is pumped at a constant flow rate, the water level drawdown is measured in the observation wells. The drawdown difference in these two wells and their distances from the pumping well are used to calculate permeability (see Chapter 11).

Well preparation is very important to obtain representative results. The region close to the well has the greatest impact on measured permeability. Improper drilling muds can plug voids and channels near the borehole and lower the measured permeability. Test wells should be cleaned on completion, and wells in loosely consolidated formations must be cased to prevent excessive sloughing of the formation rock. Permeability tests can be run through perforations in the casing in the test interval.

Permeabilities should be measured at different intervals in the formation by moving the packers and test interval. Considerable variation in permeability often occurs, particularly in sedimentary formations. An example with several permeability measurements, showing a permeability depth profile in a 78 m (260 ft) deep well, is shown in Fig. 10.16, (Chamberlain, 1979).

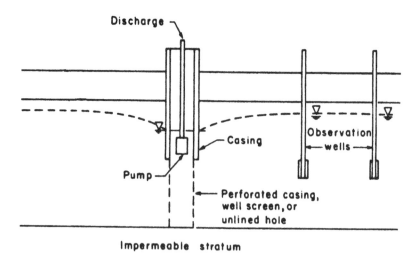

Figure 10.15. Pumping drawdown test with observation wells to determine permeability (Chamberlain, 1979).

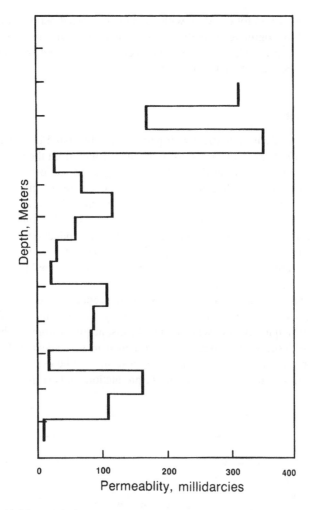

Figure 10.16. Typical measured permeability log from well testing (Chamberlain, 1979).

WELL PATTERNS

Several well patterns have been used for *in situ* solution mining, each with advantages and disadvantages. A single well can be used for injection of lixiviant followed by withdrawal of the injected solution, with repeated cycles. Each cycle usually involves a fixed amount of solution injected over a fixed period followed by withdrawal over a fixed period. Although rarely used for commercial solution mining, this "huff/puff" or "push/pull" technique is an excellent low-cost approach to field testing of a solution mining prospect in a "pilot plant" mode.

Vertical Wells

Three commonly used vertical well patterns are illustrated in a plan view in Fig. 10.17. These wells are most frequently drilled from the earth's surface, but they also can be drilled from underground platforms. The five spot pattern with a single central injector well is one of the most efficient patterns for only a few wells. Again, it is used primarily for field testing. Often a successful five spot field test will be expanded into a larger commercial array of alternate injection wells and production wells. The *extended* five spot pattern, illustrated in the center of Fig. 10.17, is the most frequently used commercial wellfield array, and it provides an equal number of injection and production wells.

Sometimes it is more efficient to have more production wells than injection wells, or vice versa, in a continuous array. For example, production wells may frequently plug with saturated salts, acquired from leaching, or it may be desirable to produce more liquid than was injected. In either case, more production wells than injection wells may be desired. A seven spot pattern is a hexagonal array of six production wells surrounding one

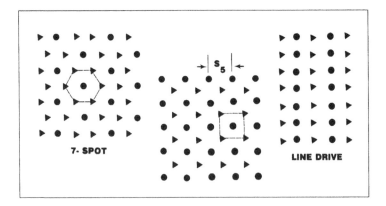

Figure 10.17. Well patterns in plan view, all at the same wellfield density (wells/unit area), used for solution mining.

injection well. An extension of this pattern will provide two production wells for each injection well.

A line drive is another well arrangement. It usually consists of alternating parallel rows of injection wells and production wells.

Deviated and Horizontal Wells

New well drilling technology, developed in the petroleum industry, permits the drilling of several wells in the ore zone that deviate from a single pilot well. This can be a cost advantage for deep ore deposits because the deviation can start at considerable depth and thereby reduce the total drilling length. Wells that run within and parallel to the bedding plane of tabular ore deposits, rather than across it, can provide more efficient access to the ore. Horizontal wells are being used in coal beds for methane production as well as for oil and gas production from stratabound formations.

WELL COMPLETION

Well drilling and completion is usually performed by contractors with specialized equipment and experience, but well completion is expensive and critical to a successful *in situ* leaching operation. Materials of construction must be compatible with the injected leach solution and pregnant liquor. This often requires acid resistant cements, plastic piping (usually PVC or fiberglass) and stainless steel pumps. Although injection wells may be completed somewhat differently than production wells, it is often a good idea to complete them identically so they can be used interchangeably, reversing flow of lixiviant and oxidant occasionally to obtain better oxidant utilization and improved overall metal recovery.

Boreholes drilled in incompetent rock must be cased, with steel or plastic pipe, to prevent borehole collapse. Nearly all solution mining wells will require cementing for a distance above the top of the ore zone. Often the initial drilling at the surface is in soil or unconsolidated alluvium, requiring the borehole to be cased down to bed rock. After the initial drilling and surface casing emplacement, a smaller bit that will pass through the surface casing must be used. For deep wells a series of casings with successively smaller bore may be required. The higher strength of steel pipe is needed for deep hole casings.

Cementing the space between the casing and wall of the borehole helps bond and support the casing and prevents solution and groundwater flow in the borehole outside the casing. This primary cementing prevents both loss of pregnant liquor and contamination of adjacent or penetrated aquifers.

Cement slurry is usually injected down the inside of the casing to the bottom and returns in the outer annulus. A wiper plug separates the slurry from a displacement fluid, usually water, that pushes the wiper plug to the bottom of the casing. Spaced centralizers center the casing so the cement will uniformly surround the casing after injection. It is often unnecessary to extend cementing back to the surface.

For surface casing and shallow wells, cement slurry can be injected into the outer annulus from the surface. When the borehole has been drilled through the producing formation (ore deposit), a packer or cement basket can be used at the top of the producing formation to prevent cement from entering the formation. A packer expands between the borehole and outer wall of the casing providing a seal that prevents vertical movement of fluids at the packer location.

Squeeze cementing is the injection of a cement slurry into the void space surrounding a borehole, using hydraulic pressure. Squeeze cementing is used to seal porous formations, usually adjacent to or above the ore zone.

Casing is usually considered to be permanently installed. Consequently, a separate corrosion resistant well string of pipe, that can be inserted inside the casing and removed as needed, is usually used. When injecting corrosive solutions at a high pressure, a stainless steel injector pipe sealed with a packer between casing and injector pipe at the top of the producing horizon is required. Both the injection and production well solution pipes must be perforated or slotted in the production zone.

For production wells at high pumping rates, a downhole multistage turbine pump (shown in Fig. 10.18) can be located at the end of the production pipe string. In shallow solution mining wells, these pumps are usually driven from the surface by a drive shaft, analogous with irrigation wells. A sealed electric motor drive in the borehole is also possible and often used for deep solution mining. Each turbine pump stage is essentially a single centrifugal pump, and as a centrifugal pump's diameter decreases, its lifting capacity or head also decreases for the same speed of rotation. Solution mining production wells are often located in tight formations that involve small solution flow rates relative to agricultural irrigation wells. Consequently, small well bore diameters are used to save well drilling and well completion costs. As a result, solution mining turbine pumps for deep production wells will usually contain many stages (10–50, or more)

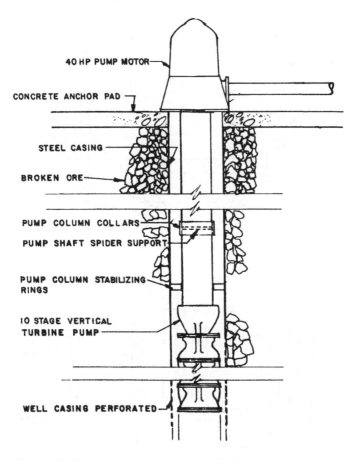

40 HP PUMP MOTOR

CONCRETE ANCHOR PAD

STEEL CASING

BROKEN ORE

PUMP COLUMN COLLARS
PUMP SHAFT SPIDER SUPPORT

PUMP COLUMN STABILIZING
RINGS

10 STAGE VERTICAL
TURBINE PUMP

WELL CASING PERFORATED

Figure 10.18. Large flow production well with turbine pump.

constructed of stainless steel or other corrosion resistant alloys and driven by a downhole motor. These pumps can be very expensive.

Pumping can also be done with a sucker rod and downhole plunger extended from a rocker arm pumping mechanism. This is well-matched to the small solution flow rates that are often necessary when solution mining a tight ore formation.

Well completion for *in situ* solution mining varies considerably with project circumstances, but one example of well completion is shown in Fig. 10.19.

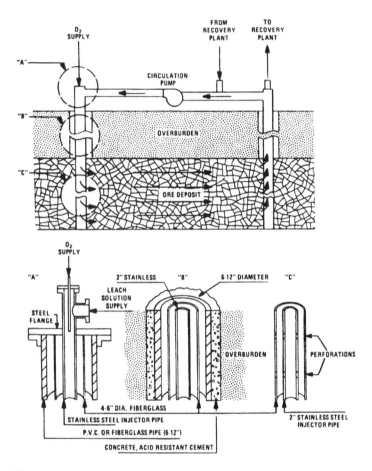

Figure 10.19. An example of well completion for *in situ* solution mining.

WELL STIMULATION

"Well stimulation" is action taken to increase the ore deposit permeability adjacent to the well borehole. Well stimulation methods include (1) acidifying, which is often used when the host rock is limestone or dolomite; (2) hydraulic fracturing the formation, with or without proppants; and (3) explosive fracturing. Well stimulation is particularly important in hard rock ore deposits that have not been rubblized. The intrinsic permeability of hard rock ore deposits may be very low (millidarcy) and prevent or limit many applications of *in situ* solution mining. Because solution flow from (and to) wells drilled from the surface is radial, velocity and velocity gradients are highest at the borehole and decrease rapidly with distance from the well. Therefore, low permeability in the ore formation adjacent to the well is particularly adverse. Effectively increasing the well diameter or radically lowering the permeability in a cylinder of rock surrounding the well will greatly accelerate solution flow rates. Hence, effective well stimulation can be extremely beneficial.

Hydraulic Fracturing

Hydraulic fracturing has been a well-established method of stimulating oil and gas wells and has been extended to solution mining applications. Fracture is initiated by pumping fluid into the well at pressures that exceed the resistance of the formation, usually related to the rock tensile strength. At great depths this is often slightly above the lithostatic pressure. In sedimentary formations, rock is usually weakest along bedding planes and will separate there. Consequently, the cracks are usually parallel with bedding.

During the process, fracture initiation is identified by operators at the surface from either a sudden drop in pressure or an increased rate of pumping of the fluid. Fluid flows into the initiated fracture and extends it. Preventing the opened fractures from closing on release of pressure can be accomplished with a **proppant**, pumped into the crack as a slurry. Proppants are typically microspheres of a ceramic sufficiently strong to support the lithostatic load without failure. Dense fired alumina is often used as a proppant for oil and gas wells. Crushed quartzite sand is fairly resistant to most lixiviants and would be adequate for shallow solution mining wells.

Over the past fifty years, nearly one million hydrofracture treatments of oil and gas wells have been completed, and about 1/3 of all oil and gas wells are hydrofractured. Hydraulic fracture fluids are viscous to minimize loss to the formation and to suspend proppant particles, which are carried with the fluid into the cracks that are formed. The high viscosity of the fracturing fluid also causes a fracture to be wider than it would otherwise be. Cross-linked guar is often used to obtain high viscosity in an aqueous hydraulic fracturing fluid. Inflatable packers are used to isolate the section of the mineral formation in which fracturing is desired.

Explosive Fracturing

Explosive fracturing is caused by detonation of a high explosive filling the borehole over the thickness of the ore zone. Detonation creates a pulverized region near the borehole and tensile fractures beyond the crushed region. Since there is no way to prop open the tensile fractures, they do not significantly contribute to increased permeability. The major cause of improved permeability is cavity formation in the pulverized region. Drilling below the formation prior to explosive emplacement provides room for expansion and deposition of pulverized rock. High energy explosives of the type that release energy and gas over a relatively long period provide the maximum pulverizing of rock. High density (aluminized) ammonium nitrate based explosives work well.

ECONOMICS OF *IN SITU* LEACHING ORE DEPOSITS

Metal price and ore grade are always major determinants of economic success, but neither of these can be controlled by the operator. The economic performance of an *in situ* solution mining operation depends on (1) the daily volume of solution processed, (2) the rate of fluid injection and production per well, (3) the rate of mineral solubilization in the leaching solution, and (4) the volume of ore swept (contacted) by the leaching solution (Davidson, 1988). Pregnant liquor concentration has a major impact on economics, since the multiple of concentration and solution volume is the amount of product and determines revenue.

The major solution mining capital investments are (1) the surface recovery facility, (2) wells, (3) pumps, and (4) the chemical cost of solution inventory that must be charged into the ore deposit. Related investment costs usually include property acquisition, ancillary facilities, and off-site facilities such as roads, offices, warehouses, laboratories and utilities.

The amount and the cost of chemicals that must be precharged before dissolved metal is obtained can be very large. This chemical inventory includes filling the void space in the ore deposit under the wellfield and, possibly, secondary enrichment reactions that rob dissolved metal from the pregnant liquor while consumption of lixiviant chemicals continues. During *in situ* leaching of copper sulfide ore deposits, large amounts of oxygen may be consumed and large amounts of acid or ammonium sulfate may be generated without much copper being recovered.

For copper recovery by SX/EW, the solvent extraction capital expense is proportional to the pregnant liquor flow volume while the electrowinning capital expense is proportional to the metal production rate.

Wellfield costs increase with well depth and the number of wells. It is desirable to have a high metal concentration in the pregnant liquor to achieve a high level of individual well capacity and minimize the number of wells. Pumping costs increase with the volume of solution, number of wells and pressure (the head for production wells and the injection pressure for injection wells).

Inventory costs are related to displacing water in the production zone of the wellfield with leaching solution. On start-up, the solubilized metal is diluted by the existing pore water and the initial pregnant liquor produced will be very weak. At least one exchange of the volume of water in the wellfield producing zone (one pore volume) will be required to reach full (steady state) metal production capacity from the wellfield, and several pore volumes may be required.

An economic dilemma occurs because higher steady state pregnant liquor grades invariably require higher chemical concentrations (solvent ions, oxidants, etc.) in the lixiviant and higher chemical inventory costs (Davidson, 1988). Achieving higher steady state pregnant liquor grades usually requires a longer solution residence time in the production zone, slower solution flow and a longer time to achieve the exchange of one pore volume. Hence, the negative cash flow connected with start-up is usually greater with higher pregnant liquor grades, but the direct costs and capital amortization are generally lower with higher pregnant liquor grades, once steady state production has been reached.

The rate of extraction of metal will decline as *in situ* ore leaching proceeds because of both mineral dissolution kinetics and chemical diffusion effects, as previously discussed. Typically, individual well production will gradually decline after reaching peak capacity. Consequently, the initial wellfield should not cover all of the ore deposit. Additional wells must be added periodically to compensate for declining production from older wells and to maintain a nearly constant total volumetric flow rate and a nearly constant average metal concentration in the aggregate pregnant liquor needed by the metal recovery plant.

Eventually, older wells produce insufficient metal and revenue to carry their operating and maintenance cost and must be shut-in, either terminated or rested. Well lifetime usually has a very significant impact on solution mining profitability. Proper selection of well spacing and both initial well stimulation and subsequent episodes of well stimulation can have a major impact on well lifetime and on solution mining project profitability.

With copper *in situ* solution mining, the maximum production rate (peak capacity) that is selected should not exceed the total *recoverable* copper divided by the project lifetime. Production beyond 20 years has little impact on profitability and 20 years is a viable SX/EW life expectancy.

Determining the net present value of a solution mining production well is difficult because of the uncertain, declining rate of metal production.

When well production declines logarithmically, following the quasi-first order dependence on the remaining (unextracted) mineral, and the rate constant, K_m, is known, then it is possible to express the gross present value of the resulting revenue, $PV_{(t_p)}$ at the time of peak concentration, t_p, by the following integral involving the time dependent cash flow, $CF_{(t)}$:

$$PV_{(t_p)} = \int \frac{d(CF_{(t)})}{(1+i)^t}, \qquad (10.13)$$

where i is the well operator's discount rate. This present value is not a *net* present value because it excludes taxes and the capital cost of well drilling and completion.

At a steady production well flow rate, the differential cash flow is related to the flow rate, Q, current metal concentration in the well fluid, c, and the metal price, P_m. The metal price may be adjusted downward to include deductions for production costs that are *variable* with the amount of metal produced, e.g., metal refining charges. The differential cash flow is:

$$\frac{d(CF_{(t)})}{dt} = P_m Q c - C_{op}, \qquad (10.14)$$

where C_{op} is all of the *fixed* operating costs of the production well, including pumping and processing liquors and associated injection wells, usually one per production well in a five spot pattern. C_{op} is expressed in dollars per unit of time, typically annually.

The concentration, c, may be expressed in terms of the rate constant, K_m, through eqn 10.9, yielding:

$$\frac{d(CF_{(t)})}{dt} = P_m Q c_p e^{-K_m t} - C_{op}. \qquad (10.15)$$

Substituting into the gross present value of the revenue integral, eqn 10.13, yields,

$$PV_{(t_p)} = P_m Q c_p \int_0^{t_f} e^{-[K_m + \ln(1+i)]t} \, dt - \int_0^{t_f} \frac{C_{op}}{(1+i)^t} \, dt. \qquad (10.16)$$

Integrating between zero time at the time of peak dissolved metal concentration from the production well, $t_p = 0$, and the time at the end of leaching, t_f, yields the following expression for the gross present value of the revenues generated by the producing well, or wellfield, being evaluated:

$$PV_{(t_p)} = \frac{P_m Q c_p}{K_m + \ln(1+i)} (1 - e^{-[K_m + \ln(1+i)]t_f}) - C_{op} \left[\frac{1 - (1+i)^{-t_f}}{\ln(1+i)} \right], \quad (10.17)$$

This present value can be further discounted back from the well's peak concentration and breakout times to the time that the well is drilled or first put into production. Subtraction of the well drilling and completion costs from the gross present value of the revenues, $PV_{(t_p)}$, will provide a useful profitability index for *in situ* solution mining, when it follows the logarithmic solution grade decline model.

In applying eqn 10.17, parameters involving time, i.e., t, K_m, i, C_{op} and Q, must be expressed in the same time units. For example, if time is expressed in months, then the fractional discount rate must be the annual interest rate divided by 12, and a 12% annual discount rate would be $i = 0.01$.

EXAMPLE PROBLEM

Given the following production well data, compute for each year out to seven years: (1) the ratio of c_F/c_p using eqn 10.10, and (2) the gross present value of the revenues using eqn 10.17. Also, plot $PV_{(t_p)}$ versus years of production on graph paper and find the year nearest the maximum in the gross revenue present value. The data are:

$$P_m = \$1.00/lb = \$0.002203/g$$

$$Q = 120 \, l/min \quad (\text{well availability} = 90\%)$$

$$= 56{,}765{,}000 \, l/yr$$

$$c_p = 10 \, g/l$$

$$K_m = 0.5 \, yr^{-1}$$

$$i = 15\%/yr = 0.15/yr$$

$$C_{op} = \$200{,}000/yr$$

ANSWER

Inserting these data into eqns 10.10 and 10.17 at each of seven years yields the following table:

Year	c_F/c_p	PV(R)	PV(C_{op})	$PV_{(t_p)}$
1	0.606	$923,596	−$186,660	**$736,936**
2	0.370	1,410,717	−348,000	**1,062,717**
3	0.223	1,607,633	−490,000	**1,177,633**
4	0.135	1,803,136	−612,400	**1,190,736**
5	0.082	1,874,602	−719,600	**1,155,002**
6	0.05	1,912,295	−812,800	**1,099,495**
7	0.03	1,932,174	−893,000	**1,039,174**

Using computed results in the table, the gross production present value, $PV_{(t_p)}$, is plotted versus years of production in Fig. 10.20. This plot shows that the maximum value of the $PV_{(t_p)}$ occurs near year four, and is declining by year five because the incremental production is less than the annual fixed cost in the out years. It is interesting, however, that little change in $PV_{(t_p)}$ occurs after year three.

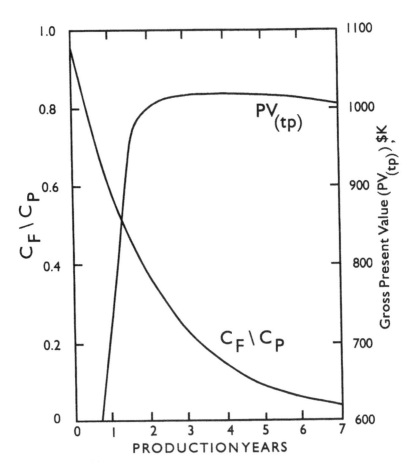

Figure 10.20. Plot of relative pregnant liquor concentration, c_F/c_p, and gross production present value, $PV_{(t_p)}$, for example problem.

WASTE WATER TREATMENT AND AQUIFER RESTORATION

The major environmental goals for *in situ* leaching are (1) containing leaching solutions during solution mining and (2) treating waste water streams during operations and later during aquifer restoration. Containment can be verified through data from samples obtained from monitor wells surrounding the mining site or downstream with respect to groundwater flow.

Waste water treatment options are listed in Fig. 10.21 and include several possible methods for precipitating, concentrating or otherwise removing

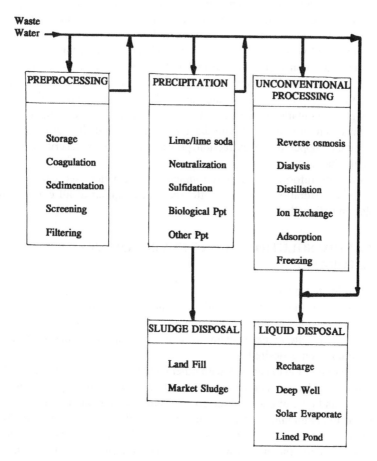

Figure 10.21. Waste water treatment options for bleed streams and restoration following solution mining.

dissolved solids. This list is not exhaustive. The result of waste water processing is usually a sludge or concentrated liquid for further processing or environmentally acceptable disposal.

Restoration aims at returning the mined deposit to the geochemical conditions that existed before mining. This may include (1) *in situ* treatment, (2) removal of contaminated groundwater followed by surface treatment of this waste water to remove contaminating dissolved solids, or (3) a combination of *in situ* and surface treatments.

Both *in situ* restoration and removal of contaminated water will utilize the existing injection and production wells that were established for solution mining. *In situ* restoration methods may include oxidation (e.g., by air injection), reduction (e.g., with H_2S), chemical injection and neutralization, and injection of anaerobic bacteria to destroy the contaminant.

Removal of volatile contaminants, (not related to mining) by air ventilation is being considered for ground polluted with organics. Under this plan, air will be blown into the vadose zone through boreholes drilled from the surface down to the water table. The injected air will pass upward through the contaminated ground, sweeping out the volatile contaminants. This is essentially identical with the approach to forced air ventilation as a secondary copper dump extraction process, which was described at the end of Chapter 8.

For more information on environmental remediation, see Chapter 15. The final restoration activity is often cement plugging of all exploration holes and all wells.

URANIUM SOLUTION MINING ENVIRONMENTAL RESTORATION

The ultimate decomposition of alkali carbonate lixiviant and hydrogen peroxide is water, carbon dioxide and the alkali cation; only the alkali cation is of concern. Since ammonium ion has been the dominant cation used in carbonate *in situ* leaching of uranium, removal or modification of ammonium ion is desired.

Ammonium ion exchanges with other cations in clays that may be in the formation. The amount and tenacity of this exchange depends on the amount and type of clay and the concentration of ammonium ions in solution. If the ammonium ions were not bound to the clay, flushing with 3–5 volumes of fresh water per volume of contaminated water would usually suffice to return the ground water to baseline concentrations. The replaced water would be removed by pumping and either treated on the surface or used for its irrigation and fertilization value.

Results of both experiments and numerical simulation show that ammonium ion migration behaves analogously to movement in an ion chromatograph. In the absence of net groundwater lateral flow, this migration is very slow, particularly after the carbonate and bicarbonate anions have been removed from solution. Ammonia breaks down to nitrates under aerobic conditions but does not do so readily under anaerobic conditions. However, ammonia is not particularly toxic and continuance of ammonia adsorbed on clays underground does not appear to be very harmful.

PROBLEMS

1. A typical South Texas uranium mine covers 20 acres, has a 4 m thick ore zone and has an effective microporosity of 25%. How much groundwater is likely to be affected by the solution mining process if lateral migration is prevented by using guard wells?

2. Solution mining a deep copper ore deposit is being planned using an equal number of injection and production wells in an extended five spot wellfield. 22,700 tonne of copper will be produced each year once full capacity is reached. The design basis is 0.5 wt pct copper ore grade, 5% formation microporosity, 50% ultimate copper extraction, 20 yr mine life, 3 yr well lifetime, and a flow rate of 3,000 cm^3/s per well.

   Two lixiviant composition options are being considered. The first option provides pregnant liquor copper concentrations at full production capacity of 2 g/l while the second lixiviant option provides 8 g/l. The ore volume initially leached is 10×10^6 m^3. For each lixiviant, compute:

   (a) total solution flow rate (pregnant liquor production rate) in cm^3/s;
   (b) number of initial injection and production wells;
   (c) total number of injection and production wells over 20 yr;
   (d) volume, m^3, of solution stored in the **initial** wellfield;
   (e) time (days) to displace one pore volume in the initial wellfield;
   (f) percent of well lifetime used during the start-up period (less than full capacity);
   (g) approximate cumulative copper production per production well during its lifetime, in tonnes.

3. Consider two wellfield cases, case A with a one year well life and an initial wellfield of 20 wells at a total wellfield cost of $2,000,000 and case B with a 3 year well life and 42 wells required in the initial wellfield. Well cost is the same in each case. The planned mine life is 20 yr. At a 20% return on investment (ROI) the capital cost of case A

including the subsequent replacement wells (19 times the original wells) has a present value of $-\$11,600,000$. The present value of case B wellfield investments over 20 yr is 2.3 times the capital cost of the initial case B wellfield. Which case will have the smaller investment negative present value? You don't need the NPV formula or eqn 10.17 to do this problem.

4. It is planned to replace old declining injection and production wells in a uranium leaching wellfield. The new wells will be a step out from the present wells in the same ore deposit aquifer. Using eqn 10.17, estimate the gross present value of the expected revenue at the time the new wells reach their peak uranium concentration if the planned well service life is three years beyond the production peak. What do you expect the Uranium concentration in the pregnant liquor to be at the end of this three-year service life?

   The present wellfield parameters are:

$$c_p = 2.0 \, g/l \text{ of } U_3O_8 \text{ equivalent,}$$

$$P_m = \$12/lb \text{ of } U_3O_8,$$

$$Q = 500 \, gal/min,$$

$$K_m = 0.05 \, month^{-1},$$

$$i = 18\%/yr,$$

$$C_{op} = \$1,000,000/yr.$$

REFERENCES AND SUGGESTED FURTHER READING

Aplan, F.F., McKinney, W.A. and Pernichele, A.D. (Eds.) (1974). Chap. 5, *Solution Mining Symposium*, Society of Mining Engineers, Littleton, CO.

Braun, R.L., Lewis, A.E. and Wadsworth, M.E. (1974). In-place leaching of primary sulfide ores: Laboratory leaching data and kinetics model. Aplan, F.F. et al. (Eds.). *Solution Mining Symposium*, AIME, New York. pp. 295.

Braun, R.L., Lewis, A.E. and Wadsworth, M.E. (1974). *Met. Trans.*, 5, p. 1717.

Chamberlain, P.G. (1979). Evaluating ore bodies for leaching and permeability measurements. *In Situ uranium Mining and Ground Water Restoration*, Society of Mining Engineers, Littleton, CO. pp. 7–22.

Davidson, D.H. (1988). Method of estimating optimum economics for *in situ* leaching (ISL) of copper. *Preprint 88–4*, Society of Mining Engineers, Littleton, CO.

Davis, G.R. and Mays, W.M. (Eds.) (1978). *South Texas Uranium Seminar*. Society of Mining Engineers, Littleton, CO.

George, C. and Faul, R. (1985). Cementing technologies for solution mining wells and salt storage domes. Schlitt, W.J. and Larson, W.C. (Eds.), *Salts & Brines '85*, Society of Mining Engineers, Littleton, CO, pp. 11–23.

Jacobson, R.H. and Murphy, J.W. (1988). A review of solution mining—Part 1, *Min. Res. Eng.*, I, pp. 67–83.

Merritt, R.C. (1971). *The Extractive Metallurgy of Uranium*, Colorado School of Mines Research Institute, Golden, CO.

Murphy, J.W., McGrew, K.J. and Jacobson, R.H. (May 1984). Application of laboratory stream tube testing to economic evaluations of solution mining. *Mineral and Met. Processing*, I, pp. 49–56.

O'Neil, T. (1991, Aug). *In situ* copper mining at Santa Cruz: a project update, *Mining Engineering*, pp. 1031–1034.

Richner, D.R., Shock, D.A. and Ahles, J.K. (1992). Hartman, H.L. (Ed.), *Mining Engineering Handbook*, Society of Mining, Metallurgy and Exploration, Littleton, CO, Chap. 153.

Schlitt, W.J. and Hiskey, J.B. (Eds.) (1981). *Interfacing Technologies in Solution Mining*, Society of Mining Engineers & Society of Petroleum Engineers, Littleton, CO.

Schlitt, W.J. and Shock, D.A. (Eds.) (1979). *In Situ Uranium Mining and Ground Water Restoration*, Society of Mining Engineers, Littleton, CO.

Shock, D.A. and Conley, F.R. (1974). Solution mining—Its promise and its problems. Aplan, F.F. et al. (Eds.). *Solution Mining Symposium*, AIME, New York. pp. 79–97.

ELEVEN

In Situ *Leaching Hydrology*

This chapter will begin with a few simple examples of the mechanics of radial flow to and from wells penetrating a confined aquifer, and will be limited to steady flow. These examples can be applied to *in situ* solution mining of flooded horizontal ore deposits confined by impermeable strata above and below the ore deposit. Although this is a rather restrictive geometry, it is an adequate approximation for many sedimentary ore deposits that are industrially important. Extensive literature in hydrology, petroleum engineering and porous media flow generally treat many other examples of both steady and unsteady flow. A few of these that are also useful for solution mining wellfield configurations are presented.

When well pumping, or injection, begins, the flow will not be steady and will only approach a steady state asymptotically after prolonged pumping or injection. Consequently, pumping or injection tests conducted to evaluate an aquifer are usually not carried out over enough time to reasonably approximate steady flow. Under these conditions, non-steady flow methods of evaluation must be used. These methods invariably involve measurements made at various time intervals, for example the drawdown of groundwater at an observation well in the vicinity of the well being pumped. The complex solution flow mathematics and associated graphical methods used to interpret test data during non-steady flow are beyond the scope of this text but can be found in groundwater texts, e.g., Fetter (1988) and DeWeist (1965), as well as petroleum engineering texts.

STEADY STATE FLOW ASSUMPTIONS AND DEFINITIONS

The limiting assumptions (DeWiest, 1965) are: (1) the fluid is incompressible, (2) the ground (ore zone) is saturated (flooded), and (3) the permeability is isotropic and uniform. These assumptions are often valid for aqueous solutions and solution mining, at least at a macroscale. Steady flow relations in linear coordinates were presented in Chapter 7. Recall that the superficial flow velocity through porous media in linear coordinates is

$$u_l[Lt^{-1}] = \frac{Q_l}{A}\left[\frac{L^3t^{-1}}{L^2}\right] = \left(\frac{k_i}{\mu}\right)\frac{dp}{dx}, \tag{7.9}$$

where k_i is the intrinsic permeability and is a property of the porous media independent of the fluid. The hydraulic conductivity, K, is also used in hydrology and is a function of both the porous media and the fluid,

$$K[Lt^{-1}] = \frac{k_i g \rho_l}{\mu}. \tag{7.8}$$

FLOW FROM A SINGLE WELL IN A HORIZONTAL AQUIFER

The following discussion (DeWiest, 1965), pertains to steady flow from a single well being pumped (produced) at a constant rate, Q. If the well is used for fluid injection, the flow is equal and opposite using the same equations and in effect is the production flow's "mirror image." The aquifer is unlimited in extent in all lateral directions. Under the assumptions of isotropic, homogeneous permeability, the flow to the well will have radial symmetry and the head (hydrostatic pressure) will be constant along the perimeter of any circle concentric with the well.

Confined Ore-Bearing Aquifer

If complete penetration of the aquifer by the well is assumed, the flow is parallel to the horizontal formation and to the confining horizontal upper and lower strata. The formation is also assumed to be of constant thickness, infinite extent and negligible deviation from horizontal. Flow follows streamlines, s', in opposite direction to the rays, r, emanating from the center of the well, taken as the origin of plane polar coordinates, as shown in Fig. 11.1.

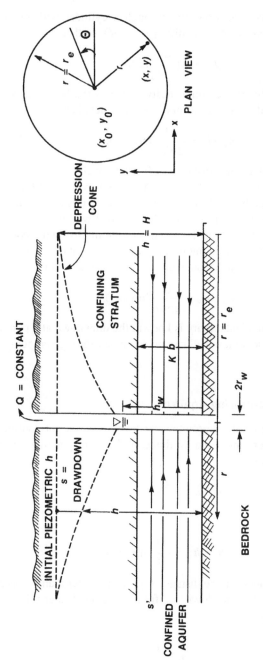

Figure 11.1. Radial flow to a well penetrating an aquifer vertically confined by the upper and lower strata; section view (DeWiest, 1965).

Hence, Q equals the flow rate through a cylinder with radius r and height b, the thickness of the aquifer (ore zone), so that

$$Q = -K\left(\frac{dh}{dr}\right)2\pi rb,\qquad(11.1)$$

where K is the hydraulic conductivity and h is the head. Equation 11.1 may be integrated after separation of the variables; hence,

$$h_2 - h_1 = \frac{Q}{2\pi Kb}\ln(r_2/r_1),\qquad(11.2)$$

where h_1 is the head at a distance r_1 from the well, and h_2 is the head at a greater distance, r_2, from the well. As shown by Fig. 10.15, measuring the depth to groundwater from two observation wells penetrating the formation at distances, r_1 and r_2, from the pumped well will give h_1 and h_2, which can be used with eqn 11.2 to determine the hydraulic conductivity, K, of the formation. Furthermore, the nearer observation wells can be the pumped well itself, with the radial distance, r_1, the radius of the well bore, r_w:

$$h_2 - h_w = \frac{Q}{2\pi Kb}\ln(r_2/r_w).\qquad(11.3)$$

If the hydraulic conductivity is known, the drawdown at a pumped well relative to an observation well can be estimated for a given pumping rate, Q, using eqn 11.3.

Unconfined Ore-Bearing Aquifer

The piezometric head, or groundwater level, is often not confined by an impermeable overlying strata. In this case the corresponding steady radial flow equations are:

$$h_2^2 - h_1^2 = \frac{Q}{\pi K}\ln(r_2/r_1)\qquad(11.4)$$

and

$$h_2^2 - h_w^2 = \frac{Q}{\pi K}\ln(r_2/r_w).\qquad(11.5a)$$

But, for a sufficiently large distance (r_2 much greater than r_w) h_2 approximates H_{GW} and the drawdown, s_w, from the groundwater reference plane is approximately

$$s_w^2 \approx \frac{Q}{\pi K}\ln(r_2/r_w).\qquad(11.5b)$$

EXAMPLE PROBLEM

Using three cased exploration drill holes in a confined but poorly consolidated sandy ore deposit, drawdown measurements were made on two holes, while the third hole was pumped at a constant rate of 100 gal/min. The drawdowns from the surface, measured after no further detectable drawdown occurred, were 430 and 440 ft for drill holes located 150 and 60 ft, respectively, from the pumped hole. The ore bearing aquifer thickness is 20 ft. Estimate the hydraulic conductivity in ft/min and cm/s and compare the calculated results with Table 7.5.

ANSWER
The difference in depth from the surface to the two measurement wells is 10 ft and equal to $h_2 - h_1$. Rearranging eqn 11.2, which describes single well flow for a confined aquifer, yields:

$$K = \frac{Q}{2\pi(h_2 - h_1)b} \ln(r_2/r_1),$$

$$\ln(r_2/r_1) = \ln(150/60) = 0.91629,$$

$$K = \frac{100 \, \text{GPM}(2.23 \times 10^{-3} \, \text{ft/s})(60 \, \text{s/min})(0.91629)}{2\pi(10 \, \text{ft})(20 \, \text{ft})},$$

$$K = 9.756 \times 10^{-3} \, \text{ft/min},$$

$$= 0.0495 \, \text{mm/s},$$

$$\approx 5 \times 10^{-2} \, \text{mm/s}.$$

Because this answer is within the expected range of a well-sorted sand, according to Table 7.5, it is reasonable.

STEADY FLOW FROM AN INJECTION WELL TO A PRODUCTION WELL IN A CONFINED AQUIFER

Fig. 11.2 illustrates flow from a single injection well (left) to a single producing well (right) in a **confined** horizontal aquifer, with both well flows equal (DeWeist, 1965). The upper view is a vertical section through the two wells showing the head (piezometric surface or water table) near the wells. The lower plan view shows the flow streamlines from the injection well to the production well and the equipotential lines (equal hydraulic head). Flow streamlines are always perpendicular to the equipotential lines.

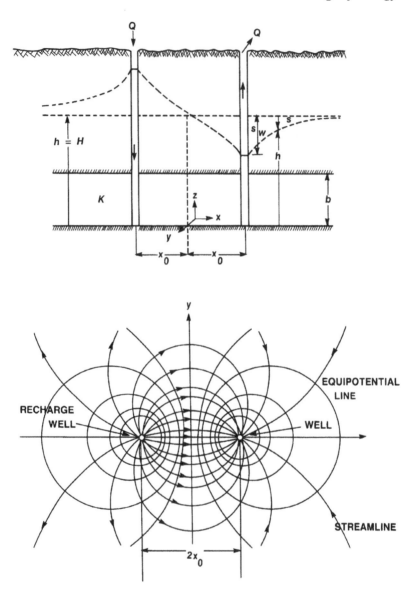

Figure 11.2. Injection well and production well in a vertically confined aquifer; section and plan views (DeWiest, 1965).

The drawdown, s_w, is equal to the increase in head above the water table in the injection well. A steady injection pressure can be achieved by continuously filling the injection tube to maintain the appropriate height. Using a packer for sealing off flow into the confining formations, the injection well can be pressured to heads above the ground surface with an injection pump of sufficient pressure rating and capacity. The drawdown, s_w, at the producing well is related to the distance between wells, $2x_0$, by the following equation:

$$s_w = \left(\frac{Q}{4\pi Kb} \right) \ln \left[\frac{(2x_0 - r_w)^2}{r_w^2} \right] \tag{11.6}$$

and when $x_0 \gg r_w$, then:

$$s_w = \left(\frac{Q}{2\pi Kb} \right) \ln \left[\frac{2x_0}{r_w} \right]. \tag{11.7}$$

However, it is worth noting that if a tracer is injected, less than the amount of tracer injected will return in a finite period because of some lateral flow outside the region following streamlines that loop away over great distances. This loss is indicated in Fig. 11.2 by the streamlines that do not converge within the illustrated area of the figure, and it was discussed in the preceding chapter.

WELL RADIUS AND SKIN EFFECTS IN HORIZONTALLY CONFINED AQUIFERS

The well radius, r_w, has a major influence on the flow rate that can be achieved with the same head. As an example, consider an injection well and a production well, each with a 40 mm well radius, separated by 64 m. If the well radius is expanded, for example, by well stimulation or by drilling the hole at a larger diameter, then the drawdown, s_w, at constant flow rate, Q (given by eqn 11.7) will decrease. The resulting drawdown change as a function of well radius change is plotted in Fig. 11.3. The drawdown change is expressed as a percentage of the drawdown at $r_{w(0)} = 40$ mm. Doubling the well radius provides a 10% decrease in drawdown.

Conversely, if the drawdown, s_w, is held constant the flow rate, Q, will increase with borehole radius, as shown in Fig. 11.4.

The "skin effect" is a zone surrounding the well with a substantially different permeability than the aquifer permeability. In many cases it has been found that the permeability of the formation near the well bore is reduced as a result of drilling and well completion. Invasion by drilling

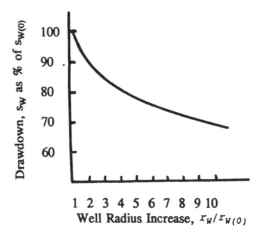

Figure 11.3. Drawdown lowering, at constant Q, by increasing the well bore radius, $r_{w(0)} = 40\,\text{mm}$, $x_0 = 32\,\text{m}$.

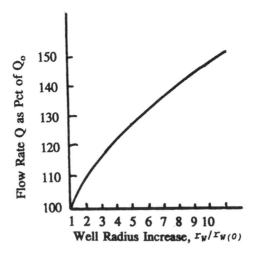

Figure 11.4. Increase in flow rate at constant drawdown, s_w, with increasing well bore radius, $r_{w(0)} = 40\,\text{mm}$, $x_0 = 32\,\text{m}$.

fluids, dispersion of clays, presence of mud cake and cement, partial well penetration, limited perforation of the injection/production casing, and plugging of the perforations are some of the occurrences that can lower permeability adjacent to the borehole.

The effect of a reduction in permeability near the well is illustrated in Fig. 11.5. The skin effect is usually accounted for as an additional pressure drop that is proportional to the flow rate, Q. Consequently, for a horizontally confined aquifer,

$$s_w = \left(\frac{Q}{4\pi Kb}\right)\left[\ln\left(\frac{(2x_0 - r_w)^2}{r_w^2}\right) + 2s_s\right] \tag{11.8}$$

and when $x_0 \gg r_w$, then,

$$s_w \simeq \left(\frac{Q}{2\pi Kb}\right)\left[\ln\left(\frac{2x_0}{r_w}\right) + s_s\right]. \tag{11.9}$$

The radius, r_s, of the skin zone around the well and the hydraulic conductivity, K_s, in this zone are related to the skin factor s_s by:

$$s_s = \left(\frac{K}{K_s} - 1\right)\ln\left(\frac{r_s}{r_w}\right). \tag{11.10}$$

Thus, if the skin hydraulic conductivity is less than in the formation, s_s will be positive, and if the hydraulic conductivities are equal, $s_s = 0$. But, if the hydraulic conductivity in the skin is greater than that in the formation (for example, if caused by well stimulation), then s_s is negative. Hydraulically fractured petroleum wells often show values of s_s ranging from -3 to -5.

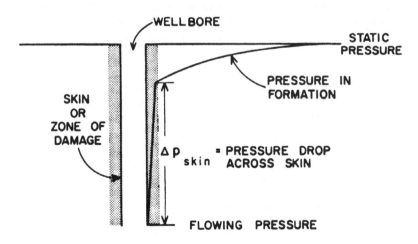

Figure 11.5. Pressure distribution near a well with a skin effect, $s_s > 0$.

STEADY FLOW FOR WELLFIELD CONFIGURATIONS

Fluid flow equations, taken from Davidson (1988), for three wellfield geometric configurations that may be encountered in solution mining follow: (1) radial flow in the horizontal direction from an *extended* five spot pattern of vertical wells drilled from the surface, or from stations at an underground mine level, (2) vertical flow between two sets of horizontal fractures separated by a fixed distance in the vertical direction, and (3) flow between two fan patterns of short wells that have been drilled perpendicular to underground mine crosscuts. In this last configuration, which pertains to solution mining wall rock in a pre-existing conventional mine, flow is parallel with the crosscut. The five spot pattern was adequately described in the previous chapter, but the geometric configurations of the other wellfields will be explained further.

Expressions for the solution flow rates, Q, from injection wells to production wells for each of these well patterns are shown in eqns 11.11–11.13, using SI units. Before using these equations, hydraulic conductivities must be converted to intrinsic permeabilities. The pressure difference, ΔP_{IP}, between the injection and production wells in a five spot pattern, is twice the drawdown, s_w, at the production well, which is illustrated in Fig. 11.2.

Table 11.1 shows the SI units used in these flow equations along with three other commonly used measurement systems and the conversion factors, C.F., that must be multiplied against the right side of the equations to obtain correct values of Q, in the units of Q that are shown in Table 11.1. However, to reduce the chance of error, it is recommended that all parameters be converted to SI units before working problems. Refer to the appendix for conversions of parameter units to SI units.

Table 11.1 Dimensions Used with Flow Eqns 11.11–11.13.
C.F. is a Parameter Conversion Factor Appearing in the
Numerator of the Right Side of Each Equation.

Parameter	SI	Metric	Common USA	Common Hydraulic
Q	m³/s	l/s	gal/min	gal/min
ΔP_{IP}	Pa	kPa	psi	inches of water
S_5, b, H_c, h_f, x_f	m	m	ft	ft
k_i	m²	m²	darcy	darcy
μ	Pa s	cP	cP	cP
C.F.	1	1.0×10^9	0.3277	1.183×10^{-3}

Flow in a Five Spot Extended Pattern of Vertical Wells

This pattern was shown in Fig. 10.17. Equally spaced vertical wells penetrate the producing aquifer with alternate injection and production wells. Production wells are separated from each other by a distance S_5 and injection wells are also separated from each other by a distance S_5. The aquifer is confined at the top (roof) and bottom (floor) by impermeable strata, but it is unbounded in the lateral direction. The flow for each production well is Q, and each injection well flow is equal and opposite.

The extended five spot flow equation expressed with the intrinsic permeability and SI units is:

$$Q = \frac{k_i \Delta P_{IP}}{\mu} \left[\frac{\pi b}{\ln(S_5/\sqrt{2}r_w) - 0.619} \right]. \qquad (11.11)$$

Flow Between Two Levels of Horizontal Fractures

This flow pattern is illustrated in Fig. 11.6. Two plane parallel fracture patterns are established at each of two horizontal levels in the mine. The horizontal fractures are created by drilling a pattern of vertical wells from the surface from which horizontal fractures are introduced at specified depths. The horizontal fractures at different levels are alternately connected to injection wells and production wells. Leach solution is injected into horizontal fractures from the injection wells; then it flows in the vertical direction to horizontal fractures at the next level (over the intervening distance, H_c); and then the solution flows to the production wells. An example of this flow pattern occurs at the Van Dyke Mine at Miami, Arizona, that was discussed in Chapter 10.

Usually the well pattern is identical with an extended five spot pattern. Alternate wells are used to generate the injection horizontal fractures while the remaining wells are used to generate the production horizontal fractures. Horizontal fractures at the different level are assumed to overlap completely, and the pressure drop for leach solution flowing through the horizontal fractures is, ideally, negligible. Hence, the flow is linear between two parallel equipotential planes (horizontal) separated by distance H_c over an area of $(S_5)^2$.

A suitable fracturing technique such as hydrofracturing is used. Hydrofracturing at moderate depth will usually create horizontal fractures, rather than other fracture orientations, because of the limited lithostatic pressure. Fracture is favored along rock formation bedding planes, so this *in situ* leaching pattern works well in sedimentary rock that remains nearly horizontal when mining begins.

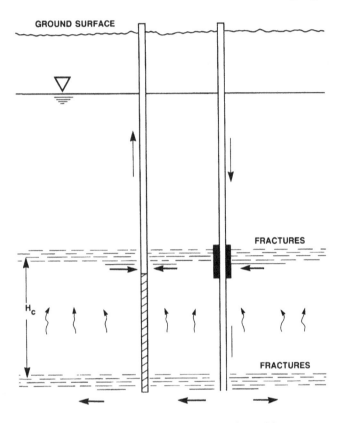

Figure 11.6.   Solution flow between parallel, horizontal fracture patterns.

Care in drilling, well completion, and fracturing must be taken to ensure that solution does not bypass the ore zone and transfer directly through one of the fracture patterns from an injection well to a production well. Wells connected to the lower fracture horizon must be sealed from the upper fracture horizon. Wells connected to the upper fracture level can be plugged below that level.

For a given injection pressure difference between the two fracture horizons, the solution flow rate will vary inversely with H_c. Consequently, it may be necessary to limit H_c and to have several injection and production levels in a thick orebody. This can be accomplished by drilling the wells to the bottom of the orebody and establishing leaching layer 1 (the bottom layer) with height H_c, which can be thought of as the orebody's first *in situ* mining "stope." Successive layers can be *in situ* mined sequentially upward, as shown in Fig. 11.7.

The equation for flow from each production well between horizontal fracture patterns in terms of the intrinsic permeability, k_i, using SI units is:

$$Q = \frac{k_i \Delta P_{IP}}{\mu} \left(\frac{S_5^2}{H_c} \right). \tag{11.12}$$

Flow Between Fan Well Patterns

This *in situ* mining method is well adapted to a mine with pre-existing underground workings that include many crosscuts and drifts, which are horizontal openings similar to tunnels, that can be used to access the drilling of injection and production wells. One-half of a fan pattern of short wells is shown in Fig. 11.8. Holes are drilled from an underground drill station located in a crosscut or drift in a direction normal to the axis of the crosscut. These wells radiate in a fan pattern with each well equally spaced at θ radians apart. Wells are drilled to a depth of h_f, or somewhat greater distances when directed toward corners of the ore block, as shown in Fig. 11.8. Generally the well depth, h_f, will not exceed 30 m.

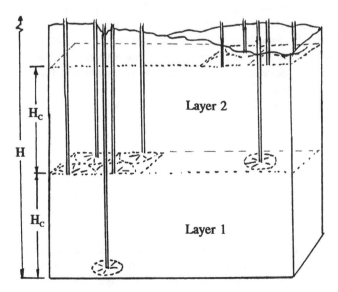

Figure 11.7. Sequential layer leaching of a thick orebody with multiple horizontal fracture patterns (Davidson, 1988).

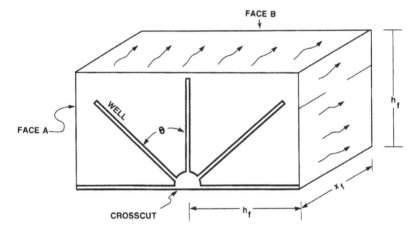

Figure 11.8. Cutaway of half of a fan pattern used for *in situ* leaching with underground crosscuts and drifts (Davidson, 1988).

Leach solution is injected in all of the wells in the fan located at each alternate crosscut station, e.g., Face A in Fig. 11.8. Solution migrates parallel with the crosscut, a distance x_f, in both the forward and backward directions to adjacent production well fans. A production well fan station is indicated by Face B in Fig. 11.8. The volume of the leached ore block between an injection fan pattern of wells and a production fan pattern of wells is $4h_f^2x_f$. Several alternating injection and production fans will occur down a long drift.

The equation for fan pattern flow to a crosscut station in terms of the intrinsic permeability, k_i, using SI units is:

$$Q = \frac{k_i \Delta P_{IP}}{\mu} \left(\frac{\pi h_f}{(\pi x_f/2h_f) + (2\Theta/\pi)\ln(h_f\Theta/4r_w)} \right). \qquad (11.13)$$

Equations 11.11–11.13 for the three wellfield geometric configurations were given for the intrinsic permeability at the standard condition of 20°C. Note that higher operating temperatures and high concentrations of dissolved salts may change the leaching solution viscosity from that of water at 20°C (1.0 cP).

EXAMPLE PROBLEM

For an extended five-spot wellfield in a confined aquifer, estimate the pumping rate, Q, in gallons per minute at each production well, if the drawdown, s_w, is limited to 15 ft. The wellfield conditions are: $K = 5 \times 10^{-3}$ mm/s, the cased well diameters are 8 in, the separation distance between production wells is 50 ft, and the formation thickness is 40 ft.

ANSWER
Q is determined by using eqn 11.11:

$$Q = \frac{k_i \Delta P_{IP}}{\mu} \left[\frac{\pi b}{\ln(S_5/\sqrt{2}r_w - 0.619} \right].$$

The intrinsic permeability is obtained from the hydraulic conductivity using eqn 7.10b:

$$k_i \, [m^2] = 5 \times 10^{-3} \, mm/s \, (1.02 \times 10^{-10}) = 5.1 \times 10^{-13} \, m^2.$$

The dilute solution is assumed to have the same viscosity of water at 20°C, 1 cP, which is 1×10^{-3} Pa s. The pressure difference from production well to injection well is twice the drawdown, or 30 ft (360 in), which converts to 89,568 Pa. The 8 in borehole of the well is $r_w = 4$ in, and the distance between wells in the same units is 600 in. Inserting these in eqn 11.11, and using SI units, yields:

$$Q \, [m^3/s] = \frac{5.1 \times 10^{-13}(89568)}{1 \times 10^{-3}} \left[\frac{(40 \, ft)\pi(0.3048 \, m/ft)}{\ln(600/1.414(4)) - 0.619} \right]$$

$$Q = 4.57 \times 10^{-5} \, m^2/s \, [9.47 \, m]$$

$$= 4.33 \times 10^{-4} \, m^{-3}/s$$

$$= 4.33 \times 10^{-4} \, m^{-3}/s \times 1.58 \times 10^4 \, gal/min \, per \, m^3/s$$

$$= 6.84 \, gal/min.$$

A sucker rod pump would be appropriate for this low flow rate.

ENVIRONMENTAL CONTAINMENT

Many *in situ* mined ore deposits are small in area compared to the aquifers in which they occur. This is particularly true for uranium deposits, which tend to occur in pods. From both an economic and environmental perspective, minimizing lixiviant escape from the solution mining zone is necessary. It is possible to surround the mining area with "guard wells" that recharge relatively small amounts of groundwater in these additional wells to obtain a no-flow boundary around the perimeter of the ore zone.

In the absence of guard wells injecting groundwater, using production wells and cones of depression in the piezometric surface at the perimeter of the ore body is desirable. In either of these cases there will be a net flow of groundwater into the ore zone. In aggregate, more liquid will be produced (pumped) than injected. After recovery of the valuable metal or mineral, the solution in excess of that needed to regenerate the lixiviant for reinjection must be treated by methods described in Chapter 10. Then, it must be disposed of in an environmentally acceptable manner, which may include reinjection as restored groundwater at a different location in the aquifer.

In some ore deposits, groundwater flows into and out of the ore zone under a hydraulic gradient. This must be considered in designing an environmental containment system for the lixiviant. If the natural hydraulic gradient is significant, it may be necessary to place a series of guard wells down stream with sufficient groundwater recharge, as a line drive, to stop continued downstream flow of contaminated groundwater. See Chapter 15 for a more detailed discussion of groundwater remediation.

PROBLEMS

1. Preparatory to solution mining in a **confined** aquifer, estimate the *average* hydraulic conductivity, in cm/s, and the intrinsic permeability in darcys, from the following six sets of drawdown data for three wells penetrating the formation. Wells A and B are separated by 60 m. Wells B and C are separated by 50 m. Wells C and A are separated by 20 m. Each well, in turn, is pumped at a steady flow of 120 l/min while the drawdown is measured, in meters, at the other two wells. The wellbore radii are 4 cm and the formation thickness is 5.0 m.

DRAWDOWNS MEASURED AT THESE WELLS (in meters)

Pumping Well	Well A	Well B	Well C
A		1.54	3.10
B	1.60		2.30
C	2.70	1.55	

Which well, if any, has the greater skin effect, s_s, positive? Explain your answers. Hint: Develop an equation using eqn 11.3 and two drawdowns (s_1 and s_2) while pumping Well 3.

2. An **extended** five spot wellfield is planned for a uranium solution mining project with well spacing at $S_5 = 64$ m, which is approximately one production well per acre. The vertically **confined** ore deposit aquifer is 5 m thick with an average intrinsic permeability of 1 darcy. Well bore radius is 40 mm (about 1.5 in). Estimate (compute) the drawdown in meters at the production wells relative to the injection wells if the production wells are pumped at 60 l/min.

3. An *in situ* copper porphyry solution mining **extended** five spot wellfield is being planned, but the intrinsic permeability is only 5 millidarcy. The orebody begins at a depth of 500 m and extends down another 500 m. The ore grade is 0.7% Cu and laboratory tests indicate that 50% is extractable. With a well radius of 40 mm and spacing, S_5, of 64 m, estimate the flow rate from each production well in liters per minute if the production well drawdown is limited to 50 m (a total head difference between production and injection wells equal to 100 m).

4. Having installed and operated the solution mine of Problem 3, it has been suggested that the production well flow rate an be increased by installing horizontal hydrofractures at the top of the 500 m thick ore zone connected to the injection wells, which would be plugged below this level, and connecting the production wells to horizontal hydrofractures at the bottom of the ore zone. The wells would be operated in the same pumping manner with a total head difference between production and injection wells of 100 m. Evaluate this suggestion and recommend whether it should, or should not, be adopted. Show calculations to justify your recommendation.

5. The Daisy Mae Mine is a steep dipping copper vein deposit with oxide copper minerals that was worked as an underground mine using hand-mining methods with drifts running in the ore vein and separated 70 ft apart along the vein's dip. Solution mining the lower grade disseminated copper surrounding the mined-out richer veins with a lixiviant solution containing sulfuric acid is being considered. The existing drifts will be reopened and fan well leaching with 40 ft wells drilled from drift stations will be used. Eight wells, separated by 45 degrees, will be drilled at each drift station. These wells will have 1-inch diameter pipe cemented in at the collar and pressurized to 75 psi during leaching. Well stations along each drift will be 50 ft apart, with alternating production and injection fan wells; i.e., production fan wells will be 100 ft apart.

The hydraulic conductivity of the heavily altered wall rock around the original veins is not known, but it is likely to vary between 10^{-7} and 10^{-4} mm/s. Compute the following expected production well flow rates for each of the listed hydraulic conductivities:

Hydraulic conductivity mm/s	Flow rate per single well gal/min
1×10^{-7}	?
1×10^{-6}	?
1×10^{-5}	?
1×10^{-4}	?

REFERENCES AND SUGGESTED FURTHER READING

Bommer, P.M. and Schecter, R.S. (1979). Mathematical Modeling of *In Situ* Uranium Leaching, W.J. Schlitt and D.A. Shock (Eds.), *In Situ Uranium Mining & Ground Water Restoration*, Society of Mining Engineers, Littleton, CO.

Carslaw, H.S. and Jaeger, J.C. (1959). *Conduction of Heat in Solids*, Oxford and Clarendon Press.

Davidson, D.H. (Apr 1988). Generic *In Situ* Copper Mine Design Manual, Vol. II, Science Applications International Corp., Contract J0267001 for the U.S. Bureau of Mines.

DeNevers, N. (1970). *Fluid Mechanics*, Addison-Wesley, Menlo Park, CA.

DeWiest, R.J.M. (1965). *Geohydrology*, J. Wiley & Sons, New York.

Fetter, C.W. (1988). *Applied Hydrogeology*, MacMillan Publishing, New York, NY, ISBN 0-675-20887-4.

TWELVE

Numerical Simulation of Fragmented Rock Leaching

INTRODUCTION

The primary goals of solution mining computer simulation are (1) to predict with reasonable accuracy results from leaching, or other solution mining activity and (2) to gain insight into significant phenomena affecting solution mining.

Often the critical leaching result to be predicted by the model is the amount of metal extraction over time (the metal extraction curve), at realistic mine conditions. Since all numerical simulations are only approximations to complex situations, a model is considered adequate or good if it predicts with reasonable accuracy actual mine or experimental results, preferably without any adjustable parameters. In the author's view, simulation programs that fit metal extraction curves using adjustable parameters suffer on two important points: (1) loss of quantitative credibility and (2) loss of prediction capability for extrapolation to another case because of the inability to determine the adjustable parameters except *a posteriori*. Hence, they may lack engineering utility.

Predicting metal extraction curves with sufficient accuracy and credibility to provide reliable engineering predictions has not always been successful. Consequently, solution mining investment analyses have not been based on numerical simulation. However, verification of many significant phenomena determining the outcome of solution mining has been good. Occasionally interesting and important new insights have come from the results of applying numerical simulation models.

A major problem in model verification has been obtaining accurate and unambiguous experimental results for comparison. Many aspects of solution mining cannot be readily replicated in the laboratory. Among these are the effects of large rock sizes, rock size segregation (classification) resulting from dumping, solution channeling during percolation and high heaps and dumps. Experimental evaluation of a 50-meter mine waste dump in a column test would require a 15-story tower to contain the column, which would be very expensive.

Several models have been developed in response to large scale, relatively well-controlled experiments. At least one large scale controlled experiment was conducted, in part, to test a simulation model, Cathles and Murr (1980) and Murr et al. (1982). However, large scale experiments are very expensive and, consequently, there have been few of them.

Numerical simulation has generally been divided into two classes of models: empirical and deterministic.

Empirical models, based on extensive operating history and data, can be useful at a specific solution mining site for continued operational planning and control. Often they involve the application of statistics to operating data. Operating rules of thumb can be very useful and "expert systems," combining both a computer based empirical model with proven rules of thumb, may eventually be employed.

Deterministic models, using what are believed to be the most significant physicochemical factors, have predominated in the solution mining literature. However, obtaining representative physicochemical characteristics of the ore, needed as input parameters to the chosen model, is a major problem in verifying and using deterministic models. For example, it is easy to measure the rock internal microporosity from a sample representing a 1 kg laboratory leaching experiment, but it becomes progressively more difficult to obtain the **representative**, or average, microporosity for the rocks as the experiment scale increases to a 1 tonne column test, to a 10,000 tonne test heap and, finally, to a 1,000,000 tonne commercial heap or dump. Another major problem is determining the percentage of ore mineral grains that are accessible to the lixiviant solution absorbed in open micropores. Because of the absence of definitive information, it is usually implicitly assumed that *all* of the mineral is accessible to the solution penetrating micropores, but this is probably always wrong to some degree.

The purpose of this chapter is to focus on deterministic models and illustrate how physicochemical factors, primarily chemical diffusion, chemical kinetics of mineral grain oxidation and dissolution, and fluid flow are incorporated in constructing numerical simulation models of leaching. The discussion will focus on concepts that have been used, and, importantly, on some of the limitations encountered in formulating and applying these models to solution mining. For full descriptions of models the reader should review the original literature.

In formulating a deterministic model of ore fragment leaching, consideration must be given to many different phenomena occurring within the entire leaching system. Model development generally begins on a microscopic scale and proceeds incrementally to the macroscopic scale representing actual mining conditions.

It is worth emphasizing that, when coupled sequential phenomena are involved, the slowest step will control the overall rate of the process if it is significantly slower than the other processes. Because of the larger rock sizes encountered in solution mining, where rock grinding for mineral liberation is not involved, chemical diffusion within the solution filled rock micropores always affects extraction rates. However, mixed leaching kinetics occur when slow reacting mineral grains (e.g., chalcopyrite) or very large mineral grains are present within the rock. Then the mineral dissolution rate must be considered along with diffusion.

A REVIEW OF DIFFUSION CONTROLLED
ROCK (ORE FRAGMENT) LEACHING

Two important examples of diffusion controlled rock leaching have already been described for quasi-spherical rocks and discussed to some extent in previous chapters.

In applying these simulation models to an aggregation of wide ranging rock sizes, the rocks are divided into several discrete narrow size range groups, and each rock size group in the resulting histogram is simulated separately, using its average size or the largest size in the group. An example of this was given in Chapter 2, and the Gates–Gaudin–Schuhmann rock size distribution appeared in previous chapters. This practice is normally used in formulating deterministic ore fragment leaching models for diffusion controlled kinetics and for mixed kinetics.

Extraction (Washing) Governed by Diffusion
in Solution-Filled Micropores

Leaching oxidized gold ore with submicron gold particles disseminated uniformly throughout the rock, but accessible to solution filled rock micropores, was described in Chapter 2. The gold particles are quickly dissolved by the intruding cyanide solution. The dissolution time is insignificant compared with the total time to complete leaching and the dissolved gold does not reach the saturation limit of the pore liquid. Simulation theory for this case was described in Chapter 1. Rock leaching is adequately described for quasi-spherical rocks by the diffusion equation for spheres, eqn 1.1. Dissolved species migrate out of the rock by ordinary chemical diffusion in the pore liquid under a concentration gradient. But this is **not** a case of steady state diffusion.

The fractional extraction, F_{t,r_0}, is provided by an analytical solution of the diffusion equation, which is eqn 1.2 repeated here:

$$F_{t,r_0} = 1 - \frac{6}{\pi^2} \sum_{n=1}^{\infty} \frac{1}{n^2} \exp\left(-\frac{D_{\text{eff}} n^2 \pi^2 t}{r_0^2}\right). \tag{1.2}$$

With the dimensionless fractional extraction curves (plotted in Fig. 1.1, based on eqn 1.2) effective simulation of this diffusion controlled leaching can be done without a computer.

Fractional extraction curves plotted from eqn 1.2 for specific gold leaching cases were shown in Figs. 2.3–2.5. Concentration profiles of the remaining mineral within the rock, computed from eqn 1.3, were shown in Fig. 2.2. Using this approach, the agreement between actual.gold extraction curves observed in heap leaching practice and simulated extraction curves (e.g., Figs. 2.3 and 2.4) is excellent over a very wide ore fragment size range. Furthermore, the simulated results correlate very well with the common mining industry practice of crushing ore to less than about 20 mm diameter to obtain good gold extractions during only a few weeks of heap leaching, before removing the spent ore.

Pseudo-Steady State Diffusion Coupled with Fast Mineral Dissolution

The second important example is the shrinking core model, which was described in Chapter 5 in connection with oxide copper ore leaching. In this case the mineral dissolves rapidly and its rate of dissolution need not be considered. However, dissolution of only a small amount of the mineral saturates the pore solution and, therefore, a significant amount of mineral dissolution cannot occur as the leach solution fills the rock micropores. Mineral dissolution occurs over time within the rock at the slowly retreating interface (discontinuity) between a completely leached rim and the unleached core of the rock. The dissolved metal migrates out of the rock under a solution concentration gradient from the higher metal solute concentration at the interface to the lower metal solute concentration at the rock surface. This is an example of pseudo-steady state diffusion between the rock's internal interface and its surface—"pseudo"—because the distance between the interface and surface is expanding so slowly that its movement can be ignored as a simplifying mathematical approximation of the situation at any instant.

In the oxide copper ore leaching example used in Chapter 5, the solution at the internal rock interface is depleted in acid compared with the external solution, and saturated with dissolved cupric ions. Equilibrium occurs for the concentration of acid and the oxide copper minerals, (e.g., chrysocolla) that are present at the internal interface. More dissolution will only occur as cupric ions migrate away (out) from the internal interface.

The geometry of the diffusion controlled rock leaching case was illustrated in Fig. 5.7 and the fractional extraction, F_t, was given by eqn 5.17 in terms of the effective diffusion coefficient, rock particle radius, leaching time and a stoichiometric factor:

$$1 - \frac{2}{3} F_{t,r_0} - (1 - F_{t,r_0})^{2/3} = \left(\frac{2 V_{Cu} D_{eff} A_0}{B r_0^2} \right) t. \tag{5.17}$$

A comparison was made by Roman et al. (1974) of the experimental copper extraction curve from oxide ore leached in columns with sulfuric acid, with simulated copper extractions; see Fig. 5.10.

In Chapter 9 the biooxidation rate was controlled only by the intrinsic mineral (pyrite) oxidation kinetics, provided that the ore was first crushed to a top size sufficiently small that diffusion of the oxidant (primarily ferric ions) to the minerals was not affecting the overall biooxidation rate. This requires that the oxidant concentration at the center of the largest rock be little reduced from the external oxidant concentration. All of the minerals in all of the rocks oxidize at about the same rate, governed by the heterogeneous mineral reaction rate kinetics. Dixon and Hendrix (1993) have designated this to be "homogeneous rock leaching." For typical refractory gold ores and sulfide copper ores, generally this requires crushing the ore to a top size of about 19 mm (3/4 in). For ores of very low internal microporosity, ε, or larger amounts of sulfides than a few percent, finer crushing will be required to eliminate an effect from oxidant diffusion within the ore particles.

ROCK (ORE FRAGMENT) LEACHING WITH MIXED KINETICS

Two simulation approaches were used in the 1970s to describe the leaching of primary chalcopyrite copper ore that include slow mineral dissolution kinetics (Bartlett, 1973; Braun et al., 1974). Most of the subsequent mixed kinetics models have been extensions of these two approaches.

Formulation of both of these models was stimulated by experimental research (Braun et al., 1974; Leah and Braun, 1976) on oxygen pressurized leaching of primary copper ore at Lawrence Livermore Laboratory (LLL), described in Chapter 10. However, these mixed kinetics models can also be used for percolation leaching of copper mine waste, where the dominant oxidant is ferric ion, rather than dissolved oxygen.

The mixed kinetics rock leaching process is illustrated by the microview of an ore fragment illustrated in Fig. 12.1, after Auck and Wadsworth (1973). Chalcopyrite grains become progressively smaller as they are dissolved and a sharp discontinuity (interface) between the fully leached rock rim and the unleached core of the rock does **not** occur. Rather, the sulfide grains gradually diminish in size over a reaction zone as they are slowly oxidized and dissolved. The reaction zone thickness will increase for the slower leaching sulfide minerals such as chalcopyrite and pyrite. As sulfide grain size becomes larger, the reaction zone will increase.

The concentration profiles of diffusing oxidant and dissolved copper will cross each other within the transition region rather than converging to zero at the core/rim interface of the shrinking core model. This is illustrated in Fig. 12.2.

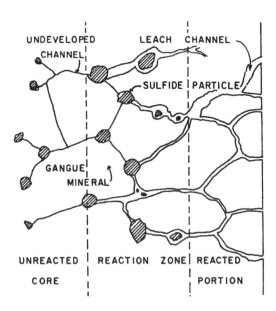

Figure 12.1. Microview of channel micropores and partial oxidation reactions of minerals in a primary copper ore fragment (Auck and Wadsworth, 1973).

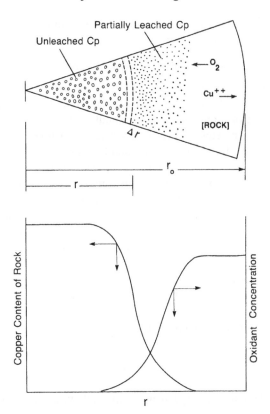

Figure 12.2. Illustration of nonsteady state mixed kinetics model of primary copper ore leaching (without a sharp interface between leached rim and unleached core) (Bartlett model, 1973).

NONSTEADY STATE MIXED KINETICS MODEL

The diffusion equation for a porous solid in spherical coordinates can be modified to include a term, R_{ox}, to account for the chemical generation of the rate limiting diffusing species, which is dissolved oxygen in this particular example of copper leaching. Because oxygen is being consumed R_{ox} is always negative

$$\varepsilon \frac{\partial c}{\partial t} = R_{ox} + \frac{D\varepsilon}{\tau} \left(\frac{\partial^2 c}{\partial r^2} + \frac{2}{r} \frac{\partial c}{\partial r} \right). \tag{12.1}$$

The diffusion coefficient in liquids normally follows a temperature dependence of the following form,

$$D = D_0 T \qquad \text{and} \qquad D_{\text{eff}} = D_{\text{eff}(0)} T \qquad (12.2)$$

which was used in applying the model.

After substitution for R_{ox}, eqn 12.1, solved numerically in finite difference form, is the ore fragment leaching simulation model (Bartlett, 1973).

Note that R_{ox} in eqn 12.1 is a volumetric reaction rate term. It is obtained by converting the rate expression for a heterogeneous reaction on the surface of many depleting mineral grains into the pseudo-homogeneous chemical reaction rate, R_{ox}. This approximation assumes that the local oxygen consumption rate, $-R_{\text{ox}(t,r)}$, at any time, t, is proportional to the local oxygen concentration, $c_{t,r}$, and the local remaining unreacted mass fraction of the original chalcopyrite mineral, $1 - q_{t,r}$, where $q_{t,r}$ is the local mass fraction of original chalcopyrite mineral, which has already been oxidized and dissolved. This approximation for the rate, R_{ox}, amounts to a local first order reaction in the remaining chalcopyrite mass that is being depleted. For more details supporting this argument see Bartlett's 1973 paper.

Thus, the fractional extraction rate is

$$\frac{\partial q_{t,r}}{\partial t} = k_1 (1 - q_{t,r}) c_{t,r}, \qquad (12.3)$$

where k_1 is the temperature dependent mineral surface reaction rate constant,

$$k_1 = \left(\frac{k_0}{d_{\text{m(avg)}}} \right) \exp\left(-\frac{E_{\text{act}}}{RT} \right), \qquad (12.4)$$

where $d_{\text{m(avg)}}$ is the volume mean diameter of the reacting mineral (chalcopyrite) grain. The activation energy for chalcopyrite oxidation by dissolved oxygen determined by Braun et al. (1974) was:

$$E_{\text{act}} = 73.2 \, \text{kJ/mol}.$$

Consequently,

$$-R_{\text{ox}} = B \frac{\partial q_{t,r}}{\partial t} \qquad (12.5)$$

and, the chemical conversion approximation is:

$$R_{\text{ox}} = -k_1 B (1 - q_{t,r}) c_{t,r}. \qquad (12.6)$$

The stoichiometric proportionality constant, B, is determined from the mols of oxygen required per mol of copper dissolved. This must also include oxygen used to oxidize pyrite that is present in the ore and that is being oxidized at about the same rate, or a little faster, than chalcopyrite. Hence,

$$B = \frac{G\rho_{ore}}{MW_{Cu}} \left(2.75 + 2.58 \frac{Py}{Cp} \right), \tag{12.7}$$

which is based on the net chemical oxidation reactions discussed in Chapter 6, and where Py/Cp is the molar ratio of pyrite to chalcopyrite. The ore grade (copper mass fraction) is G and ρ_{ore} is the ore rock specific gravity.

The boundary conditions of the ore fragment being leached are quite simple. The oxygen (oxidant species) concentration outside the rock, c_0, is constant at the solubility corresponding with the test temperature and oxygen pressure. Computations are made at isothermal and isobaric conditions throughout any particular iteration in space and time. Oxygen (oxidant) is not transferred through the center of the sphere (inner boundary), in agreement with physical reality.

The initial oxygen (oxidant) concentration in the solution within micropores of the rock is zero and the copper distribution within the rock is macroscopically uniform.

A summary of the variable parameters and derived constants needed to solve the leaching equation is given in Table 12.1. Values of these parameters used to generate extraction curves (that will be shown in subsequent figures) are also listed in Table 12.1. With the exception of the unknown rock microporosity, these are the actual parameters of the LLL experiment.

Table 12.1 Summary of Variable Parameters and Constants Used in the Leaching Equation (Bartlett, 1973).

Primary Variable Parameters	Dependent Constants
Temperature, T [90°C]	c_0 (T, P_{O_2})
Oxygen Pressure, P_{O_2} [2.75 MPa]	$D(T)$ [2.0×10^{-3} mm^2/s at 90°C]
Ore Grade, G [0.007 (0.7 wt pct)]	k_1 (T, $d_{m(avg)}$), eqn 12.4
Ore Porosity, ε [varied]	B (Py/Cp, G, ρ_{ore}), eqn 12.7

Secondary Parameters Fixed for study
$\rho_{ore} = 2.9$ g/cm^3
Py/Cp = 2.0
$\tau = 2.0$

Numerical solutions of the leaching rate equation (eqn 12.1) were computed using both explicit and implicit finite difference techniques (Bartlett, 1973). The rock was divided into J concentric shells of uniform thickness, Δr, and I increments of time, Δt. For most of the runs, $\Delta r = 1$ mm. The general approach was to solve the leaching equation for oxygen concentrations corresponding to the next time increment, c_j^{i+1}, then solve for the new fractional extraction values, q_j^{i+1}, for each radial location, j, using the leaching rate equation:

$$q_j^{i+1} = q_j^i + k_1(1 - q_{i,j})\left[\frac{c_j^{i+1} + c_j^i}{2}\right]\Delta t. \tag{12.8}$$

At the completion of all J iterations through the rock, the net extraction from the rock at that particular time, F_{t,r_0}, was calculated as follows:

$$F_{t,r_0}^{i+1} = \frac{\sum_{j=2}^{J} q_j^{i+1} 4\pi r_j^2 \Delta r}{\sum_{j=2}^{J} 4\pi r_j^2 \Delta r}. \tag{12.9}$$

The innermost core, $j = 1$, is not included in the analysis but this loss is a negligible volume fraction of the total rock.

Finally, the net average extraction for the ore, F_t, was determined by a weighted sum over the different rock sizes of the rock size histogram, following the mode of eqn 2.3,

$$FF^{i+1} = \sum_{y=1}^{Y} F_{r_0}^{i+1}(N_y) \tag{12.10}$$

which is F_t for $t = i + 1$.

The boundary and initial conditions are taken care of by specifying the following:

$$c_{J+1} = c_0, \tag{12.11}$$

$$c_1^i = c_2^i \quad \text{for all } i; \tag{12.12}$$

$$c_j^0 = 0, \quad j = 1, J. \tag{12.13}$$

The Crank–Nicolson method with centered time difference was used for the implicit solution of the leaching equation. The essence of this method is that functions are evaluated at $(c^{i+1} + c^i)/2$. It can be shown that the solution is always stable regardless of the size of the time difference, Δt. However, significant accuracy will be lost if Δt or Δr are too large.

The finite difference approximation of the leaching equation is given by eqn 12.14:

$$c_j^{i+1} = c_j^i - \frac{k_1 B \Delta t}{\varepsilon}(1-q_j^i)\frac{(c_j^i + c_j^{i+1})}{2}$$

$$+ \frac{D\Delta t}{2\tau(\Delta r)^2}[(c_{j+1}^i - 2c_j^i + c_{j-1}^i) + (c_{j+1}^{i+1} - 2c_j^{i+1} + c_{j+1}^{i+1})$$

$$+ \frac{\Delta r}{r_j}\{(c_{j+1}^i - c_{j-1}^i) + (c_{j+1}^{i+1} - c_{j-1}^{i+1})\}]. \tag{12.14}$$

Early in the leaching period, the oxygen concentrations and copper fractional extractions change more rapidly than they do later. Consequently, the time difference Δt was expanded as follows to save computation time:

Δt	Leaching Time Span
4 hours	0–160 days
1 day (24 hours)	160–736 days (2 years)
7 days	2–15 years

REACTION ZONE MODEL: PSEUDO-STEADY STATE MIXED KINETICS

This simulation model also begins with the diffusion equation modified with the homogeneous chemical reaction rate term, eqn 12.1. The equation is simplified by assuming the pseudo-steady state approximation,

$$\frac{\partial c}{\partial t} = 0 \tag{12.15}$$

which yields:

$$0 = R_{ox} + \frac{D\varepsilon}{\tau}\left(\frac{\partial^2 c}{\partial r^2} + \frac{2}{r}\frac{\partial c}{\partial r}\right). \tag{12.16}$$

This approximation is justified if the region of reacting and partially leached mineral grains within the rock (shown in Fig. 12.2) is fairly narrow. Hence, simulation by this approach is often called the *narrow* reaction zone model. This condition is shown in Fig. 12.3 where a reaction zone of thickness, δ, moves topo-chemically inward during the course of leaching. It occurs when the mineral grains react fast enough so that their

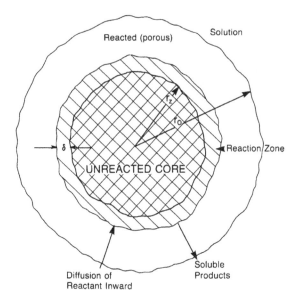

Figure 12.3. Illustration of reaction zone mixed kinetics model of primary copper ore leaching (Braun et al., 1974).

reacting period, from beginning to complete dissolution, is relatively short with respect to the overall leaching period.

Sometimes examination of partially leached and sectioned rocks obtained from copper waste dump, shows that the reaction zone is narrow and the approximation is valid, but this is not always observed with partially leached ore fragments from copper mine waste.

According to the reaction zone model, steady state diffusion occurs through the reacted outer region (rock rim) and the diffusing flux is equal to the oxidation chemical rate of reaction within the reaction zone. This model was developed by Braun et al. (1974) and has been used by Wadsworth and others at the University of Utah in subsequent copper leaching papers, Madsen et al. (1975) and Gao et al. (1981). This simulation method is reviewed by Sohn and Wadsworth (1979).

The surface area of reacting mineral particles within the moving reaction zone is assumed to be constant and independent of the mineral particle size distribution. New particles will begin to leach at the leading edge of the slowly moving reaction zone and replace particles completely dissolved at the tail of the reaction zone.

The rate of the leaching reaction within the narrow reaction zone is given by the following equation:

$$\frac{dn}{dt} = (4\pi r_z^2 \delta N_p A_p) c_z k_z. \tag{12.17}$$

With the inclusion of a sphericity factor, ψ, to account for changes in the geometry of the particle (deviations from sphericity), eqn 12.17 becomes

$$\frac{dn}{dt} = \psi^{-1}(4\pi r_z^2 \delta N_p A_p) c_z k_z, \tag{12.18}$$

where n is the moles of leachable mineral, N_p is the number of mineral grains per unit volume of rock, A_p is the average area per mineral grain in the reaction zone, c_z is the average solution concentration in the reaction zone, and k_z is the temperature dependent heterogeneous reaction rate constant for the mineral.

Under steady state conditions, diffusion through micropores to the reaction zone may be expressed by the following flux equation:

$$\frac{dn}{dt} = \psi^{-1}B^{-1}(4\pi r_z^2 D_{\text{eff}} K_h)\left(\frac{dc}{dr}\right), \tag{12.19}$$

where B is a stoichiometry factor (number of moles of reactant required per mole of metal released in dissolving the mineral), and K_h is the Henry's Law constant for solubility.

The right-hand side of eqn 12.19 may be integrated, for steady state transport for all values of r between r_z and r_0:

$$\frac{dn}{dt} = -\frac{4\pi D_{\text{eff}} K_h}{\psi B}\left(\frac{r_z r_0}{r_0 - r_z}\right)(c - c_z). \tag{12.20}$$

Equations 12.19 and 12.20 are similar to eqns 5.11 and 5.14, respectively, which describe a diffusion controlled leaching rate where the mineral reaction is rapid and the interface between the fully leached rim and the unreacted core is a sharp discontinuity. As before, these equations may be combined under the steady state approximation, giving for an ore fragment of radius, r_0, the following expression:

$$\frac{dn}{dt} = -\frac{4\pi r_0^2}{\psi}\left[\frac{1}{G\beta} + \left(\frac{B}{D_{\text{eff}}}\right)\left(\frac{r_z}{r_0}\right)(r_0 - r_z)\right]^{-1} \tag{12.21}$$

with the following groups, where G is the ore grade (mass fraction of leached mineral), and ρ_m is the mineral grain density:

$$G = \frac{\delta A_p d_{m(avg)} \rho_m}{6 \rho_{ore}},$$ (12.22)

$$\beta = \frac{6 \rho_{ore} \delta k_z}{d_{m(avg)} \rho_m}.$$ (12.23)

Equation 12.21 may be combined with the volume relation,

$$F_{t,r_0} = 1 - \left(\frac{r}{r_0} \right)^3$$ (12.24)

to yield the following expression:

$$\left(\frac{dr}{dt} \right)_z = - \frac{MW_m c}{\rho_{ore} G} \left[\frac{1}{G\beta} + \left(\frac{B}{D_{eff}} \right) \left(\frac{r_z}{r_0} \right) (r_0 - r_z) \right]^{-1}.$$ (12.25)

Integrating numerically yields the following equation, for the fractional extraction, F_{t,r_0}, of ore fragment r_0 in time t:

$$1 - \frac{2}{3} F_{t,r_0} - (1 - F_{t,r_0})^{2/3} + \frac{\beta'}{Gr_0} [1 - (1 - F_{t,r_0})^{1/3}] = \frac{\gamma t}{Gr_0^2}$$ (12.26)

where

$$\beta' = \frac{2D_{eff}}{B\beta},$$ (12.27)

$$\gamma = \frac{2MW_m D_{eff} c}{\rho_{ore} B \psi}.$$ (12.28)

Equation 12.26 is somewhat similar to eqn 5.17. Using a summation similar to eqn 12.10, the values for F_{t,r_0} for each ore fragment size can be aggregated to obtain F_t.

This reaction zone model was later extended to the leaching of rocks containing more than one copper sulfide mineral where the total copper extraction results from a contribution from several mineral leaching processes, each with different specific rate constants for the mineral oxidation surface reaction, Lin and Sohn (1987) and Paul et al. (1988).

IN SITU LEACHING OF PRIMARY COPPER ORE WITH SPARGED OXYGEN

The Lawrence Livermore Laboratory research, referred to earlier, used heated pressure vessels to study the underground solution mining system illustrated in Fig. 12.4. A rubble chamber or "chimney" is formed in ore.

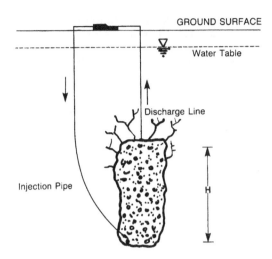

Figure 12.4. Section view of a flooded primary copper ore rubble chamber for *in situ* solution mining.

Oxygen is injected into the bottom of the flooded chamber and bubbles upward. This continues until the pregnant liquor copper concentration is adequate for recovery. Then the solution is circulated through a copper separation plant and reinjected into the ore chamber. Oxidation of chalcopyrite begins when oxygen is injected, heat is autogenously generated and the chamber temperature slowly rises. The oxygen pressure is equal to the hydrostatic pressure of water and the oxygen solubility, c_{O_2}, is a function of the depth below the water table,

$$c_{O_2} = S_0 \left[\frac{P_{O_2} - P_{H_2O}}{P_{O_2,std}} \right]. \tag{12.29}$$

Because of the low thermal conductivity of the wall rock temperature will rise to a steady state value, depending on conditions. The temperature of the large experimental study was 90°C.

An interesting part of this research project was a major experiment conducted using 6 tonne of −280 mm (−11 in) ore obtained from the San Manuel Mine in Arizona. The rock size distribution of this block-caved ore sample is shown in Table 12.2. The major experiment was conducted in a large, lead-lined pressure vessel at an oxygen pressure of 2.75 MPa (400 psi). At these temperatures and oxygen pressures the solubility of dissolved oxygen was much higher than that of ferric ion. Consequently,

Table 12.2   Rock Size Distribution of
San Manuel Primary Copper Ore.

Rock Diameter (mm)	Distribution (wt pct)
7	9.5
21	12.0
35	12.0
49	12.5
63	12.5
77	11.5
91	11.0
105	8.5
119	6.5
133	4.0
	100.0

the oxidation mechanism occurs by direct oxygen reaction rather than through the intermediary of the ferric/ferrous couple.

The large scale experiment was run continuously for two years while solutions were monitored for copper concentration and pH. Ultimately, 42% of the copper was extracted.

Although the 3 m high pressure vessel was large for a laboratory experiment, it was not large enough to cause a significant variation from top to bottom in oxygen pressure and solubility. However, for a 100 + m rubblized ore chamber there will be large differences from bottom to top in the oxygen pressure, solubility and rate of leaching. Consequently, both simulation models represent only a thin horizontal slice from a mine-sized ore chamber.

MIXED KINETICS MODELS COMPARED WITH LIVERMORE COPPER LEACHING EXPERIMENTS

Initial comparisons with the LLL flooded leaching results using mine-sized ore in the 6 tonne experiment were made with both the nonsteady state model and the reaction zone (pseudo-steady state) model.

Nonsteady State Model Comparative Results

The measured ore characteristics and the experimental parameters (listed in Table 12.1), were used in the simulation computations made by Bartlett

(1973) and reproduced in several following figures. There were no adjustable parameters except the average rock microporosity for the 6 tonne sample. The rock microporosity was not known with any reasonable degree of accuracy. The standard value for the oxygen diffusion coefficient in water at 20°C is 1.8×10^{-3} mm^3/s (Perry, 1950). A diffusion coefficient of 2.0×10^{-3} mm^3/s was estimated for 90°C and used in these simulations. Both experimental determinations and various theories yield a tortuosity near 2 (Satterfield, 1970), which was the value used in these simulations.

The computed copper extraction, carried out to one year, is shown in Fig. 12.5, along with all the experimental extraction data available at the time the simulation was made. The experimental data and simulation match at 4% rock microporosity for about two months, then trend along the 3% microporosity simulated extraction curve to one year and then, in

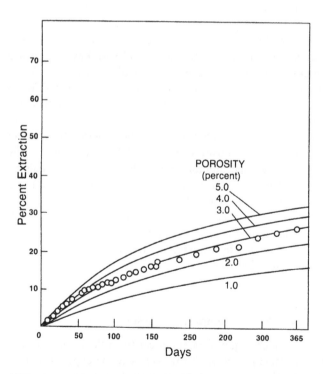

Figure 12.5. Experimental data (LLL 6 tonne rubblized ore) and simulated copper ore extraction curves at different ore fragment porosities using the nonsteady state mixed kinetics model (Bartlett, 1973).

data not shown, accelerate back to the 5% line at the end of two years. Thus, the predicted copper extractions are good, but simulation didn't predict these minor deviations.

It has been postulated by Braun et al. (1974) that the rate is initially fast because of a greater concentration of chalcopyrite mineral on the external fracture surfaces. Breakage usually occurs along pre-existing fractures where the mineral is deposited.

Rapid dissolution of small amounts of oxide copper and secondary sulfides can also account for the fast start. This usually happens in copper mine dump leaching, where acid soluble copper is nearly always present to some extent and often accounts for up to 20% of the contained copper, even in primary mine waste.

The faster than expected extraction rate during the second year could be caused by an increase in microporosity from leaching that removed minerals originally in the micropores. The leached minerals may include acid soluble gangue minerals, such as carbonates, that dissolve rapidly. Combined CaO and MgO was over 1 wt pct in the ore sample. And over half of these carbonates were extracted in 6 months in the large scale Livermore test as reported by Braun et al. (1974). Physical disintegration of the rocks, because of acid weathering of the gangue minerals, also normally occurs during copper mine waste leaching and shifts the rock size distribution to smaller sizes with an accelerating effect on copper extraction.

Estimated extraction curves from simulation computations extended out to 15 years for rock porosities of 3% and 5% are shown in Fig. 12.6. Computed relative concentration profiles for the copper remaining in the rock (as unleached mineral) at different times are plotted in Fig. 12.7 for a 15 mm ore fragment ($-3/4 +1/2$ in diameter) with 6% internal microporosity. Note the absence of a sharp discontinuity or interface between leached and unleached copper. The solid copper (mineral) concentration profiles shift gradually over a *wide* reaction zone (δ) rather than abruptly over a *narrow* reaction zone.

Reaction Zone Model Comparative Results

This model, which requires an empirical fit to the 6 tonne major experiment to determine β' and γ, also fit well with the experimental results for the first year. The extraction curve resulting from the major 6 tonne experiment, which was shown in Fig. 10.4, deviated above the simulated copper extraction curve during the second year. However, decreasing the sphericity

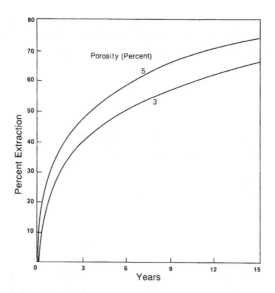

Figure 12.6. Simulated copper ore extraction curves extended to 15 years for conditions of the LLL 6 tonne test (Bartlett, 1973).

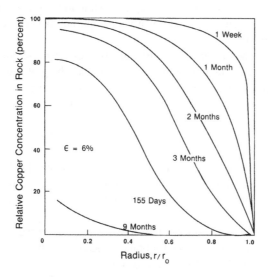

Figure 12.7. Computed residual copper mineral concentrations $[c/c_0]$ in a 15 mm ore fragment after various leaching periods; nonsteady state mixed kinetics model (Bartlett, 1973).

factor as leaching proceeded allowed fitting the copper extraction to the full two-year period of this major 6 tonne experiment. Braun et al. (1974) assumed an empirical formula in which the sphericity factor was proportional to the ever-shrinking radius of the reaction zone.

SIMULATED PARAMETRIC LEACHING OF FRAGMENTED PRIMARY COPPER ORE

Neither of the preceding two models could predict extraction curves with engineering design quality without using *a posteriori* adjustable parameters. But the observed deviations are believed to be caused by real leaching phenomena: uncertainty in the ore characteristics and variations in those ore characteristics with leaching, principally internal rock microporosity and changes in the rock size distribution due to acid weathering. Nevertheless, both models were close enough to the results of the major experiment to indicate that mixed kinetics (involving both pore diffusion and mineral chemical reaction kinetics) is the governing leaching mechanism. Consequently, it is worth examining the effects caused by variations in several leaching parameters on the copper extraction curve to determine the most sensitive parameters affecting the extraction curve and long time results. The understanding gained is a very important benefit of computer leaching simulations.

The results shown in the following figures were simulated with Bartlett's nonsteady state model for the flooded leaching of primary ore at elevated temperatures and pressures, using a standard set of leaching conditions and rock characteristics (see Tables 12.1 and 12.2). In the following figures, one parameter was varied while the other parameters were fixed at their standard condition in plotting computed F_t versus leaching time, t.

Somewhat surprisingly, variations in temperature within the available temperature range from 50°C to 110°C caused little variation in the copper extraction curves for *in situ* flooded leaching of primary copper ore (see Fig. 12.8).

The ore fragment size distribution has a very pronounced affect on the extraction curves. Two additional rubblized rock size distributions, reported by Lewis et al. (1974) and reproduced in Fig. 12.9, were used to compute estimated copper extraction curves out to 15 years. The results, plotted in Fig. 12.10, show wide differences between the extraction curves for each of these three rubble size distributions.

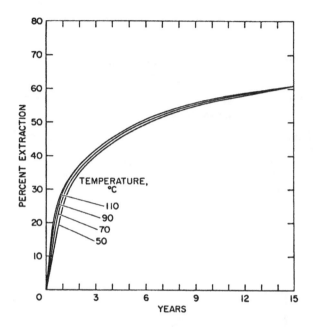

Figure 12.8. Simulated copper ore extraction curves for LLL 6 tonne rubble at different ore chamber temperatures (Bartlett, 1973).

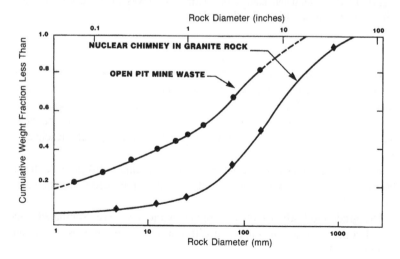

Figure 12.9. Alternative rock size distributions obtained from rubblization (Lewis et al., 1974).

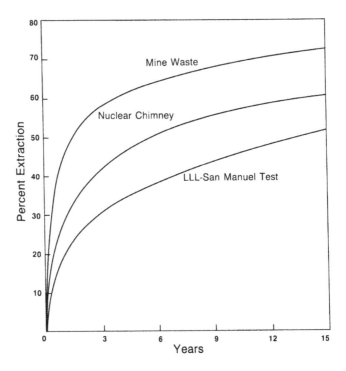

Figure 12.10. Simulated copper ore extraction curves for three selected rub-blized rock size distributions (Bartlett, 1974).

The critical role of ore fragment size on the copper extraction rate is further illustrated by Fig. 12.11, where the simulated copper extraction curves for seven different individual ore fragment sizes are shown. Ore fragments below 10 mm diameter are completely leached in one year, but ore boulders greater than 150 mm are very incompletely leached in 15 years.

Increasing the oxygen pressure in flooded leaching also gives a strong boost to the copper extraction. Simulated extraction curves are shown in Fig. 12.12.

Ore grade modestly affects the copper extraction curve. Simulated fractional copper extractions curves at different copper ore grades are shown in Fig. 12.13.

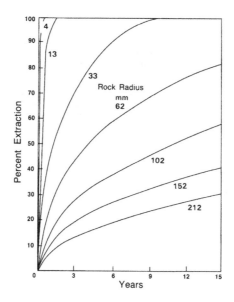

Figure 12.11. Simulated copper ore extraction curves for several monosize ore fragments, standard conditions but 6% microporosity (Bartlett, 1974).

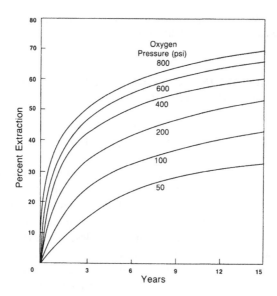

Figure 12.12. Simulated copper ore extraction curves for LLL 6 tonne rubble at different oxygen pressures in the ore chamber (Bartlett, 1973).

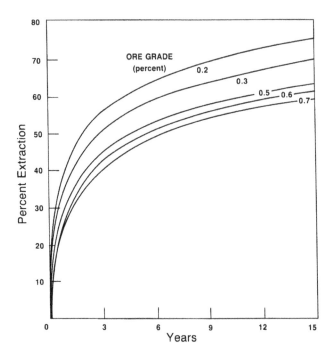

Figure 12.13. Simulated copper ore extraction curves for LLL 6 tonne rubble at different ore grades (wt pct Cu) (Bartlett, 1973).

FURTHER MIXED KINETICS ORE LEACHING MODELS WITH MACRO-EXTENSIONS TO ORE HEAPS

Madsen and Wadsworth (1981) used the reaction zone ore particle leaching model to describe the leaching of copper sulfide minerals in an ore by ferric ions, as would occur in a copper ore heap or in leaching copper mine waste dumps. This model included separate rate terms for different copper sulfide minerals. Dixon and Hendrix (1993) developed a nonsteady state model for the leaching of several mineral species within the ore with different, variable order rate expressions. Dimensionless model equations were developed and used to identify important design and scale-up factors. Depending on the parameter values selected the results can devolve to a reaction zone in the ore particle (equivalent to the reaction zone model), or to a single process step controlling the ore leaching rate, either pore diffusion within the rocks or the chemical reaction rate at the mineral surface. It also accounted for mineral grains attached to the surface of ore particles.

This is important for crushed ores, because microcrystalline ore particles generally break along preexisting microfractures and micropores where the minerals are also located. Model results confirmed the fact that minerals located at the rock surfaces are important to enhancing the overall metal extraction rate when the diffusion controlled reaction regime is important but not important when the mineral reaction rate is the slower process step dominating the overall kinetics. This model also quantitatively demonstrated that a competing second mineral will retard the extraction of a primary mineral when the kinetics are chemical reaction controlled, which is what pyrite does in retarding the rate of chalcopyrite oxidation and copper extraction in the leaching of primary copper ores and mine waste.

Using the reaction zone ore fragment leaching model, a one-dimensional macromodel developed by Gao et al. (1981) extended the simulation effort of flooded leaching at an *in situ* primary copper deposit to a full-sized ore rubble chamber. Temperature rise and leaching effects were evaluated along the vertical axis within the ore chamber over time. Additional processes incorporated in this macromodel were (vertical) axial convective transport of both mass and heat, axial dispersion of mass, and mass transfer between the bulk liquid and solid surfaces. The results of this study indicated that temperature was relatively insensitive to vertical position within the ore chamber. Temperature in the chamber increased over time, reaching a maximum of 95°C in about 200 days. Afterwards, temperature declined very slowly, reaching 90°C in 1,000 days. As expected, the higher oxygen pressure and solubility at the bottom of the ore chamber caused copper leaching there to be faster than at the top of the chamber.

Prosser and coworkers in Australia have published more than a dozen papers on heap leaching simulation. Comparisons were made with the Livermore test results already described, and with the earlier models of that test (Roach and Prosser, 1976). Box and Prosser (1986) developed a general model to handle several minerals and multiple reagents. Their procedure models sequential and complementary reactions, including those which involve substances generated within the heap. However, it is assumed that all reactions are controlled by diffusion through solution filled micropores of the ore particles.

REFERENCES AND SUGGESTED FURTHER READING

Auck, Y.T. and Wadsworth, M.E. (1973). Physical and chemical factors in copper dump leaching. Evans, D.J.I. and Shoemaker, R.S. (Eds.), *Proc. Intl. Symposium on Hydrometallurgy*, AIME, New York, pp. 645–700.

Bartlett, R.W. (1973). A combined pore diffusion and chalcopyrite dissolution kinetics model for *in situ* leaching of a fragmented copper porphyry. Evans, D.J.I. and Shoemaker, R.S. (Eds.), *Proc. Intl. Symposium on Hydrometallurgy*, AIME, New York, pp. 331–372.

Box, J.C. and Prosser, A.P. (1986). A general model for the reaction of several minerals and several reagents in heap and dump leaching. *Hydrometallurgy*, 16, pp. 77–92.

Braun, R.L., Lewis, A.E. and Wadsworth, M.E. (1974). In-place leaching of primary sulfide ores: laboratory leaching data and kinetics model. *Met. Trans.*, 5, pp. 1717–1729.

Cathles, L.M. and Murr, L.E. (1980). Evaluation of an experiment involving large column leaching of low-grade copper sulfide waste: a critical test of a model of the waste leaching process. Schlitt, W.J. (Ed.), *Leaching and Recovery of Copper from As-Mined Materials*, Society of Mining Engineers of AIME, pp. 29–48.

Dixon, D.G. and Hendrix, J.L. (1993). A general model from leaching of one or more solid reactants from porous ore particles. *Met. Trans.*, 24B, pp. 157–169.

Gao, H.W., Sohn, H.Y. and Wadsworth, M.E. (1981). A mathematical model for the *in situ* leaching of primary copper ore. Schlitt, W.J. (Ed.), *Interfacing Technologies in Solution Mining*, AIME, pp. 197–208.

Leah, D.L. and Braun, R.L. (1976). Leaching of primary sulfide ores in sulfuric acid solutions at elevated temperatures and pressures. *Trans. Soc. Min. Engrs.*, 260, pp. 41–48.

Lewis, A.E. et al. (1974). Nuclear solution mining—breaking and leaching considerations. Aplan, F.F. et al. (Eds.), *Solution Mining Symposium*, AIME, pp. 56–75.

Lin, H.K. and Sohn, H.Y. (1987). Mixed control kinetics of oxygen leaching of chalcopyrite and pyrite from porous primary ore fragments. *Met. Trans.*, 18B, pp. 497–504.

Madsen, B.W., Wadsworth, M.E. and Groves, R.D. (Mar 1975). Application of a mixed kinetics model of the leaching of low grade copper sulfide ores. *Trans. Soc. of Mining Engrs.*, AIME, pp. 69–74.

Madsen, B.W. and Wadsworth, M.E. (1981). U.S. Bureau of Mines Report of Investigations 8547.

Murr, L.E., Schlitt, W.J. and Cathles, L.M. (1982). Experimental observations of solution flow in leaching copper bearing waste. Schlitt, W.J. and Hiskey, J.B. (Eds.), *Interfacing Technologies in Solution Mining*, SME and SPE of AIME, pp. 271–290.

Paul, B.C., Sohn, H.Y. and McCarter, M.K. (1988). Model for bacterial leaching of copper ores containing a variety of sulfides. Sohn, H.Y. and Geskar, E.S. (Eds.), *Metallurgical Processes for the Year 2000 and Beyond*, The Minerals, Metals and Materials Society (TMS), pp. 451–464.

Perry, J.H. (1950). *Chemical Engineers Handbook, 3rd ed.*, p. 540.

Roach, G.I.D. and Prosser, A.P. (1976). Predicting the rates of recovery from selective chemical reaction of ores—a specific example: leaching of chalcopyrite in acid with oxygen. *Hydrometallurgy*, 2, pp. 211–218.

Roman, R.J., Benner, B.R. and Becker, G.W. (1974). A diffusion model for heap leaching and its application to scale-up. *Trans. Soc. Min. Engrs.*, pp. 247–256.

Satterfield, C.N. (1970). *Mass Transfer in Heterogeneous Catalysis*, M.I.T. Press, Cambridge, MA.

Sohn, H.Y. and Wadsworth, M.E. (1979). *Rate Processes of Extractive Metallurgy*, Plenum Press, New York, pp. 191–197.

THIRTEEN

Evaporites, Brine and Sulfur

Evaporites and brines are the sources of several soluble salts of considerable industrial importance, including sodium chloride (halite), soda ash (sodium carbonate), borates, sodium sulfate, and potash (KCl). Magnesia (MgO) and the metals, magnesium and lithium, are derived from brines.

Caverns excavated in salt deposits, particularly salt domes, are increasingly used as storage chambers because of the extremely low permeability of salt. Petroleum reserves, large amounts of natural gas, low level nuclear waste, and hazardous chemical waste are being stored in salt cavities. The excavation of these storage cavities by solution mining is practiced.

Sulfur is extracted by the Frasch solution mining method from both salt domes and tabular evaporite deposits.

EVAPORITE DEPOSITS AND MINERALOGY

Both marine evaporites and alkali evaporites, derived from inland brines, are important sources of mineral salts.

Marine Evaporite Deposits

The most abundant salt, by far, is halite (NaCl), which is found in about half of the states in the U.S. Most halite is deposited in beds that resulted from sea water entrapment in shallow bays, coupled with evaporation. These beds cover immense areas. The composition of sea water, shown in Table 13.1, is believed to have been nearly constant over geological time. Because of the many ionic species dissolved in sea water, several salts

Table 13.1 Dissolved
Constituents of Sea Water.

Cations	(%)	Anions	(%)
Na	30.61	Cl	55.04
Mg	3.69	SO_4	7.64
Ca	1.16	HCO_3	0.41
K	1.10		

may be precipitated as their saturation points are reached during solar concentration of sea water. As brines concentrate, they become more dense and migrate to depressions in the evaporite basins, where further concentration can occur during evaporation in the surrounding shallow flanks of the basin. The evaporation sequence also depends on temperature.

Generally, the insoluble carbonates precipitate first followed by gypsum $(CaSO_4 \cdot 2H_2O)$, polyhalite $(K_2Ca_2Mg(SO_4)_4)$ and halite. After 91.7% of sea water has evaporated, halite crystallizes. Gypsum can be dehydrated to anhydrite $(CaSO_4)$, which is often an encasing bed in a marine evaporite sequence. Other salts are formed after halite saturation, and include the potassium salts sylvite (KCl), carnallite $(KMgCl_3 \cdot 6H_2O)$ and langbeinite $(K_2Mg_2(SO_4)_3)$. Further brine concentration can yield bischofite $(MgCl_2 \cdot 6H_2O)$, but this mineral is rarely found in marine evaporites because it is readily resolubilized to form secondary carnallite. The mineral langbeinite is only favored at temperatures above 40°C, indicating a high temperature during crystallization in the New Mexico deposits where it occurs. Conversely, the rarely observed low temperature mineral, tachyhydrite $(CaCl_2 \cdot MgCl_2 \cdot 12H_2O)$ is abundant in Brazil and Africa. This salt is precipitated below −15°C from saturated calcium and magnesium chloride brines, indicating an arctic climate during its deposition (Schlitt and Larson, 1985).

The depositional sequence is often episodic, causing a series of horizontal beds, often with different compositions and varying intrusion of interbedded silts and clays. Subsequently, tilting, faulting and deformation of the evaporite beds may occur. Nevertheless, several large marine evaporite basins in the United States remain nearly horizontal.

Marine evaporites are commercially important mineral sources for halite, sylvite (potash) and, occasionally, sodium sulfate. Both potassium chloride minerals, sylvite and carnallite, precipitate into a halite mush. Commercial grade sylvite beds are formed, eventually, by secondary alteration from percolating sulfate-rich brines, most commonly in evaporites of Permian age.

Marine Salt Deposit Structures

The major evaporite basins in the United States are shown in Fig. 13.1. Salt domes and anticline deposits, though less common, are commercially important. Solid halite and other chlorides easily deform plastically under depth at modest heat and pressure, and because their densities are lower than the surrounding rocks, deep salt deposits tend to buoy upward, occasionally punching through overlaying strata. Thus, the salt may accumulate into immense salt domes and anticlines, especially when the source salt bed is thick. This extrusion process tends to remove impurities and inclusions and produce a more fully dense salt mass with negligible permeability.

Over 500 **salt domes** from 1 to 6 km in diameter exist in the Gulf Coast Basin of the United States where they are derived from the deep Louan salt bed (Schlitt and Larson, 1985). These salt domes are important traps for petroleum, natural gas and sulfur. The Strategic Petroleum Reserve is stored in approximately 100 million cubic meters of cavities in Gulf Coast salt domes, and the major portion of this volume has been constructed by solution mining. Natural gas is also being stored, under pressure, in solution mined salt dome cavities.

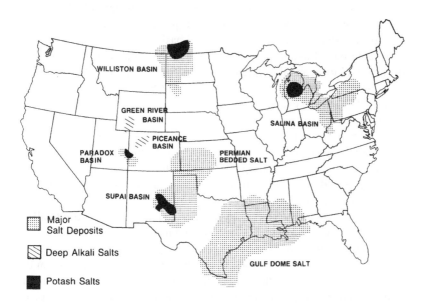

Figure 13.1. Major USA subsurface soluble salt deposits.

Anticline deposits are common in the Paradox Basin of Utah and Colorado. Salt extrusion and segregation of sylvite from halite into commercial potash deposits occurs. Because anticlines and tilted deposits are difficult to mine by conventional mining excavation methods, they are good candidates for solution mining. A potash mine in the Paradox Basin near Moab, Utah uses solution mining for extraction.

Although halite deposits are common, commercial grade deposits of the other salts are much less frequently encountered in marine evaporites. Nevertheless, Potash is an important mineral produced primarily from marine evaporites. Potash deposits are extensively mined in the Permian Basin of New Mexico and the Williston Basin in Canada.

Magnesium chloride can be separated from marine evaporites. But since this salt is not usually significantly segregated during sea water evaporation, it is usually produced directly from brine or sea water by precipitation as $Mg(OH)_2$, using calcined dolomite ($CaO \cdot MgO$).

Alkali Evaporite Deposits

These deposits are derived from inland seas like the Great Salt Lake in Utah, Searles Lake in California, and the Dead Sea in Israel and Jordan. The evaporites lying under these contemporary inland seas are geologically very recent. The Searles Lake evaporite sequence is only of Pleistocene age. Evaporation leads to formation of a playa or salt bed. Subsequent geological burial can cause a buried alkali basin.

Notable buried alkali evaporites in the United States are the Piceance Basin in Colorado and the Green River Basin in Wyoming. Although these latter basins are impressively large, alkali salt basins are generally much smaller than marine evaporite basins.

Inland sea brines also contain NaCl as their major dissolved salt. The sodium chloride is primarily oceanic in origin being carried by winds and deposited by rain and snow in the drainage basin. However, dissolved alkali salts are formed during the weathering of rocks in the drainage basin, so that the composition of dissolved salts entering the inland sea differs from the composition of sea water.

The alkali salts have limited solubilities and are precipitated into alkali evaporites, often interbedded with clays and muds that flow into the inland sea during extreme floods.

Alkali evaporites and their associated brines are sources of halite in the western United States and provide potash in competition with marine deposits. Alkali evaporites are the only important sources of trona (soda ash), borates, sodium sulfate and lithium (LiCl). Utah's Great Salt Lake

Table 13.2 Principal Boron Minerals.

Mineral	Chemical Composition	B_2O_3, wt %
Sassolite	$B(OH)_3$	56.4
Borax (Tincal)	$Na_2B_4O_7 \cdot 10H_2O$	36.5
Tincalconite	$Na_2B_4O_7 \cdot 5H_2O$	47.8
Kernite	$Na_2B_4O_7 \cdot 4H_2O$	51.0
Ulexite	$NaCaB_5O_9 \cdot 8H_2O$	43.0
Probertite	$NaCaB_5O_9 \cdot 5H_2O$	49.6
Priceite	$Ca_4B_{10}O_{19} \cdot 7H_2O$	49.8
Inyoite	$Ca_2B_6O_{11} \cdot 13H_2O$	37.6
Meyerhofferite	$Ca_2B_6O_{11} \cdot 7H_2O$	46.7
Colemanite	$Ca_2B_6O_{11} \cdot 5H_2O$	50.8
Hydroboracite	$CaMgB_6O_{11} \cdot 6H_2O$	50.5
Kurnakovite	$Mg_2B_6O_{11} \cdot 15H_2O$	37.3
Szaibelyite	$MgBO_2(OH)$	41.4
Boracite	$MgB_7O_{13} \cdot Cl$	62.2

brine is used to produce salt (NaCl), potash, magnesium chloride, and magnesium metal. Trona is mined from nearly horizontal beds by both solution mining and conventional mining methods in the Green River Basin of Wyoming. Solution mining is preceded by conventional room, and pillar mining that develops passages through which the solutions flow while leaching pillars, walls and rubble.

Major oil shale deposits are found associated with nahcolite ($NaHCO_3$) in both the Piceance and Green River alkali evaporite basins. Nahcolite is commercially recovered by *in situ* mining.

Many molecular combinations of salts are possible and do occur in nature as minerals. As an example of this, fourteen principal boron minerals are listed in Table 13.2. Only minerals with a substantial boron content, above 35%, were listed. Borax (tincal), colemanite, kernite and ulexite are commercially important, but usually only one of these is dominant in any one deposit. These differences lead to many corresponding differences in brine recovery processes, which tend to be site specific.

SOLUTION MINING PERMEABLE EVAPORITES

Evaporites can be divided into impermeable and permeable classes. Both types are mined by conventional underground excavation methods, as well as by solution mining. Because salt minerals are fairly soft and the deposits are often bedded and nearly horizontal, equipment and methods adapted from underground coal mining are often used. This includes both room and pillar mining with continuous (mechanical) miners and longwall mining. Thick deposits and salt domes can be mined underground by excavating very large stopes or rooms using blast hole bench methods similar to those used for surface mining.

Permeable evaporites are solution mined using flooded solution methods similar to those described in Chapter 10, and the considerations of hydrology and environmental containment that were discussed in Chapter 11 pertain. The goal of permeable evaporite solution mining is to extract soluble salts, usually from fairly horizontal beds using a field of injection and production wells. The lixiviant is normally water, without any other chemical reagent, but there are exceptions where chemical reagents are used to obtain or enhance solubilization. Permeability must be adequate. The evaporite bed must be bound by formations with low permeability to contain the solutions.

In fairly tight evaporites such as trona, solution mining has been recently used after conventional mining as a secondary recovery method. Brine slowly flows through the preexisting excavations.

An important additional consideration for evaporite solution mining is the high density of the brine and the variability in brine density that can occur naturally or be caused by brine processing prior to reinjection. Variable density can lead to brine aquifer stratification or other forms of poor mixing even to the extent that different brine fluids behave somewhat like immiscible liquids in the underground evaporite reservoir.

Salting out, crystallization of salts, usually induced by temperature changes in the brine, can severely affect wells, pipelines and process equipment. The use of fresh or brackish water to control salting out is nearly always required in one or more areas of a brine solution mining system.

SOLUTION MINING IMPERMEABLE EVAPORITES

Evaporite deposits are highly porous and permeable when originally deposited, but with burial they become consolidated because of the viscoplastic nature of salt coupled with lithostatic pressure. Although some trapped voids containing brine or gas may occur in salt domes, these formations are highly impermeable.

Because of the high NaCl concentration in saturated brine, halite is rapidly dissolved by water and large amounts of salt can be produced from a single, coaxial water injection and brine recovery well drilled into salt domes and sufficiently thick salt beds. Single wells have produced as much as 1200 tonnes of salt per day.

The roof of a salt cavity dissolves much faster than its side walls. Consequently, solution mining brine begins by forming a small cavity at the bottom of the evaporite zone. Then the cavity is enlarged and extended upward gradually. This sequence also allows nonsoluble minerals, usually anhydrite, to settle to the bottom of the cavity where the resulting sediment cannot impede continued solution mining.

When using this method, it is necessary to prevent unsaturated brine from rising in the borehole above the cavity roof, which is the desired horizon of brine extraction. This is accomplished by using a less dense, immiscible fluid as a blanket over the brine. The blanket fluid is usually petroleum, diesel fuel or pressurized natural gas.

An example of this method of brine solution mining is illustrated in Fig. 13.2. Water is injected and brine is extracted through separate tubing. A small amount of dilution water is added to the rising brine to prevent

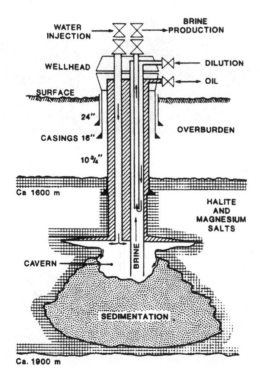

Figure 13.2. Well configuration for solution mining impermeable salt (Schlitt & Larson, 1985).

crystallization as it cools. An oil blanket is introduced through a fourth tube to control the cavity roof. Periodically, the tubing is raised a small amount, and this leads to successive solution mining cuts (illustrated in Fig. 13.3).

The size of the cavity must be governed by consideration of rock mechanics to prevent cavity collapse. This usually limits cavity diameters to about 150 m. Several salt wells and corresponding salt cavities can occur in the same salt dome, as illustrated by the map in Fig. 13.4 for the Bryan Mound salt dome near Freeport, Texas. Fluid pressure in the salt cavity may be kept near the local lithostatic pressure, rather than at the hydrostatic pressure, to prevent collapse of the cavity.

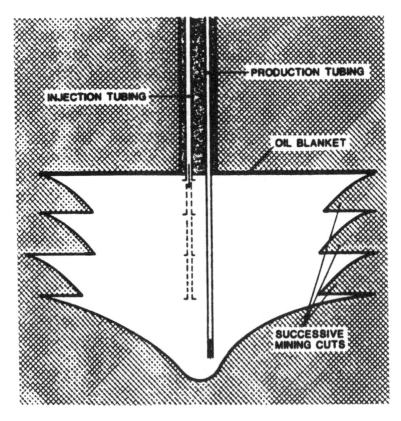

Figure 13.3. Successive impermeable salt solution mining cuts that occur as the well tubing is raised (Schlitt & Larson, 1985).

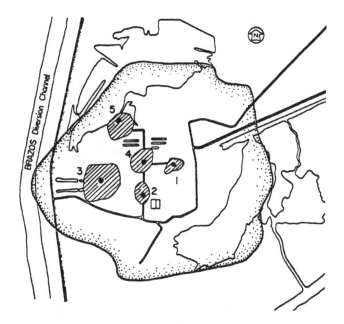

Figure 13.4. Plan view of five solution mined caverns in the Bryan Mound Salt Dome, Freeport, TX (Schlitt & Larson, 1985).

The second method of solution mining impermeable evaporites is used in layered deposits and requires two or more wells. After drilling and completing the wells, hydraulic fracturing, at a pressure above the lithostatic pressure, provides a fracture path between the wells, which are typically 150–300 m apart. Note that this is the opposite strategy of that used in metal solution mining where short circuit paths are avoided to provide good sweep efficiency through all of the ore. Water pumped from an injection well to a production well dissolves salt and opens the connecting channel further. Often one central injection well will provide water for several surrounding brine production wells.

SEARLES LAKE BRINE

Searles Lake is an important example of an inland alkali lake overlaying a sequence of evaporite beds containing three billion tonnes of soluble mineral salts. Numerous layers of salts and muds have been classified into six main stratigraphic units, which are listed in Table 13.3 (Smith, 1979). Many salt minerals are present in the Searles Lake evaporite layers

Table 13.3 Stratigraphic Units of Searles Lake.

Unit	Thickness (m)	Age (yr)
Overburden Mud	7	0–3,500
Upper Salt	15	3,500–10,500
Parting Mud	4	10,500–24,000
Lower Salt	12	24,000–32,500
Bottom Mud	30	32,500–130,000
Mixed Layer	198	130,000–Quaternary

(see Table 13.4). The dominant minerals in the lower salt are halite, trona and burkeite. Brines from this layer are solution mined principally to produce sodium carbonate. Brines occupy the extensive volume of voids, about 45%, within the permeable evaporites.

Searles Lake brines are saturated in sodium chloride, sodium sulfate, and sodium carbonate. Sodium bicarbonate and small quantities of potassium borates, organics and sulfides are also present. Trace quantities of bromine, lithium, tungsten and uranium are also in these brines. Although the concentration of the tungstate anionic complex is very low, 55 ppm in the upper salt brine, the total amount of dissolved tungsten at Searles Lake is very large and commercial tungsten extraction has been considered.

Brine extraction has been practiced at Searles Lake for over 100 yr. Over the past 70 yr brine processing has produced sodium, potassium and lithium salts of chloride, sulfate, carbonate, borate and phosphate. All salt production at Searles Lake is from brine solution mining. Evaporation of the brine and crystallizing of the contained salts in surface plants duplicate to some extent nature's crystallization sequence. However, crystallization paths in boiling brine differ from those for brine vaporizing at ambient temperatures.

Historically most of the brine production was from the upper salt. Since 1979, soda ash has been produced on a very large scale from the lower salt brine in a separate plant, using brine carbonation. Water inflow to Searles Lake is managed to maintain a fairly constant brine level, essentially at the surface of the playa. Shallow ponding of open brine expands in winter months and contracts in summer, corresponding with annual weather and solar evaporation patterns.

Huge amounts of brine at 22°C are pumped from a wellfield and sent through insulated pipes to settling ponds where gaseous sulfides are stripped before entering the processing plant. Because of the extensive

Table 13.4 Nonclastic Minerals in the Searles Lake Evaporites.

Mineral	Composition
Adularia	$KAlSi_3O_8$
Analcime	$NaAlSi_2O_6 \cdot H_2O$
Aphthitalite (glaserite)	$K_3Na(SO_4)_2$
Aragonite	$CaCO_3$
Borax	$Na_2B_4O_7 \cdot 10H_2O$
Burkeite	$2Na_2SO_4 \cdot Na_2CO_3$
Calcite	$CaCO_3$
Dolomite	$CaMg(CO_3)_2$
Galeite	$Na_2SO_4 \cdot Na(F,Cl)$
Gaylussite	$CaCO_3 \cdot Na_2CO_3 \cdot 5H_2O$
Halite	$NaCl$
Hanksite	$9Na_2SO_4 \cdot 2Na_2CO_3 \cdot KCl$
Mirabilite	$Na_2SO_4 \cdot 10H_2O$
Nahcolite	$NaHCO_3$
Northupite	$Na_2CO_3 \cdot MgCO_3 \cdot NaCl$
Phillipsite	$KCa(Al_3Si_5O_{16}) \cdot 6H_2O$
Pirssonite	$CaCO_3 \cdot Na_2CO_3 \cdot 2H_2O$
Schairerite	$Na_2SO_4 \cdot Na(F,Cl)$
Searlesite	$NaBSi_2O_6 \cdot H_2O$
Sulfohalite	$2Na_2SO_4 \cdot NaCl \cdot NaF$
Teepleite	$Na_2B_2O_4 \cdot 2NaCl \cdot 4H_2O$
Thenardite	Na_2SO_4
Tincalconite	$Na_2B_4O_7 \cdot 5H_2O$
Trona	$Na_2CO_3 \cdot NaHCO_3 \cdot H_2O$
Tychite	$2Na_2CO_3 \cdot 2MgCO_3 \cdot Na_2SO_4$

evaporite void space, the hydraulic conductivity at Searles Lake is high and pump drawdown at wells is modest. After processing, depleted brines are returned to the lake at a location far from the production wells. The return solutions, coupled with precipitation and runoff, dissolve additional salts from the vast quantity of surrounding evaporites in the underground reservoir. Nevertheless, after years of extraction, some brine components have declined in concentration. Comparative brine compositions for the upper salt at various times past are shown in Table 13.5. Note the decrease in concentration of the unsaturated components, potassium and borate, caused by prolonged brine pumping and mineral extraction.

Each area of Searles Lake experiences different mineral and brine components. Near the edges, brackish run-off water reacts with the evaporite to form new brine, while near the center of the lake residual old brines predominate. Thus, the composition of brines varies concentrically with

Table 13.5 Typical Analysis of Brines
Pumped from Searles Lake Upper Salt.

	Date		
Component	Old	1955	1968
Potassium (%)	2.61	2.0	1.33
Carbonate (%)	2.73	2.5	2.66
Sulfate (%)	4.62	4.6	4.29
Chloride	12.15	12.2	—
Borate (% B_4O_7)	1.24	1.0	0.88

the highest concentrations of unsaturated salts found near the center. However, because of the much higher salt concentrations, compared with metal concentrations typical of flooded leaching described in Chapter 10, the lateral variation of the major salt components is modest.

Equilibration of the evaporite and reservoir brine takes time. Consequently, if brine pumping rates are increased, steady state salt concentrations may decrease. These effects need to be anticipated because an unexpected lower alkali salt grade will adversely affect economic results in the surface chemical separation plant, where most of the total production costs are encountered.

Sodium chloride is produced at the south end of Searles Lake in diked solar evaporation ponds. At the end of summer evaporation the residual bittern is drained and returned to the lake. Crude salt is harvested with scrapers and stacked to drain away residual bittern.

Typically, during solar concentration (evaporation) of shallow brine layers, the brine is stagnant unless there is considerable wind and wave action. Consequently, salt crystals nucleate and grow at the air/brine surface where evaporation and supersaturation are occurring. The salt crystals are held there by surface tension until they grow to a critical size and sink. Once reaching the pond bottom the sinking crystals do not coalesce and considerable intercrystal void space results. Hence, the crystallizing salt bed is highly porous and vuggy. This salt crystallization and sinking process accounts for the average 45% volume of void space in the Searles Lake beds.

A considerable volume of residual brine is trapped in the evaporite void space. Hence, the crude salt crystals are contaminated with impure brine solution, as well as other crystallizing salts when they occur in saturated amounts. At Searles Lake sodium sulfate and sodium carbonate are also saturated and crystallize during solar evaporative concentration.

The Searles Lake solar evaporation bitterns have enhanced concentrations of potassium and borate and are potentially useful as a concentrated brine feed to a plant extracting these components, although this has not been commercially practiced. Figure 13.5 projects the change in concentration of several components with the solar brine concentration ratio, defined as the volume ratio of original brine to residual bittern. In principle, further solar concentration of the bitterns and salt precipitation could provide a complex salt mixture for harvesting and further processing. However, higher concentration ratios are more difficult to achieve because the solar evaporation rate decreases with further concentration. Also, a small rate of pond bottom leakage can severely limit ultimate brine concentration and production of end-member salts. This will be discussed in the next chapter.

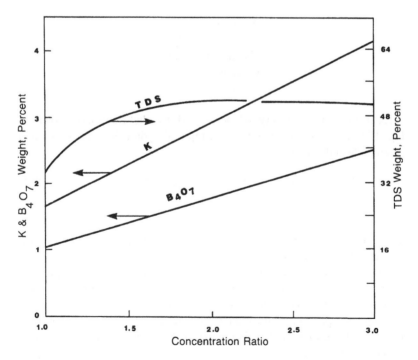

Figure 13.5. Increase in unsaturated components and total dissolved salts (TDS) during solar evaporation of Searles Lake brine.

FORT CADY BORATE *IN SITU* LEACHING

The Fort Cady colemanite ($Ca_2B_6O_{11} \cdot 5H_2O$) deposit near Barstow, California provides another and quite different example of the solution mining of an evaporite deposit. In this example the evaporite beds are deeper, at least 400 m below the ground surface, and less porous. The permeability is much lower, 2–6 md, and a lixiviant chemical, HCl, is required to dissolve the boron mineral, Wilkinson et al. (1987).

Solution mining this deposit is similar to flooded leaching of metallic minerals, such as uranium, using a field of injection and production wells. The deposit may be an attractive candidate deposit for horizontal well drilling if the wells can be accurately located within the thin colemanite beds. Leaching is conducted well below the water table, which is located at a depth of 100 m from the surface. Hydrologic control of the richer leach solutions is more critical than at Searles Lake. The ore zone is 36 m thick and composed of a sequence of numerous, thin clastic and evaporite beds. The average mineralogy of the entire ore zone is:

Anhydrite	38%
Clastics	32%
Colemanite	12%
Celestite	10%
Calcite	8%

Extraction and recovery of boric acid by the flowsheet shown in Fig. 13.6 has been piloted at the site. Injected hydrochloric acid dissolves colemanite and calcite,

$$2H_2O + 4HCl + Ca_2B_6O_{11} \cdot 5H_2O = 2CaCl_2 + 6H_3BO_3, \qquad (13.1)$$

$$2HCl + CaCO_3 = H_2O + CaCl_2 + CO_2(g). \qquad (13.2)$$

The extracted solution, containing dissolved boric acid and calcium chloride, is pumped to the surface and concentrated by solar evaporation to crystallize crude boric acid, which is eventually harvested from the evaporation pond. The solubility of H_3BO_3 is about 3 wt pct in $CaCl_2$ brine, so some boric acid must be recycled to the injection wells.

After solar pond crystallization, the solution, rich in calcium chloride, is reacted with sulfuric acid. This removes calcium as gypsum and regenerates hydrochloric acid,

$$3CaCl_2 + 3H_2SO_4 + 6H_2O = 6HCl + 3CaSO_4 \cdot 2H_2O. \qquad (13.3)$$

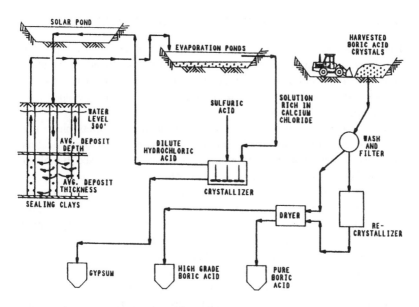

Figure 13.6. Proposed extraction and recovery of boric acid at Fort Cady deposit in California (Wilkinson et al., 1987).

Gypsum is filtered and washed to minimize soluble HCl losses and it is a potential by-product. The reactions are in balance with respect to HCl consumption and generation. However, HCl losses, especially in the evaporite formation, will require continued make-up of hydrochloric acid as well as consumption of sulfuric acid.

An extended five spot well pattern was planned following injectivity tests with three test wells. Injection pressures must be set safely below the modest fracture pressure of the formation to prevent hydrofracturing, which would allow solution short circuiting directly to production wells. Solution mining design considerations are essentially the same as those discussed in Chapters 10 and 11. Because of the low evaporite permeability and limited injection pressure, closely spaced wells (26 m apart) were planned. Thus, well drilling and completion costs are critical economic factors for this project.

A major early concern for the Fort Cady project was loss of permeability due to gypsum precipitation and plugging in the evaporite reservoir, particularly near the wells. Small amounts of sodium sulfate are present in the evaporite and will dissolve and react with calcium chloride in the brine, forming gypsum.

BRINE EVAPORATIVE CRYSTALLIZATION PATHS

Brine crystallization paths describe the sequence of salts that are precipitated as water is removed. Brine crystallization paths depend on the starting composition and solution temperature. The starting point is a sodium chloride saturated brine obtained from extraction or prior evaporation, e.g., seawater evaporation.

The usual practice is to determine ternary or more complex phase diagrams for the critical dissolved components showing the stable crystalline phases as a function of the brine solution composition, expressed in mol pct, of only those components shown on the phase diagram. A ternary phase diagram can only include three components, so water is normally excluded. Higher order phase diagrams are rare and difficult to casually interpret. Since many more components than three are usually present in complex brines, such as those at Searles Lake, interpretation of crystallization paths can be difficult and is certainly beyond the scope of this text.

Because of the complexity of salt crystallization paths, froth flotation is often a useful adjunct to separating salts. Bitterns are further crystallized either with solar ponds or by thermal evaporation. The resulting mixed crystals are milled and subjected to differential flotation in saturated brine using fatty acids or similar organic compounds as reagents. Flotation is also extensively practiced on crystalline salt mixtures from evaporites mined by conventional mining methods.

SALT RECOVERY AND SEPARATION PROCESSES

A few examples of commercial processes to recover and separate various salts from solution mined brines follow.

Sodium Chloride

Solution mined brine is normally saturated with sodium chloride, but usually contains about 0.5% $CaSO_4$ and smaller amounts of calcium and magnesium chlorides. It may be necessary to strip H_2S by aeration and chlorination and settle out solids before evaporation to crystallize the salt, usually in steam heated multiple-effect vacuum pans. Special salt grades may require removal of calcium and magnesium prior to crystallization using caustic soda or soda ash.

Salt produced by solar evaporation of brines is generally crushed and washed with water after harvesting. Some of this crude salt can be marketed directly. For higher salt grades the crude salt is redissolved, purified and recrystallized by thermal evaporation.

Soda Ash

The primary process for producing soda ash from brines is carbonation, which is extensively conducted at Searles Lake. The brine may be partially concentrated by evaporation or processed directly. Brine is heated and carbonated in large columns with sparged CO_2. Normally, the CO_2 bearing sparge gas is boiler flue gas. Sodium bicarbonate is produced and crystallizes because it is less soluble than sodium carbonate.

$$Na_2CO_3 + CO_2(g) + H_2O = 2NaHCO_3(s). \qquad (13.4)$$

The crystals are separated from the brine, which is recycled to the evaporite lake. The sodium bicarbonate filter cake is dried and calcined to sodium carbonate, releasing CO_2 for recycle to the carbonation step:

$$2NaHCO_3(s) = Na_2CO_3(s) + CO_2(g) + H_2O. \qquad (13.5)$$

The resulting sodium carbonate hot calcine, called "light ash," is bleached and quenched in water to recrystallize sodium carbonate monohydrate. The monohydrate is dried to form the final product, dense soda ash.

Potash and Borates

KCl is separated from other salts by flotation from conventionally mined sylvite ores in New Mexico and Saskatchewan, and also from mixtures of salt crystals obtained from the Great Salt Lake and Lake Bonneville, Utah by solar evaporation.

Potash is produced at Searles Lake by taking advantage of a favorable crystallization path at elevated temperature using thermal boiling in triple-effect evaporators (Mark et al., Vol. 3, 1964). Figure 13.7 represents the system, Na^+, K^+, SO_4^{2-}, CO_3^{2-}, Cl^-, and H_2O at 100°C. Searles Lake brine at composition 'B' is combined in the correct proportion with an adjacent plant mother liquor at composition 'M' to obtain an evaporator feed brine at composition 'F'. Since this composition is in the burkeite field, burkeite ($Na_2CO_3 \cdot 2Na_2SO_4$) is precipitated during evaporation, along with halite, which is not included in this ternary phase diagram. As evaporation continues, the brine composition follows the tie line from point F to point A in Fig. 13.7. When saturation with $Na_2CO_3 \cdot H_2O$ occurs at brine composition A, this phase begins to precipitate and the brine composition proceeds toward point C.

The advantage of this process is that only sodium salts have been removed up to point C while all of the potassium and boron salts have concentrated in the residual hot liquor. After removal of the sodium salt crystals, this hot

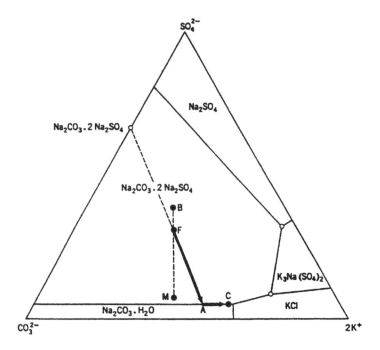

Figure 13.7. Ternary phase diagram at 100° showing evaporative crystallization path to produce a potassium and boron rich end liquor (Kirk–Othmer, 2nd ed., Vol. 3, 1964).

liquor is rapidly cooled to yield KCl crystals. Borates, highly enriched in the remaining brine, are recovered after the KCl crystals are separated.

Sodium Sulfate

Sodium sulfate is saturated in several brines including those at Searles Lake and the Great Salt Lake. Sodium sulfate solubility decreases rapidly when saturated brines cool, precipitating Glaubers salt, $Na_2SO_4 \cdot 10H_2O$, during cold months. Hence, sodium sulfate can be harvested during winter. Spraying warm brine into cold winter air has also been used to crystallize Glaubers salt.

Magnesium

Magnesium is used extensively in the form of $Mg(OH)_2$ and MgO, but $MgCl_2$ is required for fused salt electrolysis to produce magnesium metal.

Conventionally, cell grade $MgCl_2$ is produced by recovering $MgCl_2$ from brine or sea water, converting it to $Mg(OH)_2$ and then chlorinating $Mg(OH)_2$ back to anhydrous $MgCl_2$. But direct use of $MgCl_2$ crystallized from brine would eliminate these steps. Although difficult, because the $MgCl_2$ electrolysis salt must be completely anhydrous, this route is being pursued at a plant located on the rim of the Great Salt Lake using solar ponds to obtain the crude $MgCl_2$ crystals.

Lithium

Lithium carbonate is produced from brine pumped from alkali evaporites in Nevada and Chile. Economic lithium deposits are those with relatively higher lithium concentrations and lower magnesium concentrations than other alkali evaporites. Brine is pumped from wells in the evaporite, similar to the recovery of brine at Searles Lake. The brine is concentrated in dissolved lithium chloride as it passes through a series of solar evaporation ponds. Sodium chloride crystallizes and remains in the bottom of the solar ponds as an ever-thickening layer. This requires constant raising of the pond dikes. At the Silver Peak, Nevada operation there are 22 evaporation ponds covering 16 sq km (4,000 acres).

Midway through the two year solar concentration cycle, magnesium is removed from the bittern by precipitation upon adding lime. The settled precipitate, containing magnesium hydroxide and magnesium sulfate, is removed with a dredge. The remaining bittern continues through further solar evaporation ponds until its lithium concentration is about 40 times greater than in the starting brine. Then the bittern is pumped to a surface plant and heated. Lithium carbonate is precipitated by adding soda ash, and the precipitate is filtered and dried.

SULFUR DEPOSITS

Approximately 35% of the sulfur produced in the United States is derived by a solution mining method (the Frasch process). Another 45% of domestic sulfur is recovered by separating H_2S from sour natural gas. Frasch sulfur is produced from gulf coast salt domes. Significant amounts of Frasch sulfur are now being produced from tabular evaporite deposits in west Texas, as well as several other places in the world. Mexico has become a significant Frasch sulfur producer from gulf coast salt domes.

Sulfur often occurs as well-developed crystal aggregates in veins and vugs and as finer grain disseminations in the porous limestone and gypsum sections of the cap rock of a salt dome (Grayson, Vol. 22, 1983). Salt domes usually contain anhydrite ($CaSO_4$) as the source of the gypsum ($CaSO_4 \cdot 2H_2O$) in the cap rock. Salt domes are major traps for hydrocarbons and anaerobic bacteria that derive their metabolic energy from the hydrocarbons that are present in this environment. Some of the anaerobic bacteria (e.g., *Desulfovibrio desulfuricans*) reduce sulfate ions from the gypsum to form H_2S dissolved in the formation water. A net reaction for gypsum reduction is:

$$CaSO_4 \cdot 2H_2O + CH_4 = H_2S + CaCO_3 + 3H_2O. \qquad (13.6)$$

The dissolved H_2S is precipitated to sulfur by subsequent oxidation reactions, possibly after some migration into nearby regions of higher oxidation potential, such as porous limestone capping beds. Evaporite basin sulfur deposits are formed by the same process, hydrocarbon reduction of sulfates assisted by anaerobic bacteria.

THE FRASCH PROCESS

The Frasch process (Grayson, Vol. 22, 1983) utilizes the low melting point of sulfur (114.5°C for the monoclinic allotrope) to extract molten sulfur through wells drilled into the sulfur deposit. Sulfur is heated and locally melted with pressurized hot water (typically about 160°C) that is pumped into the deposit. Because the molten sulfur is more dense than water, it settles by gravity and forms a liquid pool around the bottom of the production well. In conventional Frasch technology, three concentric pipes are placed inside the well casing, as illustrated in Fig. 13.8. The outer pipe, typically about 200 mm diameter conducts the hot water, which is injected into the bottom of the sulfur bearing formation through perforations in the pipe. An inner pipe, about 100 mm diameter conducts molten sulfur up and out of the well under the hydrostatic pressure of the formation. Part way up the well, the third innermost pipe (about 25 mm diameter) injects compressed air into the well. The air bubbles rise with the molten sulfur in the sulfur conducting pipe, lowering the aggregate density of the two-phase fluid as it rises. The rising compressed air becomes an air lift pump for extracting the dense, corrosive molten sulfur. Simultaneous control of sulfur pumping, air injection, and hot water injection is necessary to keep the sulfur molten and flowing.

The well production sequence progresses from left to right in Fig. 13.8. At the beginning of each well operation, hot water is injected without an attempt at sulfur recovery for a period of time until an adequate molten

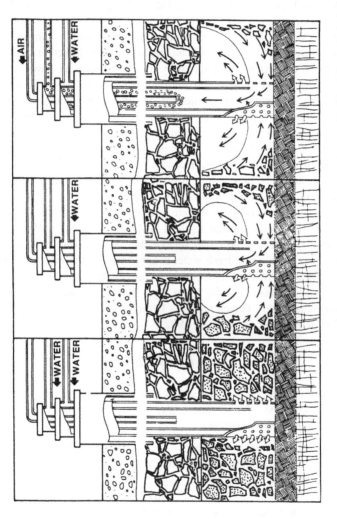

Figure 13.8.   Illustration of Frasch process molten sulfur recovery well (from Freeport Minerals in Kirk–Othmer, 3rd ed., Vol. 22, 1983).

pool of sulfur has formed. During this time upward flow of sulfur is suppressed by also forcing hot water down the sulfur production annulus. When this hot water flow ceases, liquid sulfur rises in the production pipe because of the hydrostatic pressure to an elevation about 100 m above the sulfur melt pool, where it meets the bubbling compressed air. Hot water injection continues through the outermost annulus after molten sulfur production begins. During production an excessive sulfur pumping rate may deplete the molten sulfur pool causing hot water to be produced. When this happens, the process is restarted by injecting hot water down the production pipe to melt more sulfur. Individual wells last a year or more, then, declining production or subsidence of the ground around the bottom of the well depleted of sulfur stops the production process. Recently, horizontal drilling has been used to obtain additional sulfur from depleted deposits.

On the surface molten sulfur is pumped through steam heated pipelines to an air separator. The sulfur can be cooled and solidified by spraying it onto a stockpile in a storage bin or vat, and it can also be stored and shipped in thermally insulated tank cars.

Large amounts of hot water are injected into the enclosed sulfur deposit. Bleedwater wells located a short distance from the sulfur production region are often required to relieve the excess hydrostatic pressure. In some locations where fresh water is not available, seawater is being heated and deoxygenated for injection in sulfur production wells. This occurs in the Gulf of Mexico at the Main Pass Block, southeast of New Orleans, Louisiana, where Frasch sulfur is being produced at platforms located in 63 m of seawater (Ackerman, 1992).

PROBLEMS

1. An extended 5-spot wellfield is being planned to solution mine a deep evaporite, similar to Fort Cady, to recover boron. Planned wellfield conditions are as follows:

ore deposit thickness = 36 m,
intrinsic permeability = 0.004 darcy,
well spacing, $S_5 = 26$ m,
well bore hole radius = 2.2225 cm (1–3/4 in ID pipe),
pressure drop between injection and production wells is 1500 kPa.

(a) What is the expected production rate in liters/minute from each production well?

(b) A separate economic assessment found that the production cost per well, including capital amortization, would increase by 50% if wells were drilled and completed with 4-in ID pipe. Do you recommend this? Explain your answer.

(c) A second suggestion was to double the number of wells by placing them closer together. How much closer should they be ($S_5 = ?$) to double the number of wells for the same area of deposit? With no change in the production cost per well, do you recommend that this suggestion be adopted? Explain.

2. After this boron solution mine was in production for three years, with wells spaced as originally planned at $S_5 = 26$ m, it was found that sweep efficiency was decreasing and recovery of boron had dropped off considerably. After shutting in some of the wells and surveying the permeability, it was found that there was considerable more permeability variation in the different strata, indicating that preferred leaching along the higher permeability strata had dissolved salts and made these strata even more permeable. Consequently, a switch to vertical flow between the two most open strata is being considered. These two high permeability strata are separated by 21 m, and one has a permeability of 10 darcy while the other has a permeability of 5 darcy. With packers, a pressure difference between the open strata of 1000 kPa can be maintained, but the *vertical* permeability between strata is very uncertain because of interposed thin clastic layers.

Compute the average vertical permeability required to provide the same well production rate (l/min) that you projected in Problem 1 for the original extended 5-spot wellfield.

REFERENCES AND SUGGESTED FURTHER READING

Ackerman, J.M. (1992), *Mining Engr.*, Vol. 44, No. 3, pp. 222–226.
Grayson, M. et al. (1983). *Kirk–Othmer Encyclopedia of Chemical Technology*, 3rd Edition, J. Wiley & Sons, New York, Vol. 22, p. 78.
Mark, H.F. et al. (1964). *Kirk–Othmer Encyclopedia of Chemical Technology*, 2nd Edition, J. Wiley & Sons, New York, Vol. 3, pp. 609–651.
Schlitt, W.J. and Hiskey, J.B. (1982). *Interfacing Technologies in Solution Mining*, Proceedings of Second SME-SPE International Solution Mining Symposium, American Inst. of Mining, Metallurgical and Petroleum Engineers, New York, pp. 21–54.
Schlitt, W.J. and Larson, W.C. (1985). *Salts & Brines, 85*, Society of Mining Engineers, Littleton, CO.
Smith, G.I. (1979). Subsurface stratigraphy and geochemistry of late quaternary evaporites, Searles Lake, California. USGS Professional Paper 1043.
Wilkinson, P.A.K. et al. (1987). Fort Cady in situ leach project. *Society of Mining Engineers Preprint 87–140*, Littleton, CO.

FOURTEEN

Solar Evaporation Ponds

INTRODUCTION

Solar evaporation ponds are used extensively to remove excess water from leaching solutions, to concentrate brines, and to precipitate and harvest crystalline salts from saturated solutions. These are among the largest commercial applications of solar energy. Brine sources include: sea water, brine aquifers, playas or natural salt lakes, and solution mining of evaporites and salt domes.

This chapter relates primarily to brine evaporation because of its industrial importance and because evaporation rates depend on brine concentration (brine density). This variability is not generally encountered with the more dilute solutions used in solution mining of metals. See Butts (1985) for the basis of this chapter.

A solar pond and its materials balance can be described by the flows shown in Fig. 14.1. A series of linked materials balance equations must be considered that include water and each of the salt components. Often several ponds are involved with brine processing, with the brine discharged from one pond becoming the feed for the next pond.

Because of wind action and long average residence times, brines and solutions in solar ponds tend to be fairly well mixed. Thus, the solar pond brine concentration is equal to the exit brine concentration. When crystallized salt is the desired product, the brine may be evaporated to dryness with no effluent brine (bittern). Brine occupying void space in the salt layer, **entrainment**, will remain with the salt.

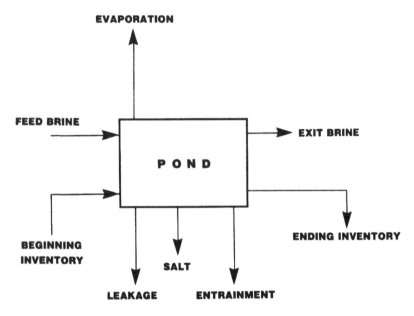

Figure 14.1. Solar evaporation pond material and brine flows.

EVAPORATION RATES

Average brine evaporation rates vary with the time of year and brine density, as illustrated in Fig. 14.2. For temperate dry climates, most of the evaporation occurs in the summer months at rates from about 3–6 mm per day. High temperatures, low humidity and wind favor high evaporation rates. Wind accelerates the evaporation rate by decreasing the vapor boundary layer over the evaporation pond's brine surface.

Effective management of solar evaporation ponds requires knowledge of the average evaporation rate as a function of brine density for each month during the summer brine evaporation season. An example is shown in Fig. 14.3. Obtaining the necessary data requires a suitably instrumented experimental program conducted over a year at the planned location for the solar ponds. Evaporation data collected in an abnormal climatological year may be in error and require judicious adjustments in planning the required pond area. Brine evaporation rates invariably decline as density increases because the chemical activity, and, therefore, the equilibrium vapor pressure of water, decreases as the dissolved salt concentration increases. For a simple (one salt) brine, density becomes constant when saturation is reached and salt crystallization begins. For more complex

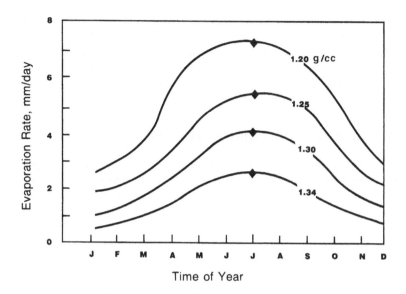

Figure 14.2. Typical brine evaporation rate curves.

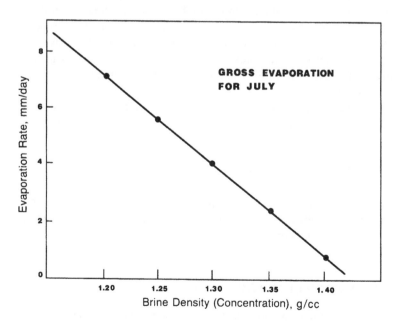

Figure 14.3. Evaporation rate dependence on brine density.

brines, density will continue to increase while one or more salts are saturated but other salts continue to concentrate during evaporation.

Solar evaporation ponds may encounter negative evaporation during winter months because of the reduced temperature and increased amounts of rain and snowfall.

SALT DEPOSITS AND BRINE ENTRAINMENT

In the first phase of concentrating ocean brine, salts often do not crystallize. Many solar pond systems crystallize only one mineral such as halite (NaCl) or sylvite (KCl). Often halite, as the dominant salt, is deposited and remains in the pond bottom while the bittern, a concentrated brine with the more valuable components, is then processed in a surface plant or by further solar evaporation in separate ponds. Each salt that is crystallized, including mixed salts, must be included in the pond materials balance.

Salt minerals crystallize sequentially, usually by nucleating at the brine/air interface where evaporation is occurring and the solution is the most supersaturated. Often these surface nuclei grow into fairly large coarse crystals that impinge on each other while still floating on the surface. Eventually the force of gravity exceeds the surface tension holding the crystals at the surface and they sink. Loose networks of massive coarse crystals or dendrites can provide very large open void spaces upon settling, with high permeability and large amounts of entrained brine in the salt sediment.

The entrainment factor is defined as the mass ratio of entrained brine to solid salt. The entrainment factor for most salts is in the range of 0.3–0.5. Selection of an entrainment factor of 0.4 is a good rule of thumb in the absence of measured data. The entrainment factor for halite evaporites is 0.35, which is also approximately its void volume fraction. However, some salt minerals crystallize from solar ponds with a much greater void fraction than solid fraction, leading to high entrainment factors. As an example, mirabilite ($Na_2SO_4 \cdot 10H_2O$) and bischoffite ($MgCl_2 \cdot 6H_2O$) have entrainment factors of 2–3.

SOLAR POND LEAKAGE

Solar ponds often extend over very large areas, and all solar ponds leak. When harvesting a crystallizing salt deposit is the goal, leakage has little impact on the results. But leakage is important when concentrated brine production is the goal. A tight solar pond is one that leaks less than 0.3 mm/day, and a leakage rate of 1 mm/day is generally not tolerable. A solar pond placed directly over a dry lake bed to obtain flat topography may be very suitable for producing a salt deposit, e.g., halite, but would not work for brine production without a more impervious bottom liner. When brine production is the objective, synthetic geomembranes such as hypalon and PVC are used (see Chapter 4). Examples of geomembrane use to deter leakage are found at the evaporation ponds at the operations of Salar de Atacama in Peru (borates), Silver Peak in Nevada (lithium), and Texas Gulf near Moab, Utah (potash).

EXAMPLE PROBLEM

Compute (1) the amount of feed brine in *tonnes per day* and (2) the solar pond area in *acres* that are required to deposit 10,000 tonnes per month of halite (NaCl) during the summer evaporation season. During this period the average evaporation rate is estimated to be 5 cm/day and the average leakage rate through the bottom of the pond is estimated to be 0.5 cm/day. The saturated brine is 78 wt pct water and the feed brine is 90 wt pct water. Salts other than NaCl in the brine are negligible, and there is no exit brine.

ANSWER

To solve this problem, we must link the *daily* water mass balance around the pond (see Fig. 14.1) with the daily salt (solids) mass balance around the pond. Note that the following calculations are made with area in cm^2, time in days and mass in grams. Mass rates are given in g/day. Conversions of answers to the desired units are made later.

$$\rho_{H_2O} = 1.0 \, g/cm^3, \quad \rho_{NaCl} = 2.16 \, g/cm^3$$

The following mass rates are defined:

$$\text{Brine feed rate, g/day} = F$$

$$\text{Evaporation rate, g/day} = E$$

$$\text{Leakage rate, g/day} = L$$

$$\text{Salt Product, g/day} = S$$

Exit brine and entrainment are negligible and beginning and ending inventory of brine are equal.

Components	ρ	Mass Fractions of Components	
		Feed Brine	Saturated Brine
NaCl	2.16 g/cc	0.10	0.22
H_2O	1.0	0.90	0.78

The evaporation and leakage *volumetric* rates, Ev and Lv, expressed in cm^3/day, are related to the pond area, A, expressed in cm^2, and the velocities of evaporation and leakage, which were given as 5 cm/day and 0.5 cm/day, respectively. Hence:

$$Ev = 5A \text{ cm}^3/\text{day},$$

$$Lv = 0.5A \text{ cm}^3/\text{day}.$$

Furthermore, the mass rates are related to the volumetric rates through the densities; hence:

$$E = Ev = 5A, \text{ and } L = Lv * (\rho_{sat}) = 0.5A * (\rho_{sat}),$$

where ρ_{sat} is the density of the saturated brine.

The volume of 1.0 g of saturated brine is related to the mass fractions and component densities. Assuming negligible volume change when NaCl is dissolved in brine:

$$\rho_{sat} = 1/\{(0.22/2.16) + (0.78/1)\}$$

$$= 1/\{0.102 + 0.78\}$$

$$= 1.134 \text{ g/cc}.$$

Using the Daily Water Mass Balance:

$$F_{H_2O} = 0.9F, \qquad E_{H_2O} = E = 5A,$$

$$L_{H_2O} = 0.78L = 0.78(0.5A)(\rho_{sat}) = 0.4226A, \qquad F_{H_2O} = E_{H_2O} + L_{H_2O}.$$

Substituting and solving for F in terms of A yields:

$$0.9F = 5A + 0.4226A, \text{ or}$$

$$F = 6.047A.$$

Using the Daily NaCl Mass Balance:

$$F_{NaCl} = 0.10F, \qquad L_{NaCl} = 0.22L = 0.22(0.5A)(\rho_{sat}) = 0.125A \text{ g/day},$$

$$F_{NaCl} = L_{NaCl} + S, \qquad S = 10,000 \text{ tonne/month} = 3.33 \times 10^8 \text{ g/day}.$$

Substituting and solving for F in terms of A yields:

$$0.1F = 0.125A + 3.33 \times 10^8, \text{ or}$$

$$F = 1.25A + 3.33 \times 10^9.$$

Equating values of F from the water and NaCl mass balances yields:

$$6.047A = 1.25A + 3.33 \times 10^9 \text{ g/day, or}$$

$$4.797A = 3.33 \times 10^9 \text{ g/day, or}$$

$$A = 69.5 \times 10^7 \text{ cm}^2, \qquad A = 69{,}486 \text{ m}^2, \qquad A = 7.48 \times 10^5 \text{ ft}^2.$$

$$A = 17.18 \text{ acres}$$

$$F = 6.047A = 6.047(69.5 \times 10^7) = 4.2 \times 10^9 \text{ g/day.}$$

$$F = 4{,}202 \text{ tonne/day}$$

TOPOGRAPHY AND ENGINEERING CONSIDERATIONS

Large areas of flat land are desirable. One firm at the Great Salt Lake has over 60 km^2 (15,000 acres) of solar ponds. The Dead Sea Works has over 130 km^2 (32,000 acres) of solar ponds. As the topography varies from horizontal, dikes must be used at ever closer intervals as the slope angle increases. Flat areas are usually at the bottom of valleys and often at nearly the same level as the brine source. Consequently, they may be prone to flooding.

Other important siting and engineering factors are the mineral reserve and its accessibility, favorable meteorology, and availability of fresh or brackish water. Water is needed to eliminate unwanted salt build-up and purify salts during processing. Brine transfer points are usually plagued with salt deposits that become serious operating problems. Salt incrustation in pumps and pipes is hard to remove mechanically, but it can be dissolved by proper use of fresh water.

Natural disasters, such as heavy rains and flooding, can upset operations and seriously damage solar ponds. These can be short-term events or extended climate changes. An increase in the level of the Great Salt Lake, following several years of increased snow and rain, flooded and ruined many solar brine evaporation ponds on the periphery of the lake during the early 1980s.

Figure 14.4.   Crawler mounted salt dredge harvesting from a Dead Sea evaporation pond (RA Hanson Co., Inc.).

HARVESTING SALT DEPOSITS

Salt deposits are usually harvested after the summer evaporation season. During the evaporation and deposition season, saturated brine is kept at a depth of only about 0.3 m (1 ft). Adequate amounts of feed brine are added as required to maintain this depth.

Upon completion of the salt making season the residual brine is drained. Drainage may include trenching a grid of ditches in the salt layer to draw down brine entrained in the deposited salt layer. Salt layer thickness varies widely, but for halite deposited from saturated brines in hot dry climates, the thickness is typically about 0.3 m (1 ft). After draining brine, the salt layer is ripped using a dozer or motor grader. The broken salt layer is either windrowed with a grader or scalped, loaded, and hauled with an elevating scraper. Haul trucks are loaded from windrows on the floor of the dry solar pond using front-end loaders. At Great Salt Lake Minerals and Chemicals Corporation in Utah, a 0.35 m (14 in) floor of halite is needed to support the weight of 45 tonne haul trucks.

Slurry harvesting of saturated brine containing suspended crystals is sometimes practiced. Solar ponds with a brine depth of about one meter are used.

Special dredges floating in brine are now being used to harvest potassium salt (carnallite) simultaneously while solar evaporation continues (see Fig. 14.4). A cutterhead with augers cuts a layer of salt from the pond bottom. Large, on-board pumps move salt slurry through a floating pipeline to a processing plant on shore. These dredges weigh up to 350 tonnes and, although most of the dredge is buoyed by the brine, four crawler tracks on the pond bottom maneuver the harvester around the solar evaporation pond. Work boats assist the harvester in making turn-arounds and relocating the slurry pipeline anchors.

PROBLEMS

1. The estimates of evaporation and leakage rates in the EXAMPLE PROBLEM may be incorrect. Redo this problem by re-computing the solar pond area in **acres** and the brine feed rate in **tonnes per day** with the following more conservative estimates: an average evaporation rate of 40 mm/day, a leakage rate of 10 mm/day and a more dilute feed brine that is 92 wt pct water.

2. Low-density brine will be pumped from an **unconfined** vuggy salt evaporite ($k_i = 1000$ darcy) into a solar evaporation pond at the rate of

1,000 gal/min from a single well with a 16 inch (400 mm) diameter perforated casing penetrating 100 ft of the evaporite. Compute the expected drawdown in feet that would be measured from an observation well located 1,000 ft away.

3. A solar evaporation pond has an area of $4 \times 10^6 \, \text{m}^2$ (about 1,000 acres) and is used to concentrate valuable unsaturated salts from an alkali brine feed saturated in NaCl (30 wt pct NaCl) with a density of 1.20 g/cc. The volumetric concentration ratio is 2.5. The required bittern production is $2.0 \times 10^6 \, \text{m}^3$ each summer season, and the bittern has a density of 1.30 g/cc. Trenches dug in the deposited salt layer are used to drain off **all** of the entrained bittern as part of the annual bittern production at the end of the season. Leakage losses are expected to be only 2.5%, because of the underlying geomembrane.

   (a) Calculate the cubic meters and tonnes of annual brine feed needed to meet the bittern production goal.
   (b) Calculate the tonnes of NaCl salt deposited each season.
   (c) With an entrainment factor of 0.35, estimate the thickness of the salt layer that is deposited each season in both centimeters and inches. The specific gravity of NaCl(s) is 2.16 g/cc.

REFERENCES AND SUGGESTED FURTHER READING

Butts, D.S. (1985). Basic solar pond modeling and material balance techniques. Schlitt, W.J. and Larson, D. (Eds.), *Salts & Brines '85*, Society of Mining Engineers, Littleton, CO, pp. 135–147.

FIFTEEN

Remediation of Contaminated Ground and Water

INTRODUCTION

Environmental control and containment to prevent contamination of earth and water is nearly always less expensive than remediation after contamination has occurred. Environmental control and containment for ore heap leaching was covered in Chapter 4. Controlling the zone of contamination during *in situ* leaching was discussed in Chapter 10, primarily using guard wells to prevent flow outside the ore deposit. Several treatment methods for the necessary excess water that is pumped out over that injected, when using guard wells, were shown in Fig. 10.21. In spite of the best practices, contamination of spent ore heaps and solution mined subsurface regions is an inevitable result of many mineral extraction processes. Generally in solution mining, the concern is with toxic metals and chemicals used in the lixiviant. Upon completion of the mining operation, reclamation must occur that includes remediation of contaminated ground, dilute solutions retained in the mine and ore heaps and draining from them, and contaminated groundwater.

Cyanide detoxification of spent ore heaps was also discussed in Chapter 4. Currently, several of these already detoxified spent ore heaps are encountering an arsenic drainage problem. As meteoric water percolates through them, pH is dropping to levels lower than the high basic pH used in cyanide leaching, and arsenopyrite and arsenic bearing pyrite are being slowly biooxidized. Consequently, arsenic is being solubilized. While the arsenic concentrations are very low, the EPA drinking water standard for arsenic is only 50 ppb. Some spent gold heaps are also encountering leachates with nitrate, from cyanide decomposition, in excess of environmental standards.

Most earth and groundwater remediation processes are divided into two approaches: (1) removal from the ground, using either water or air as a carrier, followed by treatment to remove the contaminant from the carrier fluid, and (2) *in situ* treatment to immobilize the contaminant and/or render it harmless by a suitable chemical or biochemical transformation.

CONTAMINANTS

Toxic metals are the primary contaminants resulting from mining operations. The USA maximum concentration levels (MCL) for metals and other inorganic contaminants in drinking water are listed in Table 15.1. Many of these are encountered in both solution mining leachates and conventional mine waters. Soluble metals, nitrate, arsenic and selenium are often an environmental problem at mines. Other forms of contamination, including a large assortment of organic and chloroorganic compounds, may occur from spills at various sites, whether or not these sites are mining related.

There are Federal primary and secondary standards for radionuclides, turbidity, bacteria, coliform and many organic compounds and substances,

Table 15.1 Inorganic Federal
Primary Drinking-Water Standards.

Contaminants	MCL (ppb)
Antimony	6
Arsenic	50
Asbestos	7.0 MFL
Barium	2000
Beryllium	4
Cadmium	5
Chromium	100
Cyanide	200
Fluoride	4000
Mercury	2
Nickel	100
Nitrate (as N)	10,000
Nitrite (as N)	1000
Selenium	50
Thallium	2

including several that have limited volatility. Most, but not all liquid organic substances, have little solubility in water. These are often referred to as non-aqueous phase liquids, or NAPLs. They include light organics (LNAPLs) which, upon reaching the water table, float on the groundwater. A well intersecting the groundwater will show evidence of LNAPL from the oil slick on the water in the well. Dense organics (DNAPL) have specific gravities greater than $1\,g/ml$ and pass through the groundwater to bedrock. Examples of DNAPLs are naphthalene, nitrobenzene, trichloromethane and many other chloroorganic compounds. The volatility and solubility of organic contaminants in water are important parameters in designing a remediation program.

MINE WATER AND ACID MINE DRAINAGE

The focus of this chapter is on remediation of contaminated ground and the water contained within, or draining from it. The contaminated water includes both saturated water (groundwater) and unsaturated water contained within the macropores or void space of porous earth materials. Examples of drainage from closed mining operations include acid mine water flowing from mine adits (defined as tunnels with only one opening) and from ore heaps and mine waste dumps. In the absence of an impervious liner, drainage flowing directly into the underlying ground will occur, percolating by gravity through the vadose zone and eventually reaching the water table and entering the local groundwater. Percolation of contaminated water through the vadose zone is identical with transport of a leaching solution through an ore heap, discussed in Chapter 7.

While it is natural to divide treatment methods between solid earth materials and water, in many cases they are so intimately mixed that both must logically be remediated together. An important consideration is the duration of treatment (life cycle). A long life cycle generally translates to a very high treatment cost. In most ground remediation cases, the source of contamination has ceased, but in others the source continues to produce contaminant. Abandoned mines containing sulfide minerals continue to drain acid mine water as long as air and meteoric water can enter the mine and continue mineral biooxidation. Sealing underground mines is an obvious answer to this problem, but it is often very difficult to seal large underground workings from meteoric water and air. As water accumulates above a plugged mine opening, very large hydrostatic pressures tend to force water around the plug through permeable rock. Most abandoned mines have simple pressure doors or concrete plugs that typically are less

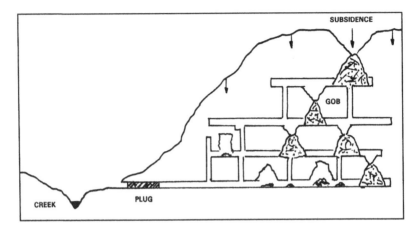

Figure 15.1. Section of an underground mine showing a thick drainage plug, mine caving to produce biooxidizable gob, ground surface subsidence, and infiltration of meteoric water and air.

than a few centimeters thick. Substitution of concrete plugs several meters in thickness, as illustrated in Fig. 15.1, combined with pressure grouting fractures in the surrounding rock are needed to have any reasonable chance of avoiding leakage.

GROUNDWATER PLUME MOVEMENT

Spills of highly insoluble substances generally do not penetrate very far through earth materials. Most heavy oils fall into this category. However, many organic compounds have high aqueous solubilities; examples are alcohols, acetone, and several chlorinated organic compounds. Metal aqueous solubility depends on pH with generally greater solubility at low pH, but some metals form soluble anion complex ions at basic pH.

Even when contaminants are soluble, passage through the vadose zone to the groundwater will usually occur over a very extended period of time. This was shown from tracer experiments for the analogous transport of water through an ore heap in Chapter 7. **Axial dispersion** of water occurs in the vertical direction because of the slower flow in smaller capillaries. Some **lateral dispersion** also occurs because of tortuous flow paths and capillary action drawing water sideways. Consequently, the area of entry into the groundwater will be somewhat greater than the area of the spill or other source of contaminant at the surface. Contaminant adsorption on the earth and diffusion into the micropores of rocks will also retard vertical transport to the water table.

Groundwater is generally in motion because of a hydraulic gradient. While groundwater flow is usually very slow, over an extended period of time a horizontal groundwater **contaminant plume** will be created that may become very large. This is illustrated in Fig. 15.2. Sedimentary hosted uranium roll front ore deposits, an example of which was shown in Fig. 10.7, are the relics of groundwater plumes bearing solubilized uranium that became precipitated and fixed in place upon reaching a reducing environment.

The processes that cause a contaminant plume to form and expand are **advection**, or flow, **dispersion** and **retardation**. Basically, these are the same processes that are operative with percolation in the vadose zone above the water table. Chemical or biochemical precipitation or transformation may also be occurring and affect plume movement, as described in the uranium ore deposition example. Advection is generally nonuniform because the ground is not homogeneous in its composition, particle size and permeability. This can lead to fingers of faster migrating contaminant in higher permeability zones separated by slower migrating contaminant in lower permeability zones that experience lower groundwater flow velocities.

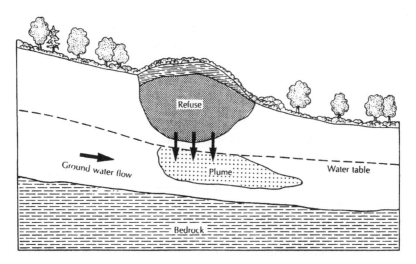

Figure 15.2. Section of a contaminant plume in groundwater (Fetter, 1988).

CONTAMINANT REMOVAL WITH A WATER CARRIER

Drainage from an ore heap, or the adit of an underground mine, falls into this category, as does pumping contaminated groundwater. Several methods of treatment of the water once pumped to the surface were cataloged in Fig. 10.21. **Pump and treat** is usually the first choice when encountering a contaminated groundwater or drainage problem.

Treatment methods to remove inorganic contaminants include oxidation and hydrolysis (lime neutralization) to precipitate metals as hydroxides from acidic solutions, sulfide precipitation, and cementation. The latter is the use of zero valent metals to reduce other metals and possibly arsenic and selenium. Several bioremediation methods can be used for pumped water treatment and will be discussed later in connection with *in situ* treatment.

Arsenate ion can be effectively removed by **co-precipitation** with amorphous ferric hydroxide at pH 4–5 (Robins et al., 1987). This has become a principal method for treating arsenic bearing metallurgical plant effluents, and there are indications that selenium is also removed by this process after the more stable selenate ion is reduced to selenite ion. However, reduction of selenate ion to selenite ion is a kinetically slow process. Experimentally, ferric hydroxide precipitation has been used to remove arsenic from gold leaching barren solutions after cyanide destruction (Bucknam, 1997).

By creating a drawdown so that all of the flow in the contaminated region is toward the pumped well, it is effective in removing the bulk of the contaminants from groundwater and controlling further spreading of plumes. However, pumping groundwater often cannot lower the contaminant concentration sufficiently to meet environmental standards. This unsatisfactory result is caused by the continued slow release of contaminant contained in the fine interstices and micropores of the earth materials. This may occur over months and sometimes years. This later period is sometimes referred to as the diffusion controlled stage of contaminant removal (Nyer et al., 1996). After the bulk of the contaminant has been removed, it becomes too expensive to continue pumping and operating a treatment plant for the carrier water, with its very low loading of contaminant. Most of the daily pump and treat costs are fixed and independent of the amount of contaminant captured. Alternative processes, either as substitutes for pump and treat or as follow-on terminal processes, are needed and will be considered in the following sections.

The high capital and operating cost of the contaminated water treatment process is also an economic problem. Aeration and neutralization with lime is the standard process for treating acid mine water. While environmentally effective, it is an example of a high operating cost process that must be conducted in perpetuity at many mine sites. The solubilities of heavy metal hydroxides as a function of pH are shown in Fig. 15.3. Consequently, as acid mine water is neutralized, several heavy metals are precipitated as hydroxides while sulfate ions are converted to gypsum ($CaSO_42H_2O$). The resulting low-density sludge must be settled, filtered and properly discarded as a potentially hazardous waste. A water treatment plant has several pumps and chemical unit operations with moving equipment that require operator attention. Generally, labor is the largest operating cost item in the process. Hence, a satisfactory *passive* process for mine drainage as it flows by gravity through the treatment "reactor" is needed. Passive systems for treating acid mine water will be discussed in a subsequent section.

IN SITU SOIL WASHING

Soil washing is any process in which the contaminated soil or ground is excavated and processed to separate the contaminants from the soil. The name implies washing the soil with a detergent or surfactant, but any of several physical separation methods common to mineral processing may be used (Bunge et al., 1995). The treatment of contaminated soil that has been excavated and placed in a heap for soil washing is essentially the same as the percolation heap leaching of ores. The information contained in previous chapters, especially Chapter 7, is highly pertinent (Bartlett, 1993).

In situ soil washing is a soil treatment process that mobilizes contaminants in the vadose zone and transports them to the groundwater where they are pumped out through a well. Generally, *in situ* soil washing involves water containing a detergent or surfactant that percolates through the contaminated soil dissolving the contaminant. An example is the use of sodium dodecylsulfate as the surfactant to remove anthracene, a highly hydrophobic contaminant (Liu and Roy, 1992).

VOLATILE CONTAMINANT REMOVAL
WITH AN AIR CARRIER

Air has some advantages as a carrier for volatile organic compounds including a faster exchange of carrier pore volumes than can be obtained with water. A brief review of air extraction methods follow, but for more details on these processes see Nyer et al., 1996. While carrying the contaminants

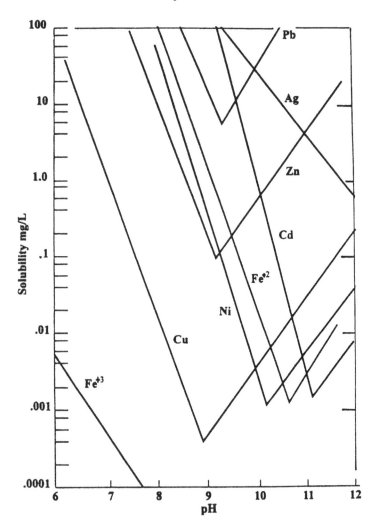

Figure 15.3.   Solubilities of metal hydroxides at various pH.

out of the contaminated ground is the main goal, biooxidation or other aerobic biodegradation may enhance remediation with these processes. **Vapor extraction** is used to extract volatile hydrocarbons from unsaturated earth, the vadose zone of contaminated ground by suction. Potentially, this can be used for ore heaps if contaminated with volatile substances. Lowering the pH in spent gold ore heaps followed by aeration to remove cyanide as HCN(g) is a variant of vapor extraction that was discussed in Chapter 4.

In vapor extraction practice, usually one or more perforated wells are attached to a vacuum pump. Air is drawn down into the contaminated zone and flows to the vapor extraction wells, see Fig. 15.4. Alternatively, there can be air injection wells and air recovery wells. As discussed in previous chapters, the flow velocity near wells will be greater than away from them, and the slow flow at locations most distant from a well must be considered in designing a vapor extraction system. Equations for fluid flow to and from wells that were developed in Chapters 9 and 11 are pertinent to vapor extraction systems.

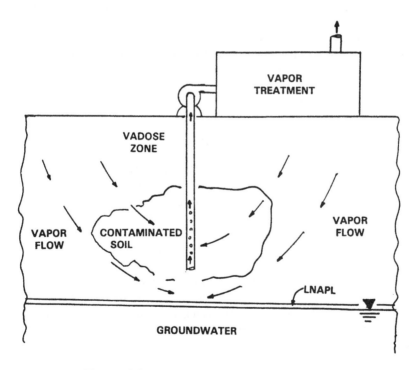

Figure 15.4. Illustration of vapor extraction system.

The collected volatile contaminants are treated by various processes that include catalytic combustion, incineration (high temperature combustion with additional fuel) and adsorption on granular activated carbon. The rate and thoroughness of contaminant removal by vapor extraction depends on the volatility and solubility of the contaminant, and in particular the equilibrium constant partitioning the contaminant between air and water. Important geological properties of the contaminated ground include microporosity, permeability to air, contaminant adsorption capacity, and moisture content.

Modifications to air vapor extraction include **flushing with hot air** and, more commonly, **steam flushing** to enhance earth washing and volatilization of the contaminants.

Air sparging is a process wherein compressed air is injected below the surface of water to strip volatile contaminants. Air bubbles rise in the water carrying volatile compounds out of it. *In situ* air sparging is the introduction of air into groundwater below the water table. It requires fairly high porosities and permeabilities in the contaminated ground, including ground above the water table where air sparging is occurring. Sometimes this can be artificially introduced by substituting a "reactive wall" of highly permeable material to capture the contaminants as groundwater flows through it. The air sparging wells can be inserted in the **"reactive wall,"** which would normally be located at the down gradient end of a contaminated plume.

IN SITU CONTAMINANT FIXATION

Because removing the last remnants of contamination, needed to meet environmental regulations, has proven difficult with either water or air stripping, *in situ* methods have been sought for stabilizing the contaminant, transforming it to a harmless product or transforming it to an insoluble product. As *in situ* processes, most of these are in the development stage, but some show considerable promise and are based on sound chemistry and biochemistry.

Precipitation of metals in acidic water, including acid mine water, by raising the pH to precipitate metal hydroxides is a well-established process that can also be conducted *in situ*. Precipitation occurs with coprecipitation of gypsum, as shown by the following generic reaction for a divalent metal cation:

$$MSO_4 + Ca(OH)_2 + 2H_2O \rightarrow M(OH)_2 + CaSO_4 2H_2O. \qquad (15.1)$$

Refer to Fig. 15.3 for metal hydroxide solubilities. These solubilities are adequate for some heavy metals, but generally inadequate for amphoteric metals and metalloids such as arsenic that form soluble oxyanion complexes. Precipitation of metals by hydrolysis is generally more effective if the solution Eh is raised (oxidation), because for metal cations with more than one valence state, the higher valence cation is generally more insoluble at the same pH. For example, ferric ions are much more insoluble than ferrous ions.

Lime pebble beds in a reactive wall, as shown in Fig. 15.5, have been used to intercept a groundwater plume and could be used to intercept acid mine water effluent from a mine adit or tunnel. However, maintaining the permeability and reactivity of the pebble bed as reaction occurs and coats particles with the hydroxy-gypsum sludge is a serious concern for a passive process. Injecting milk of lime into contaminated ground or an ore heap may be effective, especially if done with cement grouting to internally seal the contaminated region.

While hydrolysis does not work particularly well on some toxic metals, other means of inorganic precipitation can be used effectively or are readily available. Hydroapatite has been used to precipitate lead as an insoluble phosphate (Schwartz and Xu, 1992). However, sulfate anion is present in acid mine water and often in high concentrations because of the

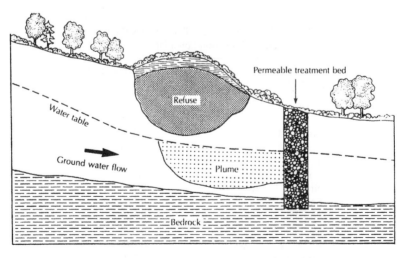

Figure 15.5. Installation of a reactive wall containing a permeable treatment bed of pebble lime (Fetter, 1988).

accumulation of aluminum, magnesium and iron counter ions. Because of the low solubility product of lead sulfate, lead is not found in acid mine water. The calculated solubility of lead in typical copper mine waste dump leach liquor, with a sulfate ion concentration of ore molar, is about 2 ppb. After acid mine water is neutralized with lime to produce calcium sulfate (gypsum), the residual sulfate concentration cannot exceed about 1 gpl, yielding a calculated maximum possible lead concentration of 0.2 ppm.

Cementation, or **zero valent metal reduction**, reactions have long been used in extractive metallurgy to precipitate dissolved metals. Iron and zinc, as scrap or metal powders, have been most frequently used to precipitate copper, nickel and cobalt. The EMF series, shown in Table 15.2, can be used as a guide to the effectiveness of this approach. Iron is preferred over zinc because of its substantially lower cost. Hexavalent chromium can be reduced to the less soluble trivalent chromium using iron powder as the reactant (Puls et al., 1995). Iron powder is also under consideration for reduction of soluble selenite ions to elemental selenium. Iron has also been used to dehalogenate chlorinated aliphatic organic compounds including PCE and TCE (Gillham et al., 1993). This was field tested with a reactive wall, containing ground iron and sand, that intercepted the contaminant plume.

Precipitation of metal sulfides can be a very effective method of removing many soluble heavy metals to very low concentrations provided the solution is in a reducing environment. Either hydrogen sulfide, sulfide

Table 15.2   Electromotive Force Series (Standard Oxidation–Reduction Potentials at 25°C).

Electrode Reaction	Standard Potential (V)
$Zn = Zn^{++} + 2e^-$	-0.762
$Cr^{++} = Cr^{+++} + e^-$	-0.41
$Fe = Fe^{++} + 2e^-$	-0.440
$Cd = Cd^{++} + 2e^-$	-0.402
$In = In^{+++} + 3e^-$	-0.340
$Tl = Tl^+ + e^-$	-0.336
$Co = Co^{++} + 2e^-$	-0.277
$Ni = Ni^{++} + 2e^-$	-0.250
$Sn = Sn^{++} + 2e^-$	-0.136
$Pb = Pb^{++} + 2e^-$	-0.126
$Cu = Cu^{++} + 2e^-$	0.345
$Fe^{++} = Fe^{+++} + e^-$	0.771
$2Hg = Hg_2^{++} + 2e^-$	0.799
$Ag = Ag^+ + e^-$	0.800

ion (e.g., Na_2S) or bisulfide ion must be supplied as a source of sulfide. Under acidic to neutral conditions, dissolved H_2S is the most prevalent soluble sulfide. At basic pH, HS^- is the predominant species. Metal precipitation with H_2S and HS^- are shown by the following typical acid generating reactions, for a divalent metal cation:

$$H_2S(aq) + M^{2+} = MS + 2H^+, \tag{15.2}$$

$$HS^- + M^{2+} = MS + H^+, \tag{15.3}$$

with metal sulfide solubility depending on pH in both cases. The thermodynamically computed solubilities of several metal sulfides, as a function of pH are shown in Fig. 15.6. Note that at neutral pH, with the exception of Fe^{3+}, these solubilities, expressed as mg/l of metal (ppm), are considerably smaller than the corresponding metal hydroxide solubilities shown in Fig. 15.3. The solubilities of the sulfides of copper, silver and mercury are off the scale (less than $0.0001\,mg/l$). In contrast with the voluminous sludge that results from oxidation and hydrolysis to precipitate gelatinous metal hydroxides that coat the lime pebbles or other basic source, sulfide precipitates are dense and dispersed. They do not block further precipitation, which is an important advantage.

In chemically treating a solution of metal sulfates, such as acid mine water, metal sulfides are precipitated, but sulfuric acid is also generated, which is usually undesirable. However, if the sulfate ions in the contaminated water are also reduced to bisulfide ions or H_2S, then *both heavy metals and sulfate ions are transformed and rendered insoluble*. This can be done biologically, as will be discussed in the next section.

BIOREMEDIATION

In situ bioremediation is increasingly being used to destroy organic compounds and cyanide, and it is the subject of considerable research (Nyer et al., 1996). Bacteria, which are ubiquitous in the earth's near-surface lithosphere, are the primary microorganisms being used. *In situ* bioremediation with bacteria generally amounts to enhancing the activity of natural bacteria at the site being remediated in order to speed up the decontamination process. The characteristics of bacteria, including cell structure and micronutrient requirements, were discussed in Chapter 6 in conjunction with biooxidation of sulfide minerals during heap leaching of ores and mine wastes. The use of bacteria to destroy cyanide was discussed in Chapter 4.

Typically, bacteria outnumber other organisms in soil and constitute the largest biomass in soil. Microorganism populations are shown in

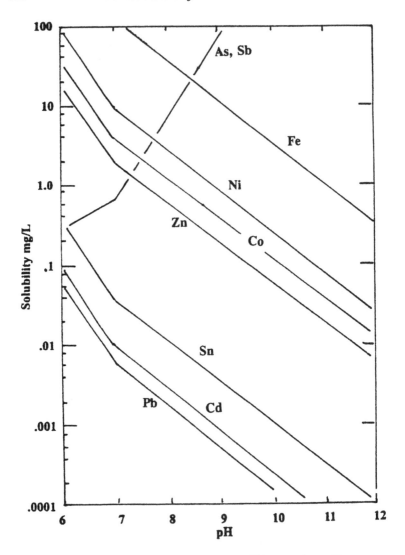

Figure 15.6. Solubilities of metal sulfides at various pH.

Table 15.3. While their population is usually highest in the root zone and declines with depth, as does organic matter, bacteria have been found at surprising depths by sampling deep wells. Fine soils generally contain more bacteria than coarse sandy or rocky soils, including ore heaps. However, the latter, with their typical higher permeabilities, can transport

Table 15.3 Microorganism Population Distribution in Soil and Ground Water (Nyer et al., 1996).

	Population Size	
Organism	Typical	Extreme
Surface Soil (cells/gram soil)		
Bacteria	0.1–1 billion	10 billion
Actinomycetes	10–100 million	100 million
Fungi	0.1–1 million	20 million
Algae	10,000–100,000	3 million
Subsoil (cells/gram soil)		
Bacteria	1000–10,000,000	200 million
Ground water (cells per mL)		
Bacteria	100–200,000	1 million

bacteria, nutrient and other reactants (including oxygen) more rapidly. Consequently, ore heaps and other coarse earth materials often can be more quickly bioremediated than fine soils.

Microorganisms, derive their energy from oxidation–reduction reactions, usually mediated with enzymes generated by the microorganisms. The energy source, for example an organic molecule or a sulfide mineral, is oxidized. Ultimate reaction products are CO_2 and SO_4^{2-}, respectively, but intermediate products usually are involved in the reaction pathway. Basically, the energy source is an electron donor transferring electrons to an accepter and releasing energy conserved in a chemical bond, e.g., a C–C bond. The intracellular mediator of redox processes in all life forms is nicotinamide adenine dinucleotide, or NAD^+. It can accept hydride ion from the electron donor species and transfer the electrons to the acceptor species, while the hydride ion is oxidized and released as water. Acceptor species are oxidized species such as dissolved oxygen, sulfate or nitrate ions and cyanide. NADH is the reduced form of NAD^+ which carries the hydride species.

The cell must have NAD^+ available so that energy can be extracted from ongoing metabolism, and cells will respire on any acceptor species that is available. If supplies of NAD^+ are low, the cell will use up excess NADH through respiration on pyruvate, a process called **fermentation**. Fermentation is operative when a human is running on a treadmill (energy outlay) without getting enough oxygen (electron acceptor), and NADH is accumulating. The muscles will attempt to regenerate NAD^+ to allow the human to continue producing energy and will place the electrons from the excess NADH on pyruvate, resulting in a buildup of lactic acid. This lactic acid will precipitate in the muscle tissue and cause the soreness that you will feel the next day as you limber up for golf or other activities demanding NAD^+.

Aerobic respiration, where oxygen is the electron acceptor, is the most common microbiological oxidation process. It is clearly operative in the biooxidation of sulfide minerals and is very energetically favored when even small amounts of oxygen are present.

Anaerobic respiration can only occur in the absence of free oxygen, and then only if other molecular species are present to provide oxygen or accept electrons. The most common of these in contaminated ground, mines, tailings, and ore heaps are soluble ferric, nitrate, carbonate and sulfate ions. In anaerobic respiration, the species being oxidized (electron donor) is usually an organic compound with some water solubility. Examples of readily soluble organic electron donors are sugar (molasses), starches, and alcohols. Anaerobic respiration can drive irreversible reactions such as nitrate reduction to nitrogen gas or cyanide reduction to ammonia, and these irreversible reactions are the easiest and most economically competitive biological processes to operate.

The order of standard reduction potentials, shown in Table 15.4, determines the predominance of competing microbial respiration reaction pathways. Dissolved (free) oxygen has the highest reduction potential, which is why aerobic respiration will be the dominant pathway if oxygen is available. Because of its strong reduction potential, no more than about 1% atmospheric oxygen, less than 1 ppm of dissolved oxygen, is generally required to prevent aerobic respiration from occurring. In the absence of oxygen, other bacteria that use one of the other respiration processes will take over if a suitable electron acceptor is present.

While the reduction potential for sulfate ion is very small, sulfate ion can be anaerobically reduced by a bacteria consortium that would likely

Table 15.4 Standard Reduction Potential at 25°C and pH 7 for Some Redox Couples that are Important Electron Acceptors in Microbial Respiration.

Oxidized Species	Reduced Species	$E°$ (V)
$O_2 + 4H^+ + 4e^-$	$= 2H_2O$	$+0.92$
$2NO_3^- + 12H^+ + 10e^-$	$= N_2 + 6H_2O$	$+0.74$
$MnO_2(s) + HCO_3^- + 3H^+ + 2e^-$	$= MnCO_3(s) + 2H_2O$	$+0.05$
$FeOOH(s) + HCO_3^- + 2H^+ + e^-$	$= FeCO_3(s) + 2H_2O$	-0.05
$SO_4^{2-} + 9H^+ + 8e^-$	$= HS^- + 4H_2O$	-0.22
$CO_2 + 8H^+ + 8e^-$	$= CH_4 + 2H_2O$	-0.24

include *Desulfovibrio desulfuricans* to produce soluble sulfide species. In the following example, sucrose is oxidized to CO_2 while sulfuric acid is reduced to H_2S, with an increase in pH.

$$12H^+ + 6SO_4^{2-} + C_{12}H_{22}O_{11} \rightarrow 12CO_2 + 6H_2S + 11H_2O. \quad (15.4)$$

The H_2S in turn will chemically react with any heavy metal cations that are present to precipitate metal sulfides, by eqn 15.2. These reactions are the basis of bioremediation of acid mine water, both as drainage and *in situ* within spent ore heaps and within the gob of underground mines. Again, using sucrose as an electron donor example, the net reaction converting dissolved metal sulfates to precipitated metal sulfides and harmless carbon dioxide is:

$$6M^{2+} + 6SO_4^{2-} + C_{12}H_{22}O_{11} \rightarrow 12CO_2 + 6MS + 11H_2O. \quad (15.5)$$

The sulfuric acid contained in acid mine water is also bioreduced. The reduction of more sulfate ions than matched with the dissolved metal sulfates also generates H_2S species in excess of that needed to stoichiometrically match the heavy metals. This drives the reaction equilibria (eqn 15.2) to precipitate more metal sulfides and lower further the concentration of dissolved metal.

Figure 15.7 shows the measured metal concentration in the drainage resulting from *in situ* biological treatment of an underground mine near Lincoln, Montana using a proprietary process biochemically similar to eqn 15.5 (Harrington, 1997).

Other species can have a favorable reduction potential over sulfate and be preferentially reduced, e.g., nitrate, arsenate, selenate and manganate. For example, precipitation of arsenic during bacterial sulfate reduction has been studied by Rittle et al. (1995). Again, using sucrose as the electron donor, the net reaction for bacteria catalyzed reduction of arsenic acid to As_2S_3 in the presence of sulfuric acid, e.g., in acid mine water is:

$$36H_2SO_4 + 24H_3AsO_4 + 7C_{12}H_{22}O_{11} \rightarrow 84CO_2 + 149H_2O + 12As_2S_3. \quad (15.6)$$

Nitrate ion concentration in a spent gold ore heap leachate was reduced from 130 ppm to a non-detectable amount. A two acre section of a heap with a height of 8 m was bioremediated. After injection of the soluble organic substance, it took 30 days to reach the analytical detection limit of 0.3 ppm nitrate in the leachate (Harrington, 1997).

For these **analogs** to be reduced and precipitated, it is necessary to have an accurate estimate of the redox titration curve that will dictate the most favorable energy pathway. For example, an excess of carbon source due to over-estimation of the reducing demand will result in formation of organic

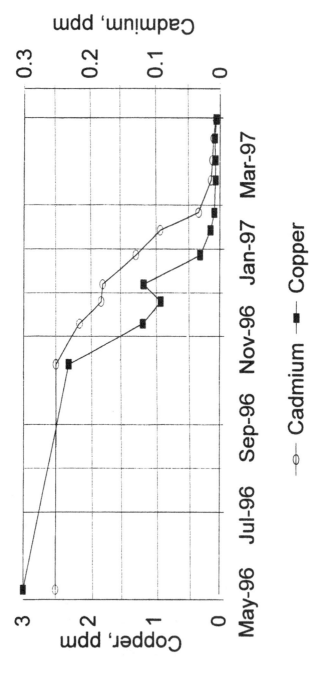

Figure 15.7. Reduction in metal concentration in the drainage from the Lincoln Mine, Montana, resulting from *in situ* bioremediation to convert soluble metal sulfates to insoluble metal sulfides (Harrington, 1997).

acids from fermentation rather than use of sulfate or its biochemical analogs. This fermentative process is one of the main reasons for past failures of laboratory and field experiments to bio-stabilize metals as sulfides. These excess organic acids will often chelate the metals ions and increase their solubility, rather than precipitate them as sulfide minerals.

In studies where carbon sources were added to a high sulfate medium, such as acid mine water, the optimal dosage and formulation of the carbon source was determined by measuring the terminal pH and the levels of accumulated organics from various levels of the added carbon source. These studies have shown that a large amount of a rich carbon source will likely be a poor choice for optimization of sulfate reduction to meet environmental standards. Fermentative byproducts can result not only in chelation, but also in lowered pH and a strong odor. These odors emanate from the reaction of sulfide ion and organic acids to yield thiols and esters. Mercaptans are examples of biologically generated odor species.

Because bioremediation processes are natural processes, they are typically unpatentable. However, some of the bioremediation processes describing optimization and formulations of carbon sources to achieve a particular result have been protected in the U.S. and elsewhere by patents. Similarly, biooxidative processes on sulfide minerals are unpatentable. Nevertheless, some processing advances have been viewed by the U.S.A. Patent Offices as novel and patents have been issued.

Toxic metals can be removed from aqueous solutions by biological mechanism other than bacteria catalyzed reduction and precipitation of sulfides. The most important of these other biological mechanisms is **bioadsorption**, wherein the soluble metal ion is either adsorbed, complexed or chelated on the cell walls or surfaces of abundant biomass. For certain biota, a large mass of metal can be adsorbed in relation to the biomass adsorbent, 10–15%. Dead biomass often works as well as living material. Bacteria, yeast, and algae are effective biosorbents for some heavy metals. **Bioaccumulation** or absorption of heavy metals involves transfer of the metal solute *through* the cell wall. Typically, the biota has limited tolerance for the metal being accumulated and, consequently, bioaccumulation usually exhibits relatively lower metal mass uptake capacity than bioadsorption.

Both bioadsorption and bioaccumulation can occur by bringing the contaminated water in contact with the biota in a tank, bed or lagoon (Rossi, 1990). A well-known example is the use of activated sludge in sewage treatment plants to also remove soluble heavy metals. Another example is the use of wetlands, particularly with cattails to adsorb metals from acid mine water as it passes through the wetlands and provide cellulose for anaerobic bioreduction. *Typha latifolia* is the predominant cattail species in wetlands used for mine water treatment with *Typha angustifolia* less commonly present. Cattails are readily available, transplant well, and are tolerant of a wide range of conditions (Sencidiver and Bhumbla, 1988). The advantage of this is that it can be a passive treatment with much lower costs than lime neutralization.

A major problem with both bioadsorption and bioaccumulation that has retarded more widespread use is disposal of the metal laden biomass following metal loading.

In a few instances, heavy metal solutes can be transformed to a volatile species and swept out of the contaminated solution. The best known example is conversion of mercuric ion to methylmercury by bacteria (Hamdy and Noyes, 1975), which can occur under both oxidizing and reducing conditions.

PASSIVE SYSTEMS FOR TREATMENT OF
ACID MINE DRAINAGE

Conventional acid mine drainage (AMD) systems, involving agitated tank reactors, thickeners and filters, require near constant operator attention and are expensive to operate. Passive treatment processes, once established, have no operating machinery and require infrequent attention, and then for only short durations. They take advantage of naturally occurring geochemical and biological processes. For a more detailed review of passive AMD treatment systems than presented here, see Gazea et al. (1996).

Natural wetlands containing *Sphagnum* moss were observed to lower iron concentrations in AMD. Consequently, the first engineered wetlands for AMD were based on using this material, but it proved to be less durable over years than cattails. The commercial applications of passive AMD systems have been on drainage from coal mines, and over 200 AMD wetlands have been installed in Appalachia (Bastian and Hammer, 1993). The heavy metal contaminants in drainage from coal mines are usually restricted to iron and manganese, and the primary metal removal method is oxidation and hydrolysis, occurring as the pH of the AMD is raised. These **aerobic wetlands** are flooded shallow basins very similar to natural wetlands, with a relatively impermeable bottom to prevent seepage. Soil in the basin supports vegetation, which controls the distribution and flow of AMD slowly through the system to provide a long residence time. The vegetation helps to promote oxygen transfer through roots into the water at depth.

Over less than a score of years, passive AMD systems have evolved and there are presently three principal passive technologies. **Anaerobic systems** use organic materials to cause bioreduction of metal sulfates to precipitated metal sulfides. **Anoxic limestone drains**, neutralize acid in the absence of ferric iron, which would otherwise precipitate and foul the system with sludge. Anoxic limestone drains are usually used in tandem with anaerobic systems to neutralize acid either before or after AMD passes through it. This saves on the consumption of organic material, that would otherwise be needed to bioreduce sulfuric acid.

Anaerobic systems consist of basins filled with compost, and often covered to prevent atmospheric oxygen from entering. The compost is porous to allow easy passage of the AMD. The compost consists of natural organic materials such as manure, hay bales, wood chips and peat. Limestone may be added to partially neutralize sulfuric acid in the AMD during treatment.

The applicability of each passive technology depends on the mine water composition. Aerobic wetlands are satisfactory when the AMD is not to acidic and toxic non-ferrous metals are not present, which is typical of coal mine drainage. Anaerobic treatment is necessary for more acidic water or AMD containing many nonferrous heavy metals, such as copper, arsenic, mercury and zinc. These contaminants are typically found in drainage from metal mines. There are complex passive treatment systems sequentially using all three technologies.

Because compost cannot readily be inserted into spent ore heaps, tailings and closed underground mines, *in situ* anaerobic systems that do not use compost are under investigation. In these systems a water soluble organic substance or organic/water emulsion is carried into the ground or mine generating the AMD (Harrington, 1997).

MISCELLANEOUS CONTAMINANT FIXATION METHODS

Encapsulation of the contaminated material in a cell, that prevents both egress of the material and passage of water through it, is a standard approach to encapsulation of earth materials. Impoundment of spent ore heap residues and mine tailings that includes covering the material during mine reclamation is one example. Unless there are unusual chemical or radioactive conditions, these mining residues are not presently classified by EPA as hazardous. Materials that are classified by EPA as hazardous must be contained in a Class II landfill with double liners for leachate collection and permanent (forever more) monitoring. A double-lined encapsulation cell (landfill) with leachate collection is illustrated in Fig. 15.8.

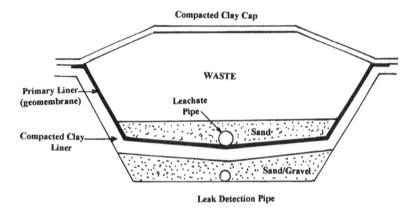

Figure 15.8. Double lined waste encapsulation cell with leachate collection.

Because of the difficulty and expense of constructing and permitting a Class II landfill, classified hazardous materials are usually shipped, at considerable expense, to an existing Class II landfill.

Stabilization of contaminated materials can be accomplished by blending the solid material with enough water, portland cement and lime to form a paste that hardens to an impervious mass. This is a relatively inexpensive process. The stabilized paste can be placed in a final lined repository or cell that is relatively impervious to meteoric water. Enough lime and cement must be added so that the hardened paste meets EPA's TCLP test for leachable toxic materials. Stabilization was used to store flue dust at the closed copper smelter in Anaconda, Montana. Several thousand tons of flue dust containing arsenic, lead, cadmium and other toxic elements was stabilized with cement and permanently encapsulated in a lined, covered landfill at the smelter site. The total cost of this remediation was about $100 per tonne of flue dust.

Vitrification is a relatively expensive process in which the material is melted in a high temperature furnace to render the material impervious to leaching. A vitreous (glassy) or semi-vitreous product is made. Thermal arc plasma furnaces have been constructed for this purpose. *In situ* vitrification has been tried at the Hanford National Laboratory using electrodes placed in the contaminated ground to melt the soil in place. It is viewed as a process for radioactive or highly toxic soil. However, results have been somewhat disappointing with respect to the depth of penetration into the ground at which vitrification can be achieved with this method.

PROBLEM

1. Write an equation similar to eqn 15.4 for the biochemical reaction between sucrose and nitrate ion to generate nitrogen and carbon dioxide. For *one million gallons* of acid mine drainage (AMD) with the listed concentrations, calculate the stoichiometric amount of sucrose required to reduce both sulfate ion and nitrate ion to the permitted concentrations as well as precipitate copper as copper sulfide. The AMD concentrations and permitted concentrations of these species are:

Species	Concentration in AMD	Permitted Concentration
SO_4^{2-}	400 ppm	250 ppm
NO_3^-	10 ppm	1 ppm
Cu^{2+}	5 ppm	0.05 ppm
dissolved oxygen	8 ppm	no limit

HINT: Don't forget the thermodynamic order of reduction of species, see Table 15.4.

REFERENCES AND SUGGESTED FURTHER READING

Bartlett, R.W. (1993). "Soil decontamination by percolation leaching." *Extraction and Processing for the Treatment and Minimization of Wastes*, J. Hager et al. (ed.), The Minerals, Metals and Materials So., Warrendale, PA, pp. 411–424.

Bastian, R.K. and Hammer, D.A. (1993). The use of constructed wetlands for wastewater treatment and recycling. *Constructed wetlands for water quality improvement*, G.A. Moshiri (ed.), Lewis Publ., ISBN 0-87371-550-0, pp. 59–68.

Bucknam, C.H. (1997). Cyanide solution detoxification jar tests. *Global Exploitation of Heap Leachable Gold Deposits*, ed. by D.M. Hausen, The Minerals Metals and Materials Soc., Warrendale, PA, pp. 191–204.

Bunge, R., Bachmann, A. and Chien, D.N. (1995). Soil-washing: mineral processing technology in environmental engineering. Proceedings of the XIX Int'l Mineral Processing Congress, Vol 4, SME, Littleton, CO. pp. 123–129.

Fetter, C.W. (1988). *Applied Hydrogeology*, MacMillan Publishing, New York, NY, Chap. 10, ISBN 0-675-20887-4.

Gazea, B., Adam, K. and Kontopoulos, A. (1996). A review of passive systems for the treatment of acid mine drainage. *Minerals Engineering, 9, No. 1*, pp. 23–42.

Gillham, R.W., O'Hannesin, S.F. and Orth, W.D. (March 1993). Metal enhanced abiotic degradation of halogenated aliphatics: Laboratory Tests and Field Trials. Proc. HazMat Central Conference, Chicago, IL.

Hamdy, M.K. and Noyes, O.R. (1975). Formation of methyl mercury by bacteria. *Appl. Microbiol. 30*, p. 424.

Harrington, J.G. (May 1997). Private Communication, GreenWorld Science Inc., Moscow, ID.

Liu, M. and Roy, D. (1992). *In situ* soil washing by surfactant. SME preprint 92–167. Littleton, CO.

Nyer, E.K. et al. (1996). *In Situ Treatment Technology*, CRC Press, Boca Raton, FL, ISBN 0-87371-995-6.

Puls, R.W., Powell, R.M. and Paul, C.J. (April 1995). *In situ* remediation of ground water contaminated with chromate and chlorinated solvents using zero-valent iron: a field study. Proc. 209th American Chemical Society National Meeting, Anaheim, CA.

Rittle, K.A. et al. (1995). Precipitation of arsenic during sulfate reduction. *Geomicrobiology Journal, 13*, pp. 1–11.

Robins, R. G., Huang, J.C.Y. and Nishimura, T. (1987). The adsorption of arsenate ion by ferric hydroxide. *Arsenic Metallurgy Fundamentals and Applications*, ed. by R.G. Reddy, The Metallurgical Soc., Warrendale, PA, pp. 99–112.

Rossi, G. (1990). *Biohydrometallurgy*, Chap. 6, McGraw-Hill, New York, NY, ISBN 0-07-053931-6.

Schwartz, F.W. and Xu, Y. (1992). Modeling the behavior of a reactive barrier system for lead. Proc. Modern Trends in Hydrogeology. Conference of the Canadian National Chapter, International Association of Hydrogeologists, Hamilton, Ontario.

Sencidiver, J.C. and Bhumbla, D.K. (1988). Effects of cattails (*Typha*) on metal removal from mine drainage, in: 1988 Mine Drainage and Surface Mine Reclamation, Vol, I: Mine Water and Mine Waste, USBM IC 9183, pp. 359–366.

APPENDIX

Table A-1 Engineering Grain-Size Classification.

Name	Size Range (mm)	Example
Boulder	>305	Basketball
Cobbles	76–305	Grapefruit
Coarse gravel	19–76	Lemon
Fine gravel	4.75–19	Pea
Coarse sand	2–4.75	Water softener salt
Medium sand	0.42–2	Table salt
Fine sand	0.075–0.42	Powdered sugar
Fines	<0.075	Talcum powder

Table A-1(B) Typical One Stage Primary Crusher Settings.

Nominal (in)	Opening (mm)	Approximate Top Size (mm)
8	200	175
6	150	130

427

Table A-2 Tyler Standard Testing-Sieve
Scale.

Mesh	Nominal (in)	(mm)	(in)
		Aperture	
	3		2.97
	2		2.10
	1/2		1.48
	1		1.050
	3/4		0.742
	1/2		0.525
	3/8		0.371
3	1/4		0.263
4		4.699	0.185
6		3.327	0.131
8		2.362	0.093
10		1.651	0.065
14	micron	1.168	0.046
20	800	0.833	0.0328
28	600	0.589	0.0232
35	400	0.417	0.0164
48	300	0.295	0.0116
65	200	0.208	0.0082
100	150	0.147	0.0058
150	100	0.104	0.0041
200	74	0.074	0.0029
270	53	0.053	0.0021
(325)*	44	0.044	0.0017

*Not in standard series.

Table A-3 Conversion Factors for Particle Size, Distance (L).

	Multiply by Table Values to Convert to these Units					
Given a Quantity in these Units	m	cm	mm	μm	in	ft
m	1	10^2	10^3	10^6	39.37	3.281
cm	10^{-2}	1	10	10^4	0.3937	0.03281
mm	10^{-3}	10^{-1}	1	10^3	3.937×10^{-2}	0.00328
μm	10^{-6}	10^{-4}	10^{-3}	1	3.937×10^{-5}	3.281×10^{-6}
in	0.0254	2.54	25.4	2.54×10^4	1	0.08333
ft	0.3048	30.48	304.8	3.048×10^5	12	1

Table A-4 Conversion Factors for Area (L^2).

	Multiply by Table Values to Convert to these Units					
Given a Quantity in these Units	m^2	cm^2	in^2	ft^2	acre	sq mile
m^2	1	10^4	1550	10.76	2.47×10^{-4}	3.86×10^{-7}
cm^2	0.0001	1	0.155	1.08×10^{-3}	2.47×10^{-8}	3.86×10^{-11}
in^2	6.45×10^{-4}	6.452	1	6.94×10^{-3}	1.59×10^{-7}	2.49×10^{-10}
ft^2	0.0929	929	144	1	2.3×10^{-5}	3.59×10^{-8}
acre	4047	4.04×10^7	6.27×10^6	43,560	1	1.56×10^{-3}
sq mile	2.59×10^6	2.59×10^{10}	4.01×10^9	2.79×10^7	640	1

Table A-5 Conversion Factors or Volumetric Flow Rate (L^3/t).

Given a Quantity in these Units	Multiply by Table Values to Convert to these Units				
	m^3/s	l/s	ft^3/s	acre-ft/day	gal/min
m^3/s	1	10^3	35.31	70.05	1.58×10^4
l/s	0.001	1	0.0353	0.070	15.85
ft^3/s	0.0283	28.32	1	1.984	448.8
acre-ft/day	0.0143	14.276	0.5042	1	226.28
gal/min	6.3×10^{-5}	0.0631	2.23×10^{-3}	4.42×10^{-3}	1

Table A-6 Conversion Factors for Force (F or ML/t^2).

Given a Quantity in these Units	Multiply by Table Values to Convert to these Units			
	$kg\,m\,s^{-2}$ (newton)	$g\,cm\,s^{-2}$ (dyne)	lb_f	$lb_m\,ft\,s^{-2}$ (poundal)
$kg\,m\,s^{-2}$ (newton)	1	10^5	2.248×10^{-1}	7.233
$g\,cm\,s^{-2}$ (dyne)	10^{-5}	1	2.248×10^{-6}	7.233×10^{-5}
lb_f	4.4482	4.4482×10^{-5}	1	32.174
$lb_m\,ft\,s^{-2}$ (poundal)	0.13826	1.3826×10^4	3.108×10^{-2}	1

Gravitational constant: $g_0 = 980.665\,cm\,s^{-2} = 9.806\,m\,s^{-2}$.

Table A-7 Conversion Factors for Pressure (F/L^2 or $ML^{-1}t^{-2}$).

Given a Quantity in these Units	Multiply by Table Values to Convert to these Units						
	Pa (N/m^2)	kPa	lb_f/in^2 (psi)	in-H_2O	m-H_2O	Atm	g-cm^{-1}s^{-2} (dyne/cm^2)
Pa	1	10^{-3}	1.45×10^{-4}	0.00402	1.02×10^{-4}	9.8692×10^{-6}	10
kPa	10^3	1	0.145	4.019	0.1021	9.8692×10^{-3}	10,000
lb_f/in^2 (psi)	6.895×10^3	6.895	1	27.68	0.703	6.805×10^{-2}	6.9×10^4
in-H_2O	248.8	0.2488	3.613×10^{-2}	1	0.0254	0.0024582	2,488
m-H_2O	9797	9.797	1.422	39.37	1	0.0968	97,970
Atm	1.013×10^5	101.3	14.696	406.8	10.33	1	1.013×10^6
g-cm^{-1}s^{-2} (dyne/cm^2)	0.1	1×10^{-4}	1.45×10^{-5}	4.02×10^{-4}	4.1×10^{-5}	9.87×10^{-7}	1

Table A-8 Conversion Factors for Velocity and Hydraulic Conductivity (L/t).

Given a Quantity in these Units	Multiply by Table Values to Convert to these Units					
	m/s	ft/day	mm/s	m/day	gal/day/ft²	gal/day/acre
m/s	1	2.83×10^5	1×10^3	86,400	2.12×10^6	9.22×10^{10}
ft/day	3.53×10^{-6}	1	3.53×10^{-3}	0.305	7.48	325,829
mm/s	1×10^{-3}	283.5	1	86.4	2120	9.235×10^7
m/day	1.16×10^{-5}	3.28	0.0116	1	24.5	1,067,220
gal/day/ft²	4.72×10^{-7}	0.134	4.72×10^{-4}	0.0408	1	43,560

Table A-9 Conversion Factors for Intrinsic Permeability (L^2).

Given a Quantity in these Units	Multiply by Table Values to Convert to these Units				
	m^2	cm^2	mm^2	darcy	md
m^2	1	10^4	10^6	1.013×10^{12}	1.013×10^{15}
cm^2	10^{-4}	1	10^2	1.013×10^8	1.013×10^{11}
mm^2	10^{-6}	10^{-2}	1	1.013×10^6	1.013×10^9
darcy	9.87×10^{-13}	9.87×10^{-9}	9.87×10^{-7}	1	10^3
millidarcy (md)	9.87×10^{-16}	9.87×10^{-12}	9.87×10^{-10}	10^{-3}	1

Table A-10 Conversion Factors for Energy, Work (FL or ML^2t^{-2}).

Given a Quantity in these Units	Multiply by Table Values to Convert to these Units					
	$kg\,m^2s^{-2}$ (joule)	$g\,cm^2s^{-2}$ (erg)	kcal	Btu	kw-hr	hp-hr
$kg\,m^2s^{-2}$ (joule)	1	10^7	2.39×10^{-4}	9.478×10^{-4}	2.7778×10^{-7}	3.725×10^{-7}
$g\,cm^2s^{-2}$ (erg)	10^{-7}	1	2.39×10^{-11}	9.478×10^{-11}	2.7778×10^{-14}	3.725×10^{-14}
kcal	4184	4.184×10^{10}	1	3.9657	1.1622×10^{-3}	1.5586×10^{-3}
Btu	1.055×10^3	1.055×10^{10}	0.25216	1	2.93×10^{-4}	3.93×10^{-4}
kw-hr	3.60×10^6	3.60×10^{13}	860.42	3412.2	1	1.3410
hp-hr	2.6845×10^6	2.6845×10^{13}	641.62	2544.5	0.7457	1

Table A-11 Conversion Factors for Dynamic Viscosity ($ML^{-1}t^{-1}$ or $FL^{-2}t$).

Given a Quantity in these Units	Multiply by Table Values to Convert to these Units		
	$kg\,m^{-1}s^{-1}$ (Pa s)	kPa s	centiPoise [cP]
$kg\,m^{-1}\,s^{-1}$ (Pa s)	1	10^{-3}	10^3
kPa s	10^3	1	1
centiPoise [cP]	10^{-3}	1	1

$\mu_{Air} = 1.8 \times 10^{-5}\,Pa\,s$ (STP), $\mu_{H_2O} = 1.0 \times 10^{-3}\,Pa\,s$ (STP).

Table A-12 Conversion Factors for Diffusivity (L^2/t).

Given a Quantity in these Units	Multiply by Table Values to Convert to these Units				
	m^2/s	cm^2/s	mm^2/s	ft^2/hr	cm^2/day
m^2/s	1	10^4	10^6	3.875×10^4	8.64×10^8
cm^2/s	10^{-4}	1	10^2	3.875	86400
mm^2/s	10^{-6}	10^{-2}	1	0.03875	864
ft^2/hr	2.581×10^{-5}	0.2581	25.81	1	22297
cm^2/day	1.16×10^{-9}	1.16×10^{-5}	0.00116	4.495×10^{-5}	1

Table A-13 Physical Constants.

Gravity Constant
$$g = 980.665\,cm\,s^{-2}$$
Gas Law Constant
$$R = 8.314 \times 10^7\,g\,cm^2\,s^{-2}\,g\,mol^{-1}\,K^{-1}$$

INDEX

For Product Safety Concerns and Information please contact our EU
representative GPSR@taylorandfrancis.com Taylor & Francis Verlag GmbH,
Kaufingerstraße 24, 80331 München, Germany

Printed and bound by CPI Group (UK) Ltd, Croydon, CR0 4YY
08/05/2025
01864327-0007